Biomass Hydrolyzing Enzymes

This reference book provides advanced knowledge about lignocellulosic biomass production and its application in biomass hydrolysis. Lignocellulosic biomass is the most abundant, ubiquitous, and renewable raw material in the world. Though biomass can be deconstructed by other means, biological ways through enzymes are eco-friendly and sustainable. *Biomass Hydrolyzing Enzymes: Basics, Advancements, and Applications* discusses the different enzymes used for degrading biomass into its monomeric components. It covers important topics including biorefineries, hydrolysis of algal mass, kinetic modeling for hydrolysis, and inhibitory effects.

Key Features

- Highlights recent developments in biorefineries, specific enzymes, inhibitor tolerance, and enhanced efficiencies
- Provides details on various kinds of biomass hydrolysis including algal biomass
- Includes the best practices for obtaining economical and efficient high conversions of biomass
- Covers strategies to be adopted for increasing the production of highly efficient enzymes
- Explores the advancements in lignocellulosic biomass hydrolysis

The book is suitable for researchers and students in biotechnology, applied microbiology, and environmental sciences.

Sustainable Industrial and Environmental Bioprocesses

This book series aims to provide a comprehensive collection of books focusing on bioprocesses in industrial and environmental biotechnology. The multidisciplinary content encompasses the chemical and biochemical engineering, industrial microbiology, and energy biosciences, all with a central theme of sustainable development and circular economy principles. The books align with the Sustainable Development Goals (SDGs) and offer state-of-the-art information and in-depth knowledge on the subject matter. The books in this series also emphasizes on the application of emerging tools, such as machine learning and artificial intelligence, for the advancement of bioprocesses. While primarily targeting academicians and researchers, the series is valuable for policy planners and industry professionals, with carefully tailored contents to cater to their specific needs and interests.

Series Editor: Dr Ashok Pandey

Professor Ashok Pandey is currently Distinguished Scientist at the Centre for Innovation and Translational Research, CSIR-Indian Institute of Toxicology Research, Lucknow, India. His major research and technological development interests are industrial & environmental biotechnology and energy biosciences, focusing on biomass to biofuels & chemicals, waste to wealth & energy, etc. He has 16 patents, 120 books, >1000 papers and book chapters, etc with h index of 140 and >82,500 citations. Professor Pandey is the recipient of many national and international awards and honours, which include Fellow, The World Academy of Sciences (TWAS), Highly Cited Researcher (Top 1% in the world), Clarivate Analytics (since 2018- till date), Rank #1 in India in Biology in India and Biochemistry and #417 in the world, Research.Com (2023), Rank #1 in India in Microbiology under Enabling and Strategic Technologies sector, Elsevier (2021), Rank #1 in India in Biotechnology and #8 in the world, Stanford University Report (2020-2021-2022), etc.

Biotic Resources: Circular Bioeconomy Perspective
Thallada Bhaskar, Sunita Varjani, Bhavya B Krishna, Ashok Pandey

Microbial Enzymes in Production of Functional Foods and Nutraceuticals
Amit Kumar Rai, Ranjna Sirohi, Luciana Porto de Souza Vandenberghe, Binod Parameswaran

Microbial Enzymes and Metabolites for Health and Well-Being
Ranjana Sirohi, Amit Kumar Rai, Luciana Porto de Souza Vandenberghe, Binod Parameswaran

Second and Third Generation Bioplastics: Production, Application, and Innovation
Luciana Porto de Souza Vandenberghe, Ashok Pandey, Ranjana Sirohi, Carlos Ricardo Soccol

Pharmaceuticals in Aquatic Environments: Toxicity, Monitoring, and Remediation Technologies
Vinod Kumar Garg, Ashok Pandey, Navish Kataria, Caterina Faggio

Diatoms: Ecology and Biotechnological Applications
Archana Tiwari, Ashok Pandey, Pau-Loke Show, Binod Parameswaran

Diatoms Biotechnology
Archana Tiwari, Ashok Pandey, Pau Loke Show, Binod Parameswaran

Sustainable Treatment and Management of Sewage Sludge
Mukesh Kumar Awasthi, Zengqiang Zhang, Ashok Pandey

Pharmaceuticals in Aquatic Environment: Remediation Technologies and Future Challenges
Vinod Kumar Garg, Ashok Pandey, Navish Kataria, Caterina Faggio

Biomass Hydrolyzing Enzymes: Basics, Advancements, and Applications
Reeta Rani Singhania, Anil Kumar Patel, Héctor A. Ruiz, Ashok Pandey

Biomass Hydrolyzing Enzymes

Basics, Advancements, and Applications

Edited by

Reeta Rani Singhania
Institute of Aquatic Science and Technology, National Kaohsiung University of Science and Technology, Kaohsiung City, Taiwan
Centre for Energy and Environmental Sustainability, Lucknow, Uttar Pradesh, India

Anil Kumar Patel
Institute of Aquatic Science and Technology, National Kaohsiung University of Science and Technology, Kaohsiung City, Taiwan;
Department of Marine Environmental Engineering, National Kaohsiung University of Science and Technology, Kaohsiung City, Taiwan;
Centre for Energy and Environmental Sustainability, Lucknow, Uttar Pradesh, India

Héctor A. Ruiz
Biorefinery Group at the Autonomous University of Coahuila, Saltillo, Coahuila, Mexico

Ashok Pandey
Centre for Innovation and Translational Research, CSIR–Indian Institute of Toxicology Research, Lucknow, Uttar Pradesh, India;
Kyung Hee University, Seoul, Republic of Korea; School of Engineering, University of Petroleum and Energy Studies, Dehradun, India;
Centre for Energy and Environmental Sustainability, Lucknow, India

CRC Press
Taylor & Francis Group
Boca Raton London New York

CRC Press is an imprint of the
Taylor & Francis Group, an **informa** business

Design cover image: Shutterstock, image numbers 1257276607, 1339565357, 2024617457 and 125422013

First edition published 2024
by CRC Press
2385 NW Executive Center Drive, Suite 320, Boca Raton FL 33431

and by CRC Press
4 Park Square, Milton Park, Abingdon, Oxon, OX14 4RN

CRC Press is an imprint of Taylor & Francis Group, LLC

ISBN: 9781032370972 (hbk)
ISBN: 9781032370989 (pbk)
ISBN: 9781003335313 (ebk)

DOI: 10.1201/9781003335313

Typeset in Times
by Deanta Global Publishing Services, Chennai, India

Contents

Preface

Biomass Hydrolyzing Enzymes: Basics, Advancements, and Applications is part of a book series published by CRC Press in partnership with Taylor & Francis Group and the Biotech Research Society, India (BRSI).

> "Nature calls for the transition to its own catalysts; driving the society towards biocatalysis."

The trend to find solutions to existing problems has changed drastically in recent years. The world has witnessed the change in finding sustainable solutions for utilizing lignocellulosic biomass, the most abundant, ubiquitous, and beneficial resource available via biocatalysts. Petrorefinery is shifting to biorefinery as ethanol from biomass has been found to be an alternative to liquid fossil fuels. However, to completely replace fossil fuels does not seem possible in the near future. Still, biomass has been accepted as an alternative fuel. Biomass hydrolyzing enzymes are perfect natural catalysts produced by microorganisms to disintegrate the compact structure of biomass, which is intricately woven by various components (cellulose, hemicellulose, and lignin) by various kinds of bonds. Microorganisms such as fungi and bacteria are known to produce hydrolytic enzymes to hydrolyze plant biomass as well as seaweed/macroalgal biomass, which is explained in the following chapters.

The book is divided into two parts: Section A includes Chapters 1–7 and presents specific biomass hydrolyzing enzymes, mechanisms of action, and their roles in biomass hydrolysis. Section B covers advancements and challenges in applications; Chapters 8–13 deal with advancements in the improvement of enzymes and challenges in applications with probable solutions.

The first chapter by Gurrani et al. introduces biomass hydrolyzing enzymes, where an account of all the biomass hydrolyzing enzymes is given. This chapter introduces the structure of biomass and the challenges that need to be tackled for hydrolyzing biomass using enzymes at high biomass loading. The synergism between various enzymes as well as their specific components are highlighted. The second chapter by Mathew et al. on fungal cellulases highlights white-rot fungi as a potential cellulase producer and developments in a bioprocess for its improved productivity to gain cost-effective cellulases. Its significance in biomass hydrolysis is highlighted. Chapter 3 by Bhaskar et al. is on bacterial cellulosomes. In this chapter, bacterial sources for cellulosomes and their potential for biomass hydrolysis applications are shown. The challenges in their production are discussed, presenting advantages of bacterial cellulosomes over fungal cellulase due to the short generation time of bacteria. Various parts of cellulosomes and synergy among various components are also discussed in detail. The next chapter by Mathew et al. is on fungal β-glucosidase (BGL), one of the major components of fungal cellulases and the rate-limiting enzyme. It has a significant role in the hydrolysis of cellulose, as the final step of hydrolysis of cellobiose into glucose is done by BGL. Glucose tolerance of BGL is highlighted as the desired property for high-solid loading biomass hydrolysis.

Chapter 5, written by Valladares-Diestra et al., is on xylanases. The chapter highlights xylanase production and its importance for the hydrolysis of biomass for biofuel applications. Chapter 6 is on lignin-degrading enzymes. Written by Kaur et al., it discusses various lignin-degrading and modifying enzymes, including laccase and its significance in biomass hydrolysis. Chapter 7 by Dixit et al. is on lytic polysaccharide monooxygenases, where the challenges in its analysis are highlighted along with its importance in the hydrolysis of biomass. It is strongly recommended that its supplementation in cellulase increases the hydrolysis of biomass twofold.

Chapter 8 is authored by Madhavan et al. and explains where genetic modification and metabolic engineering of biomass hydrolyzing enzymes are presented as a tool for improved enzyme quality and quantity. Chapter 9, authored by Sharma et al., gives an account of protein engineering for biomass hydrolyzing enzymes to enhance the desired properties. The next chapter (Chapter 10) by

Serra et al. explains the effects of inhibitors generated after the pretreatment of corn stover as well as those liberated in enzyme-based liquefaction before pretreatment. The effect of these compounds on the lignocellulolytic enzymes and the challenges of fermentation at high-solid contents are discussed. Chapter 11 by Shi et al. is on enzyme-substrate mixing by periodic forces during high-solids enzymatic hydrolysis of lignocellulose. The last two chapters cover the enzymatic hydrolysis of microalgal biomass (Chapter 12, Colla et al.) and macroalgal biomass (Chapter 13, Rocher et al.), in which seaweed is highlighted as a potential substrate to resolve energy issues in a sustainable way.

Each chapter in the book deals with state-of-the-art information and technological status of the applications and provides perspectives for the new potentials. These have been prepared by experts and highlight the use of biocatalysts for biomass utilization in order to develop a sustainable solution for environmental issues, including biofuel. We hope that graduate students and researchers broadly working in the areas of industrial and environmental biotechnology and energy biosciences with a focus on biorefinery and enzyme technology will find it immensely useful.

We sincerely thank the authors for their contributions and for sharing their knowledge on biomass hydrolyzing enzymes. We are grateful to the reviewers for their sincere efforts and acknowledge them for their contributions in their critical reviews of the chapters in order to improve them. Thanks are due to the Biotech Research Society, India (BRSI) for providing us with the opportunity to prepare this book and guiding us during the editing and publication process. We thank the team of Taylor & Francis Group comprising Dr Gagandeep Singh, Senior Publisher; Ms Madhurima Kahali, Former Editor II (Life Sciences); Ms Neha Bhatt, Editorial Assistant; and the entire team of CRC Press/Taylor & Francis Group for their consistent support during the publication process of this book.

Editors

Reeta Rani Singhania
Anil Kumar Patel
Héctor A. Ruiz
Ashok Pandey

Editors

Dr Reeta Rani Singhania is an Associate Professor at the Institute of Aquatic Science and Technology, National Kaohsiung University of Science and Technology, Kaohsiung, Taiwan. She is also Senior Scientist (Honorary) at the Centre for Energy and Environmental Sustainability – India. Her major research interests are in the areas of Microbial and Enzyme Technology and Energy Biosciences.

Dr Anil Kumar Patel is an Associate Professor at the Institute of Aquatic Science and Technology, National Kaohsiung University of Science and Technology, Kaohsiung, Taiwan. He is also Senior Scientist (Honorary) at the Centre for Energy and Environmental Sustainability–India. His major research interests are in the areas of Food and Bioprocess Technology and Energy Biosciences.

Dr Héctor A. Ruiz is Full Professor and founder of Biorefinery Group (www.biorefinerygroup.com) at the Autonomous University of Coahuila in Mexico and leader of the biomass pretreatment stage in the Bioalcohols Cluster of the Mexican Centre for Innovation in Bioenergy, Mexico. His major research interests are in the areas of Bioprocess Technology and Energy Biosciences.

Professor Ashok Pandey is Distinguished Scientist at the Centre for Innovation and Translational Research, CSIR–Indian Institute of Toxicology Research, Lucknow, India; and Distinguished Professor at the Sustainability Cluster, School of Engineering, University of Petroleum and Energy Studies, Dehradun, India. He is also Executive Director of the Centre for Energy and Environmental Sustainability–India. His major research interests are in the areas of Industrial and Environmental Biotechnology and Energy Biosciences.

List of Contributors

K. B. Arun
Department of Life Sciences, Christ (Deemed to be University), Bengaluru, Karnataka, India

Athulya
Microbial Processes and Technology Division, CSIR–National Institute for Interdisciplinary Science and Technology, Thiruvananthapuram, India

Mukesh Kumar Awasthi
College of Natural Resources and Environment, Northwest A&F University, Yangling, Shaanxi Province, China

Archana Bhaskar
Rajiv Gandhi Government Arts and Commerce College, Lormi, Mungeli, Chattisgarh, India

Parameswaran Binod
Microbial Processes and Technology Division, CSIR–National Institute for Interdisciplinary Science and Technology, Thiruvananthapuram, Kerala, India

Julia Catiane Arenhart Braun
Graduate Program in Civil and Environmental Engineering, Faculty of Engineering and Architecture, University of Passo Fundo, Passo Fundo, Brazil

Hongzhang Chen
Institute of Process Engineering, Chinese Academy of Sciences, Beijing, China; University of Chinese Academy of Sciences, Beijing, China

Meera Christopher
Microbial Processes and Technology Division, CSIR–National Institute for Interdisciplinary Science and Technology, Thiruvananthapuram, India

Luciane Maria Colla
Graduate Program in Civil and Environmental Engineering, Faculty of Engineering and Architecture, University of Passo Fundo, Passo Fundo, Brazil

Rosemary A. Cripwell
Department of Microbiology, Stellenbosch University, Matieland, South Africa

Antonio Jose Goncalves da Cruz
Laboratory of Renewable Resources Engineering (LORRE), Purdue University, West Lafayette, Indiana, USA; Department of Chemical Engineering, Federal University of de São Carlos (UFSCar), São Carlos, Brazil

Rosineide Gomes da Silva Cruz
Laboratory of Renewable Resources Engineering (LORRE), Purdue University, West Lafayette, Indiana, USA; Postgraduate Program in Chemical Engineering, Federal University of de São Carlos (UFSCar), São Carlos, Brazil

João Ricardo Moreira de Almeida
Embrapa–Agroenergia, Brasília, DF, Brazil

Ariane Fátima Murawski de Mello
Department of Bioprocess Engineering and Biotechnology, Federal University of Paraná, Centro Politécnico, Curitiba, Paraná, Brazil

Priscilla Zwiercheczewski de Oliveira
Department of Bioprocess Engineering and Biotechnology, Federal University of Paraná, Centro Politécnico, Curitiba, Paraná, Brazil

Luciana Porto de Souza Vandenberghe
Department of Bioprocess Engineering and Biotechnology, Federal University of Paraná, Centro Politécnico, Curitiba, Paraná, Brazil

Pooja Dixit
RISE Processum, Örnsköldsvik, Sweden

Siddhant Dubey
Vellore Institute of Technology, Vellore, Tamil Nadu, India; Institute of Aquatic Science and Technology, National Kaohsiung University of Science and Technology, Kaohsiung City, Taiwan

Victória Dutra Fagundes
Graduate Program in Civil and Environmental Engineering, Faculty of Engineering and Architecture, University of Passo Fundo, Passo Fundo, Brazil

Cristiane Sanchez Farinas
Embrapa–Instrumentação, São Carlos, Brazil

João Felipe Freitag
Graduate Program in Civil and Environmental Engineering, Faculty of Engineering and Architecture, University of Passo Fundo, Passo Fundo, Brazil

Swapnil Gurrani
Assistant Professor
Chemistry Department
Rajshree Group of Institutions
Bareilly, India

Diana M. R. Gutierrez
Laboratory of Renewable Resources Engineering (LORRE), Purdue University, West Lafayette, Indiana, USA

Himanshu
Department of Microbiology, Central University of Haryana, Mahendragarh, Haryana, India

Jedy Jose
Microbial Processes and Technology Division, CSIR–National Institute for Interdisciplinary Science and Technology, Thiruvananthapuram, India

Amanjot Kaur
Department of Microbiology, Central University of Haryana, Mahendragarh, Haryana, India

Michael R. Ladisch
Laboratory of Renewable Resources Engineering (LORRE), Purdue University, West Lafayette, Indiana, USA

Aravind Madhavan
School of Biotechnology, Amrita Vishwa Vidyapeetham, Kollam, Kerala, India

Abraham Mathew
Post Graduate and Research Department of Botany, St. Peters College Kolenchery, Kerala, India

Meenu Mathew
Post Graduate and Research Department of Botany, St. Peters College Kolenchery, Kerala, India

Reshma M. Mathew
Microbial Processes and Technology Division, CSIR–National Institute for Interdisciplinary Science and Technology, Thiruvananthapuram, India

Bipin G. Nair
School of Biotechnology, Amrita Vishwa Vidyapeetham, Kollam, Kerala, India

Ashok Pandey
Centre for Innovation and Translational Research, CSIR–Indian Institute of Toxicology Research, Lucknow, Uttar Pradesh, India; Kyung Hee University, Seoul, Republic of Korea; School of Engineering, University of Petroleum and Energy Studies, Dehradun, India; Centre for Energy and Environmental Sustainability, Lucknow, India

Anil Kumar Patel
Institute of Aquatic Science and Technology, National Kaohsiung University of Science and Technology, Kaohsiung City, Taiwan; Department of Marine Environmental Engineering, National Kaohsiung University of Science and Technology, Kaohsiung City, Taiwan; Centre for Energy and Environmental Sustainability, Lucknow, Uttar Pradesh, India

Adarsh Velayudhanpillai Prasannakumari
Microbial Processes and Technology Division, CSIR–National Institute for Interdisciplinary Science and Technology, Thiruvananthapuram, India

Athulya Anoop Puthiyamadam
Microbial Processes and Technology Division, CSIR–National Institute for Interdisciplinary Science and Technology, Thiruvananthapuram, India

Amit K. Rai
National Agri-Food Biotechnology Institute, SAS Nagar, India

Sindhu Raveendran
Department of Food Technology, T K M Institute of Technology, Kollam, Kerala, India

Valan Rebinro
Microbial Processes and Technology Division, CSIR–National Institute for Interdisciplinary Science and Technology, Thiruvananthapuram, India

Alan Rempel
Graduate Program in Civil and Environmental Engineering, Faculty of Engineering and Architecture, University of Passo Fundo, Passo Fundo, Brazil

Dominique F. Rocher
Department of Microbiology, Stellenbosch University, Matieland, South Africa

Héctor A. Ruiz
Department, Faculty of Chemistry Sciences, Autonomous University of Coahuila, Saltillo, Coahuila, Mexico

Jitendra Kumar Saini
Department of Microbiology, Central University of Haryana, Mahendragarh, Haryana, India

Meena Sankar
Microbial Processes and Technology Division, CSIR–National Institute for Interdisciplinary Science and Technology, Thiruvananthapuram, India

Luana Assis Serra
Laboratory of Renewable Resources Engineering (LORRE), Purdue University, West Lafayette, Indiana, USA; University of Brasília (UnB), Embrapa–Agroenergia, Brasília, Brazil

Manisha Sharma
National Agri-Food Biotechnology Institute, SAS Nagar, India

Xin Shi
Institute of Process Engineering, Chinese Academy of Sciences, Beijing, China; University of Chinese Academy of Sciences, Beijing, China

Viviane Simon
Graduate Program in Civil and Environmental Engineering, Faculty of Engineering and Architecture, University of Passo Fundo, Passo Fundo, Brazil

Sudhir P. Singh
Center of Innovative and Applied Bioprocessing, SAS Nagar, India

Reeta Rani Singhania
Institute of Aquatic Science and Technology, National Kaohsiung University of Science and Technology, Kaohsiung City, Taiwan; Centre for Energy and Environmental Sustainability, Lucknow, Uttar Pradesh, India

Carlos Ricardo Soccol
Department of Bioprocess Engineering and Biotechnology, Federal University of Paraná, Centro Politécnico, Curitiba, Paraná, Brazil

Athira Sreeja-Raju
Microbial Processes and Technology Division, CSIR–National Institute for Interdisciplinary Science and Technology, Thiruvananthapuram, India

Rajeev K. Sukumaran
Microbial Processes and Technology Division, CSIR–National Institute for Interdisciplinary Science and Technology, Thiruvananthapuram, India

Ashutosh Tiwari
Institute of Aquatic Science and Technology, National Kaohsiung University of Science and Technology, Kaohsiung City, Taiwan

Kim Kley Valladares-Diestra
Department of Bioprocess Engineering and Biotechnology, Federal University of Paraná, Centro Politécnico, Curitiba, Paraná, Brazil

Sabrina Vieira
Department of Bioprocess Engineering and Biotechnology, Federal University of Paraná, Centro Politécnico, Curitiba, Paraná, Brazil

Marinda Viljoen-Bloom
Department of Microbiology, Stellenbosch University, Private Bag X1, Matieland, South Africa

Lan Wang
Institute of Process Engineering, Chinese Academy of Sciences, Beijing, China; University of Chinese Academy of Sciences, Beijing, China

Eduardo Ximenes
Laboratory of Renewable Resources Engineering (LORRE), Purdue University, West Lafayette, Indiana, USA; Department of Environment and Occupational Health, Indiana University, Bloomington, Indiana, USA

Willem H. van Zyl
Department of Microbiology, Stellenbosch University, Private Bag X1, Matieland, South Africa

Section A

General

1 Biomass Hydrolyzing Enzymes

Swapnil Gurrani, Anil Kumar Patel, Héctor A. Ruiz, Ashok Pandey, Pooja Dixit, Ashutosh Tiwari, and Reeta Rani Singhania

1.1 INTRODUCTION

Global energy demand is increasing and is expected to continue to increase with predicted population growth in the coming decades (Duarah et al., 2022). Despite significant advances in renewable energy technology, fossil fuels still control the bulk of the energy market, which are directly linked to greenhouse gas (GHG) emissions and climate change (Tursi, 2019). As a result, the global society has been striving to find alternative routes for energy production. Considering various alternative energy sources, biomass has been an indispensable part of energy that accounts for more than one-third of primary energy (Benti et al., 2021). Hence, biorefinery has been looked upon as an alternative to petrorefinery. Though in the present scenario, it is not possible to completely replace petrorefinery, a sustainable technology is the current demand. Biomass energy, or bioenergy, is produced when biomass is converted into electricity, heat, power, or transportation fuels and a manufacturing unit that facilitates the conversion of plant biomass to energy and chemicals is known as *biorefinery*, in which plant biomass can be separated into several compounds and energy is recovered (Dong et al., 2023). Lignocellulosic biomass is considered the second-generation feedstock for biorefinery to produce biofuels and bio-based products. The most important biomass sources are agricultural and forestry residues (wastes from the wood processing industry such as shavings, sawdust, etc.), animal residues (livestock farms), sewage, algae, and aquatic crops. Municipal solid waste (MSW) and the waste streams originating from anthropogenic activities also fall in the biomass category if they are not reusable in subsequent processing (Kaltschmitt, 2019). Potential sources for producing alternative fuels, known as biofuels, can be found in biological entities such as plants and microbes. These biofuels include bioalcohols, biodiesel, algal biofuels, biohydrogen, and microbial fuel cells (Patel et al., 2014).

Due to the abundance of lignocellulosic polymers on Earth, many microorganisms have evolved strategies to consume this abundant material as an energy source. Degradation of lignocellulosic biomass is not only a survival strategy of these microbes, but this destruction also plays a pivotal role in carbon recycling of Earth through the fixation of photosynthetically fixed carbon present in plant biomass.

Microbes that can utilize lignocellulosic material as their energy source are called cellulolytic organisms and they are mainly confined to bacterial and fungal kingdoms. All of these organisms secrete a complex cellulolytic cocktail of at least 10–30 different enzymes at a time. These enzymes help microbes to access the component sugars of biomass as an energy source. Cellulolytic microbes can be split into two main classes: anaerobic microorganisms such as *Clostridia* species or the *Chytridomycetes*, which produce a bound multienzyme complex, known as a cellulosome; and aerobic microbes (fungi and bacteria) that secrete free individual cellulases and synergizing accessory enzymes.

Filamentous fungi are the predominant degrader of plant biomass and have evolved different mechanisms for attacking different lignocellulosic components in biomass. Industrially, filamentous fungi are dominant producers of industrial cellulases, due to their ability to secrete protein in

DOI: 10.1201/9781003335313-2

larger quantities, the stability of fungal cellulases, the production of cellulases along with synergistic components, etc. (Patel et al., 2020).

Cellulolytic fungi that degrade lignocellulosic biomass contain a complex suite of hydrolytic enzymes, namely, cellulases and hemicellulases. These enzymes cleave the glycosidic bonds present in polymers (cellulose and hemicellulose) through the addition of one water molecule. Due to their mode of action, these enzymes are also known as glycosyl hydrolases.

Apart from glycosyl hydrolases, cellulolytic fungi also possess non-hydrolytic enzymes that are involved in the disruption or loosening of lignocellulose structure. These enzymes do not generate soluble sugars after cellulose disruption, but they do induce structural modifications in recalcitrant cellulose like disruption, swelling, fibrillation, roughening, etc., which increase cellulose accessibility for enzymes, and this process is known as *amorphogenesis* (Arantes & Saddler, 2011). Some of the important non-hydrolytic enzymes are lytic polysaccharide monooxygenases (LPMOs) and swollenin. However, due to their industrial significance and unique degradation reaction mechanisms, LPMOs have gained significant attention worldwide, in industrial as well as academic research (Singhania et al., 2021a).

1.2 BIOMASS STRUCTURE/COMPOSITION

The major constituent of the plant cell wall is *lignocellulose*. As the name suggests, it consists of lignin (15%–20%), hemicellulose (25%–30%) and cellulose (40%–50%), as shown in Figure 1.1. These components together form a three-dimensional complex network with the help of covalent and non-covalent interactions. Cellulose is a polysaccharide of glucose monomers held together by β (1→4) linkages and is the most abundant biopolymer on Earth. These β (1→4) linkages are what make cellulose resistant to hydrolysis. Hemicellulose, which constitutes the second major component of biomass, is an amorphous heteropolymer composed of various carbohydrates, including xylose, mannose and glucose. The amorphous nature of hemicellulose renders it more susceptible to hydrolysis compared to crystalline cellulose. Cellulose and hemicellulose, along with lignin as the third major component, account for over 90% of lignocellulosic biomass and 80% of herbaceous biomass. Lignin, a complex network of aromatic alcohols, is intricately intertwined with the cellulose and hemicellulose fractions within the biomass structure. This interwoven nature of the lignin helps provide rigidity to lignocellulosic materials, such as trees (Williams et al., 2017).

Additional constituents found in biomass are extractives/volatiles and ash. Although they constitute a smaller proportion of the overall biomass composition, they can significantly impact the optimal conversion process. Extractives/volatiles encompass water and ethanol-soluble components. Water-soluble compounds consist of nonstructural sugars and proteins, while ethanol-soluble components primarily consist of chlorophyll and waxes. Ash, on the other hand, represents the inorganic content in biomass, which can either be naturally present or introduced through human activities. Naturally occurring ash includes materials such as calcium and potassium ions, while anthropogenic ash mainly consists of silica (dirt) accumulated during the harvesting process.

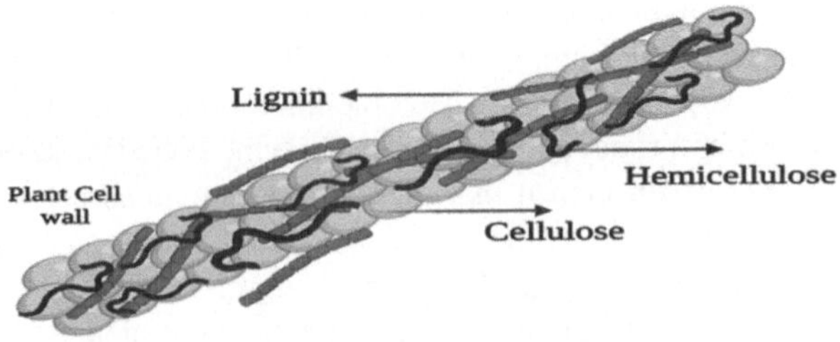

FIGURE 1.1 Structure and chemical composition of plant cell wall.

1.3 PRETREATMENT OF BIOMASS

The complex structure of biomass and lignin content is responsible for the recalcitrance of the biomass. To overcome it, pretreatment of biomass is one of the most important and required steps for biomass deconstruction. Recalcitrance of the biomass inhibits enzymatic action on the biomass unless it is converted into an amorphous structure. Pretreatment can be physical, chemical, physicochemical and biological. The choice of pretreatment is usually based on biomass and the overall process as well as the target end product. Each pretreatment has a specific action; however, the objective is similar, that is, to make the biomass amorphous to allow enzymatic action on biomass. Dilute acid pretreatment, alkali pretreatment, hydrothermal pretreatment, organosolv pretreatment and biological pretreatment are a few major and popular pretreatment methods.

In dilute acid pretreatment, acid at various concentrations is used, coupled with temperatures at 120 degrees for different durations. It usually targets xylan components and due to its liquification, the structure becomes amorphous and is the most used method. Alkali pretreatment employs NaOH or other alkali at various concentrations and it targets lignin components. However, it has a challenge of bringing the pH back to ~5.0 for enzymatic hydrolysis. Another possibility for chemical pretreatment is the use of organic solvents; in this case, an organic solvent is used in the presence of an inorganic acid catalyst to break down the bonds between hemicelluloses and lignin to bring the lignin into solution (Huang et al., 2014). The structure of lignin and the linkages between lignin and hemicellulose and cellulose can be broken by acetyl group removal as well, and the successive enzymatic hydrolysis of cellulose can be enhanced (Wang et al., 2019). Especially, for bioconversions of biomass into biochemicals, biocatalysts in comparison to solid catalysts or chemocatalysts have the excellent capability of catalyzing the reactions under mild conditions due to which not much energy is consumed. Hydrolysis of microcrystalline cellulose with solid catalysts such as Al_2O_3–B_2O_3 solid catalysts takes place at 180°C, which is energy consuming in comparison to hydrolysis in the presence of cellulase that occurs at 50°C, which is far milder than the earlier one (Singhania et al., 2022b). In the pharmaceutical industry, chemocatalysts may not be completely replaced with biocatalysts, as it is difficult sometimes for multistep complex reactions. Steam pretreatment, or hydrothermal pretreatment, has emerged as an eco-friendly and sustainable way of pretreatment where water at temperatures from 140 to 180 degrees acts as a catalyst and breaks the structure of the biomass causing it to become amorphous. It can also be coupled with a small amount of acid or alkali for better results. However, there are environmental concerns. Organosolve pretreatment is a new emerging way for the chemicals to be recovered, hence discharge in the environment can be controlled. Deep eutectic solvents have emerged as green solvents and are gaining popularity. Microwave pretreatment is an effective pretreatment, however, scale is the major issue. It is one of the fastest ways of pretreatment.

At present, there exist various approaches to depolymerize lignin, namely thermochemical, mechanical, chemical and biological methods. Among these, biological treatments are regarded as environmentally friendly catalysts due to their utilization of microorganisms or enzymes, which are typically nontoxic. Moreover, biological treatments exhibit advantages such as low energy consumption, minimal waste generation, and cost-effectiveness in downstream processes. Biological pretreatment is one of the most sustainable ways and can be done via laccase-producing fungi or using laccase itself to degrade lignin, thereby giving access to cellulolytic enzymes in subsequent steps to attack cellulosic fibers. The major limitation here is the time, which varies from days to weeks and months even.

1.4 HYDROLYSIS OF BIOMASS BY BIOCATALYSTS

Unlike solid catalysts, biocatalysts do not pose any environmental risks when they are disposed of. The recovered polysaccharides (cellulose, hemicellulose) can be degraded by mild acid or by using enzymes (Chin et al., 2020). Biomass hydrolysis efficiency increases in the presence of accessory enzymes like LPMOs. In addition to cellulase, xylanase, lacasse and pectinase, LPMOs also play an extremely important role as the biomass contains hemicellulose along with cellulose.

1.4.1 Cellulase

Cellulases, which rank as the second largest industrial enzymes in terms of dollar volume, have witnessed a surge in demand across various industrial sectors since 1995. These sectors include the detergent industry, textile industry, animal feed and food industry, paper industry, and biofuel industry (Singhania et al., 2017). Biomass hydrolyzing enzymes or biocatalysts, primarily produced by microorganisms such as bacteria and fungi, play a crucial role in the process. Filamentous fungi like *Trichoderma*, *Penicillium*, *Humicola* and *Aspergillus* are among the most studied and exploited ones for cellulase production. Cellulases are multienzyme complexes consisting of mainly three different components: endo-1,4-β-D-glucanase (EC 3.2.1.4), exoglucanase/exo-cellobiohydrolase (EC 3.2.1.91) and β-glucosidase (EC 3.2.1.21). All three components act synergistically to hydrolyze the cellulose polymer completely into its glucose monomers. Endo-1,4-β-D-glucanase acts first, randomly, between cellulose fibers to generate reducing and non-reducing ends, which are further attacked by exo-cellobiohydrolase, releasing cellobiose, which is glucose dimers linked by β-1,4 glycosidic bonds. Cellobiose is finally hydrolyzed by β-glucosidase (BGL) into glucose monomers, the final end product (Maki et al., 2009).

1.4.2 Xylanase/Hemicellulase

Hemicellulose is the second most abundant polysaccharide in plants and contains three kinds of backbones: xylose backbone in xylan, mannose in galactoglucomannan and glucose backbone in xyloglucan. These three kinds of structures are the basic framework of hemicellulose and many sugar residues are present in its different branches. These three types of hemicellulose backbone structures are hydrolyzed by a specific class of hemicellulases.

Xylanases are produced by different living organisms such as microorganisms, protozoans and mollusks, and are also found in the rumen of higher animals. They are mainly produced by microorganisms, e.g., bacteria, fungi and actinomycetes at an industrial scale (Motta et al., 2013). Xylanases are pivotal enzymes responsible for breaking down the xylosidic linkages within the xylan structure of biomass. Commercially available enzymes are classified into various glycoside hydrolase families. The enzymatic depolymerization of hemicellulose into monomeric sugars necessitates the cooperative action of multiple enzymes. These enzymes include endoxylanase (endo-1,4-β-xylanase), β-xylosidase (xylan-1,4-β-xylosidase), α-glucuronidase (α-glucosiduronase), α-arabinofuranosidase (arabinofuranosidase) and acetylxylan esterase. Among them, endoxylanases and β-xylosidases are the two key enzymes responsible for the hydrolysis of xylan. Endo-xylanases randomly cleave the xylan backbone into xylo-oligosaccharides, while β-xylosidases further hydrolyze the xylo-oligosaccharides from non-reducing end into xylose monomers. The removal of side groups is catalyzed by α-d-glucuronidases, α-l-arabinofuranosidases, ferulic acid esterases, acetylxylan esterases and p-coumaric acid esterases (Knob & Carmona, 2010).

β-1,4-endoxylanase and β-1,4-xylosidase: Fungal β-1,4-endoxylanase are the enzymes that degrade a linear chain of 1,4-linked D-xylose residues present in xylan, which belong to glycoside hydrolase families 10 and 11 (GH10 and GH11). GH10 endoxylanase has broader substrate specificity than members of the GH11 family. GH10 family enzymes are capable of degrading xylan backbones with a higher degree of substitution and smaller xylo-oligosaccharides; hence these enzymes are capable of completely degrading substituted xylan. The released xylo-oligosaccharides are subsequently degraded by β-xylosidase and they belong to the GH3 family.

Xyloglucan active endoglucanase: This endoglucanase, also known as xyloglucanase, cleaves β-1,4–linked glucose units from glucan backbone and they belong to the GH12 and GH74 families.

After the action of endoglucanases, the released glucose disaccharides are cleaved by β-1,4-glucosidase enzyme.

Mannan active hydrolases: Endomannanase degrades the hemicelluloses that have mannan backbone and they belong to GH families 5 and 26. Generally, GH5 enzymes exhibit substrate specificity toward oligosaccharides with more than three D-mannose residues. Some of these enzymes also contain carbohydrate-binding modules, mainly CBM1 like other fungal carbohydrate-active enzymes. The released mannobiose and mannotriose are further degraded by β-1,4-mannosidases, which fall under GH family 2.

1.4.3 Laccase

Laccase is an enzyme belonging to the multi-copper oxidase family, and it facilitates the oxidation of one electron in a diverse array of phenolic compounds. This enzyme is considered environmentally friendly, as it utilizes molecular oxygen as a co-substrate for catalysis and produces water as the sole by-product. In nature, fungi secrete laccase to access carbohydrates such as cellulose and hemicellulose in wood by breaking down lignin. The potential role of laccases in the production of bioethanol has been investigated through the pretreatment of ligninolytic residues. By employing laccase treatment, lignin degradation occurs, enhancing the fermentability of these residues. Studies have shown that with the assistance of mediators, laccase can degrade approximately 80%–90% of the lignin structure. The conversion of lignin components into biofuels is a significant objective in the field of biorefinery, which involves the transformation of lignocellulosic materials into bioenergy and chemicals with industrial potential (Chio et al., 2019; Ponnusamy et al., 2019).

1.4.4 Pectinase

Pectinases are thus considered to play a crucial role in the efficient hydrolysis of pectin-rich lignocellulosic biomasses. Polygalacturonases and lyases catalyze the degradation of pectin through depolymerization, while pectin methylesterases catalyze the de-esterification reaction (Li et al., 2021). There are three types of pectinases: pectin methyl esterase, which removes the methoxyl residues from pectin; polygalacturonases, which split the glycosidic bonds between galacturonic acid molecules; and pectin transeliminases, which cleave α (1→4) linkages, thus forming unsaturated derivatives of galacturonic acid (Singh et al., 2019). Polygalacturonase is the most extensively researched and widely utilized commercial pectinase. Its significance lies in the fact that pectin, a gelatinous substance, acts as the binding agent between plant cells and serves as a matrix for other cell wall constituents like cellulose fibrils. Table 1.1 provides an account of biomass hydrolyzing enzymes and their hydrolytic efficiency based on specific polymers in the biomass.

1.5 SYNERGISM BETWEEN ENZYMES AND AMONG ENZYME COMPONENTS

It is well known that due to the complexity of biomass, a plethora of enzymes is required to act synergistically to hydrolyze the biomass. Laccase, cellulase, xylanase and pectinase all play significant roles, and based on biomass components their ratio can be changed in the cocktail so as to have better hydrolysis. Xylanase and cellulase components show synergism, which is also presented in Section 1.4.2. Each of these enzymes are complex enzymes having several components such as laccase along with lignin peroxidase (Lip) and manganese peroxidase (MnP), which act together to modify or deconstruct lignin (Singhania et al., 2022a). Similarly, xylanase consists of β-xylosidase and endoxylanase majorly. Even for hydrolyzing cellulose, cellulase is required, which is not a single protein but is composed of endoglucanase, exoglucanase/cellobiohydrolase (CBH) and β-glucosidase. LPMOs have also been included as non-hydrolytic enzymes, however, they are capable of increasing the rate of hydrolysis, doubling to some extent (Singhania et al., 2021a).

TABLE 1.1
Enzymes Involved in Biomass Hydrolysis and Their Hydrolytic Efficiency on Various Biomasses

Biomass-Degrading Enzyme	Biomass Name	Hydrolysis Efficiency (Approximate)	Reference
Cellulase	Cornstalk	82%	(Peng et al., 2021)
Cellulase	Reed	91%	
Cellulase	Bamboo	97%	
Laccase	Wheat straw	70%	(Deng et al., 2019)
Xylanase	Birchwood xylan	65%	(Ostadjoo et al., 2019a)
Xylanase	Oat spelt xylan	68%	
Xylanase	Sugarcane bagasse	73%	
Xylanase	Wheat straw	84%	
Xylanase	Barley straw	75%	(García-Aparicio et al., 2007)
Pectinase	Cassava	98%	(Collares et al., 2012)
Cellulase	Cotton seed hairs	80%–95%	(Kumar et al., 2009)
Cellulase	Wastepaper	85%–99%	
Xylanase	Sugarcane bagasse	70%	(Ostadjoo et al., 2019b)
Laccase	Coffee silverskin	73%	(Giacobbe et al., 2018)
Laccase	Apple pomace	83%	
Cellulase	Wood	95%	(Álvarez et al., 2016)

Enzymatic degradation of cellulose to glucose is mainly done by the concerted action of synergism between three kinds of enzymes, namely, endocellulases, exocellulases and β-glucosidases. *Endoglucanases/endo-β-(1,4)-glucanases (EC 3.2.1.4)* degrade the glycosidic bonds in the glucan chain of cellulose at random positions, generally on the amorphous region of cellulosic polymer. The active site of these enzymes is generally located in an open cleft allowing the random binding of cellulases to the cellulose chain. The substrates of these enzymes are soluble and insoluble β-(1, 4)-glucan substrates and cleave internal β-(1, 4)-glycosidic bonds between glucose moieties of cellulosic chains. *Exo-β-(1,4)-D-glucanases/cellobiohydrolases (EC 3.2.1.74)* attack the cellulose chain from the end, hence these are called *exocellulase.* The active site of these enzymes is located within a pocket. These enzymes liberate D-cellobiose in a "processive" manner (successive cleavage of product) from the cellulosic chain. *β-D-glucosidases (EC 3.2.1.21)* release D-glucose units from cellobiose and soluble cellodextrins. After exo- and endocellulases, this class of enzymes comes into action and converts all disaccharides/cellobiose into glucose moieties. Hence, the synergistic action of cellulase is required to hydrolyze the complex polymer and is represented in Figure 1.2.

Biomass hydrolytic enzyme cocktail preparation is an outcome of this synergism only. Various enzymes at various ratios as well as various components of enzymes such as endoglucanase, exoglucanase and β-glucosidase of cellulase may also be incorporated in various ratios depending upon the biomass characteristics. Most of the enzymes available in the market for biomass hydrolyzing purposes are cocktails produced by mixing native enzymes from different sources based on synergism at different ratios or by heterologous expression of various cellulolytic components in a single host microbe (Saini et al., 2016).

Usually, few challenges arise when these enzymes are employed for biomass hydrolysis. For biomass hydrolysis, most of the research was oriented toward increasing the hydrolysis efficiency and solid loading was not so important. However, the research is now oriented toward decreasing the requirement of infrastructure as well as the number of steps. It is necessary to obtain hydrolysate

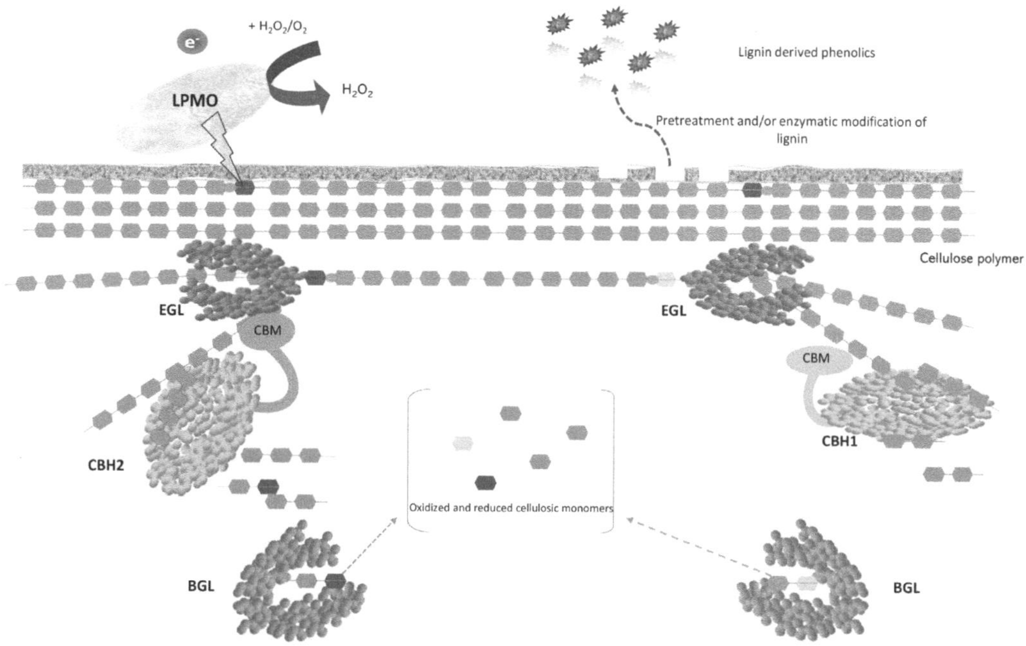

FIGURE 1.2 Synergism among cellulase components to hydrolyze cellulose microfibril.

with a higher concentration of fermentable sugars to obtain a decent concentration of product. The following section will give insight into the challenges of high biomass-loading enzymatic hydrolysis and probable solutions have also been presented.

1.6 CHALLENGES OF BIOMASS HYDROLYSIS

Enzymatic hydrolysis of biomass is one of the most strategic steps in the overall process to obtain fermentable sugars for its conversion to biofuels and biochemicals, with a more recent focus on operating at high-solids loadings. To attain a high concentration of sugars in broth, it is necessary to do enzymatic hydrolysis at high-solids loading at least higher than 20%, as then only 10% of fermentable sugars can be achieved to be converted into ethanol having about 4%–5% concentration so as to have economically feasible distillation (Singhania et al., 2021b). There have been suggestions that to enhance the economic viability of the lignocellulosic conversion process, enzymatic hydrolysis should be carried out using high-solids loadings, although it is associated with challenges. A process is considered "high solids" if the ratio of solids/liquid is such that very little to no free water is present in the slurry (Hodge et al., 2009) or roughly a solid loading of ≥15% (w/w). Increasing the solids loading in a system brings forth challenges that were previously insignificant in low-solids systems, and this observation also applies to high-solids pretreatment (Modenbach & Nokes, 2012a). Among the major challenges encountered during enzymatic hydrolysis at high-solids loading is the limited availability of water within the reactor. Water plays a crucial role in facilitating effective hydrolysis due to its contribution to mass transfer and lubricity. Water enhances the efficiency of enzymatic and chemical reactions by providing a medium for solubilizing and aiding in the mass transfer of products. Additionally, it reduces the viscosity of the slurry by increasing the lubricity of particles, thereby reducing the required shear stress for a given shear rate. Consequently, this allows for lower power input during mixing (Kristensen et al., 2009). Energy demands increase as mixing becomes more difficult, which may counteract the benefits of using high-solids loadings. Reactors suitable for effective hydrolysis of these complex networks of lignocellulosic materials are imperative, and designs implemented to overcome these limitations can be found in later chapters of the book. Material handling also becomes an issue because viscous materials are difficult to pump or pour, which may limit the pretreatments' applicability in a conventional continuous system (Modenbach & Nokes, 2012b). Other challenges specific to high-solids enzymatic hydrolysis include long hydrolysis times. Enzymatic hydrolysis is typically thought to be the bottleneck of the entire conversion process in terms of both time and money, since the reaction time needed for most enzymes to convert lignocellulose into sufficient glucose concentrations for fermentation is on the order of days (usually ≥3 days). Recent studies have suggested that enzymes can overcrowd accessible cellulose sites, thus not reaching the full hydrolytic potential for the given enzyme loading. At high-solids loading, the nonproductive binding of enzymes to lignin also increases and even inhibitors generated during pretreatments are in concentrated form, which may affect the further enzymatic hydrolysis as well as fermentation and thereby decrease the efficiency of the overall process (Singhania et al., 2022b). Also, the products are in concentrated form, which may even cause feedback inhibition of enzymes.

There are solutions to these issues such as fed-batch addition of biomass or enzymes, or both. The fed-batch addition of biomass was found to be better, which allows the biomass to easily liquefy. The addition of further biomass in liquefied hydrolysate again liquefies the biomass to obtain higher concentrations of fermentable sugars. This strategy allows the maintenance of the viscosity to an acceptable level. Enzymes may also be added in batches; however, a serious inspection is required for a particular biomass, which also depends on the type of pretreatment employed for it before hydrolysis. Also, the addition of cheaper protein may increase the rate of reaction and efficiency by inhibiting the enzyme sitting on lignin. The cheaper protein may bind on the surface of lignin to the saturation level so that cellulase or useful biomass hydrolyzing enzyme may act on cellulose without losing to nonproductive binding to lignin. Similarly, surfactants may be employed to prevent cellulase molecules from binding to the lignin.

Major challenges during high-solids loading enzymatic hydrolysis are depicted in Figure 1.3.

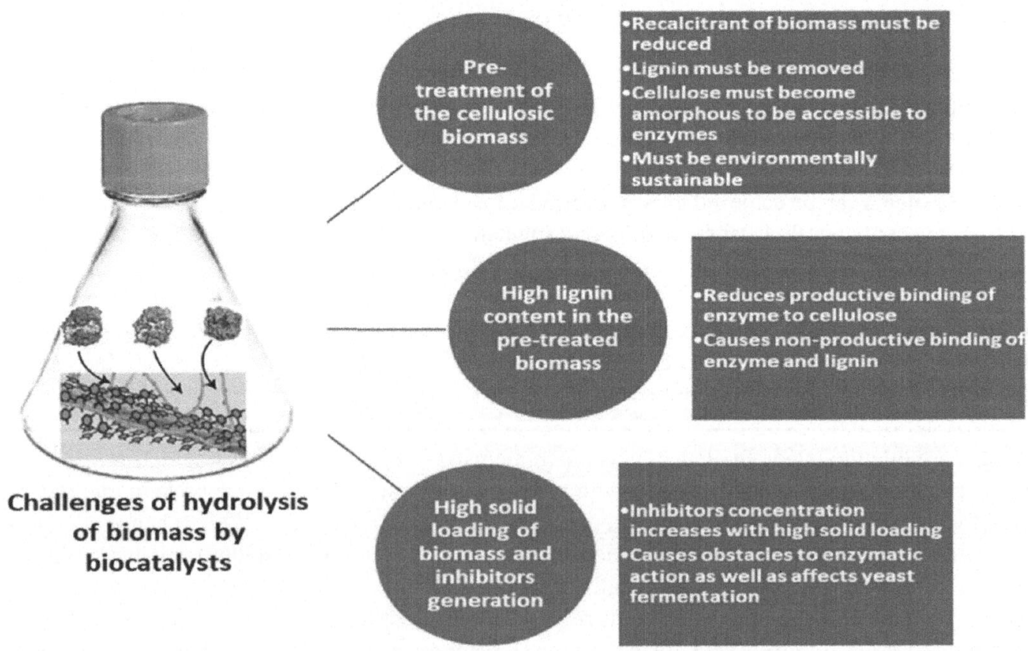

FIGURE 1.3 Challenges during biomass hydrolysis at high-solids loading. (Reproduced from Singhania et al., 2022b, under Creative Commons CC BY license.)

1.6 CONCLUSIONS AND PERSPECTIVES

Biomass hydrolyzing enzymes are excellent biocatalysts capable of degrading biomass into its monomeric compound to be able to convert it into a cascade of value-added products including biofuel, feed and platform chemicals. Robust enzymes are required for efficient hydrolysis at high-solids loading. Reactor designing coupled with chemical engineering and advances in machine learning can further lead to advancements in the field. Robust enzymes are the key to the success of biorefinery, which is capable of replacing petrorefinery to an extent for environmental sustainability.

Currently, the lignocellulose pretreatment and bioconversion strategy have achieved much advancement. The excessive cost and low efficiency of the lignocellulosic biorefinery was a limit to its industrial application, however, several industries and commercial plants are established all over the world. Biomass hydrolyzing enzymes are environmentally benevolent and sustainable, and are preferred for hydrolyzing biomass into fermentable sugars. Pretreatment technologies have advanced in recent times where green solvents, recyclability and environmental aspects have been given emphasis. In this aspect, hydrothermal pretreatment seems to be the most suitable method, as at higher temperatures the water molecule turns into hydronium ions and acts as a catalyst. Biomass hydrolyzing enzymes can finally hydrolyze the pretreated biomass, however, synergy is the most important criterion. Most of the filamentous fungi when grown on particular biomass produce enzymes required to degrade the biomass into simpler molecules as glucose and derive energy. However, most of the fungi produce enzymes in inducible conditions and stop in the presence of an easily available carbon source. Biomass components, such as cellulose, xylan and lignin, require specific enzymes like cellulase to hydrolyze cellulose. Cellulase itself is a complex enzyme, as it comprises several components including endoglucanase, exoglucanase, β-glucosidase and LPMOs, which act in synergy to convert cellulose into glucose and are similar to other polymeric components of biomass. Feedback inhibition and several other challenges have been reported, which need to be addressed at different levels. High-solids loading biomass hydrolysis requires specific reactors and strategies like the fed-batch addition of biomass to speed up the reaction time and liquefy faster to produce liquid phase–facilitating enzymes and biomass interaction. Specific bioreactor designing is required and there are advancements in this direction as well. Less free liquid medium causes nonproductive binding of enzymes on lignin, which must be avoided by using cheaper proteins as well as surfactants, or by modifying lignin or designer molecules.

Further advances in the evolution of biomass hydrolyzing enzymes, cocktail formation, and better hydrolysis strategies aided by advances in bioinformatics, reactor designing and machine learning will enable the development of a more cost-effective enzymatic hydrolysis process for biomass. In short, we believe that current and future developments in the field of high-solids loading of biomass with an excellent biomass hydrolyzing enzyme cocktail will define the future of biorefinery.

REFERENCES

Aliyah, A., Alamsyah, G., Ramadhani, R., & Hermansyah, H. (2017). Production of α-amylase and β-glucosidase from *Aspergillus niger* by solid state fermentation method on biomass waste substrates from rice husk, bagasse and corn cob. *Energy Procedia*, *136*, 418–423. https://doi.org/10.1016/j.egypro.2017.10.269

Álvarez, C., Reyes-Sosa, F. M., & Díez, B. (2016). Enzymatic hydrolysis of biomass from wood. *Microbial Biotechnology*, *9*(2), 149–156. https://doi.org/10.1111/1751-7915.12346

Arantes, V., & Saddler, J. N. (2011). Cellulose accessibility limits the effectiveness of minimum cellulase loading on the efficient hydrolysis of pretreated lignocellulosic substrates. *Biotechnol Biofuels*, *4*, 3. https://doi.org/10.1186/1754-6834-4-3

Benti, N. E., Gurmesa, G. S., Argaw, T., Aneseyee, A. B., Gunta, S., Kassahun, G. B., Aga, G. S., & Asfaw, A. A. (2021). The current status, challenges and prospects of using biomass energy in Ethiopia. *Biotechnology for Biofuels*, *14*(1), 209. https://doi.org/10.1186/s13068-021-02060-3

Bhardwaj, N., Kumar, B., & Verma, P. (2019). A detailed overview of xylanases: An emerging biomolecule for current and future prospective. *Bioresources and Bioprocessing*, *6*(1), 40. https://doi.org/10.1186/s40643-019-0276-2

Chin, S. C., Tee, K. F., Tong, F. S., Ong, H. R., & Gimbun, J. (2020). Thermal and mechanical properties of bamboo fiber reinforced composites. *Materials Today Communications*, *23*, 100876. https://doi.org/10.1016/j.mtcomm.2019.100876

Chio, C., Sain, M., & Qin, W. (2019). Lignin utilization: A review of lignin depolymerization from various aspects. *Renewable and Sustainable Energy Reviews*, *107*, 232–249.

Collares, R. M., Miklasevicius, L. V. S., Bassaco, M. M., Salau, N. P. G., Mazutti, M. A., Bisognin, D. A., & Terra, L. M. (2012). Optimization of enzymatic hydrolysis of cassava to obtain fermentable sugars. *Journal of Zhejiang University. Science. Part B*, *13*(7), 579–586. https://doi.org/10.1631/jzus.B1100297

Deng, Z., Xia, A., Liao, Q., Zhu, X., Huang, Y., & Fu, Q. (2019). Laccase pretreatment of wheat straw: Effects of the physicochemical characteristics and the kinetics of enzymatic hydrolysis. *Biotechnology for Biofuels*, *12*(1), 159. https://doi.org/10.1186/s13068-019-1499-3

Dong, C. D., Patel, A. K., Madhavan, A., Chen, C. W., & Singhania, R. R. (2023). Significance of glycans in cellulolytic enzymes for lignocellulosic biorefinery-A review. *Bioresource Technology*, 379, 128992.

Duarah, P., Haldar, D., Patel, A. K., Dong, C. D., Singhania, R. R., & Purkait, M. K. (2022). A review on global perspectives of sustainable development in bioenergy generation. *Bioresource Technology*, *348*, 126791–126791.

Escamilla-Alvarado, C., Pérez-Pimienta, J. A., Ponce-Noyola, T., & Poggi-Varaldo, H. M. (2017). An overview of the enzyme potential in bioenergy-producing biorefineries. *Journal of Chemical Technology and Biotechnology*, *92*(5), 906–924. https://doi.org/10.1002/jctb.5088

García-Aparicio, M. P., Ballesteros, M., Manzanares, P., Ballesteros, I., González, A., & José Negro, M. (2007). Xylanase contribution to the efficiency of cellulose enzymatic hydrolysis of barley straw. *Applied Biochemistry and Biotechnology*, *137*(1–12), 353–365. https://doi.org/10.1007/s12010-007-9064-0

Giacobbe, S., Pezzella, C., Lettera, V., Sannia, G., & Piscitelli, A. (2018). Laccase pretreatment for agrofood wastes valorization. *Bioresource Technology*, *265*, 59–65. https://doi.org/10.1016/j.biortech.2018.05.108

Hodge, D. B., Karim, M. N., Schell, D. J., & McMillan, J. D. (2009). Model-based fed-batch for high-solids enzymatic cellulose hydrolysis. *Applied Biochemistry and Biotechnology*, *152*(1), 88–107. https://doi.org/10.1007/s12010-008-8217-0

Hu, J., Arantes, V., & Saddler, J. N. (2011). The enhancement of enzymatic hydrolysis of lignocellulosic substrates by the addition of accessory enzymes such as xylanase: Is it an additive or synergistic effect? *Biotechnology for Biofuels*, *4*(1), 36. https://doi.org/10.1186/1754-6834-4-36

Huang, K., Xia, S., Zhang, X. M., Chen, Y. Le, Wu, Y. T., & Hu, X. B. (2014). Comparative study of the solubilities of SO2 in five low volatile organic solvents (sulfolane, ethylene glycol, propylene carbonate, N-methylimidazole, and N-methylpyrrolidone). *Journal of Chemical & Engineering Data*, *59*, 1202–1212. https://doi.org/10.1021/je4007713

Kaltschmitt, M. (2019). Renewable energy from biomass: Introduction. In *Energy from organic materials (biomass)* (pp. 1–14). Springer. https://doi.org/10.1007/978-1-4939-7813-7_924

Knob, A., & Carmona, E. C. (2010). Purification and characterization of two extracellular xylanases from *Penicillium sclerotiorum*: A novel acidophilic xylanase. *Applied Biochemistry and Biotechnology*, *162*(2), 429–443. https://doi.org/10.1007/s12010-009-8731-8

Kristensen, J. B., Felby, C., & Jørgensen, H. (2009). Yield-determining factors in high-solids enzymatic hydrolysis of lignocellulose. *Biotechnology for Biofuels*, *2*(1), 11. https://doi.org/10.1186/1754-6834-2-11

Kumar, P., Barrett, D. M., Delwiche, M. J., & Stroeve, P. (2009). Methods for pretreatment of lignocellulosic biomass for efficient hydrolysis and biofuel production. *Industrial and Engineering Chemistry Research*, *48*(8), 3713–3729. https://doi.org/10.1021/ie801542g

Li, F., Foucat, L., & Bonnin, E. (2021). Effect of solid loading on the behaviour of pectin-degrading enzymes. *Biotechnology for Biofuels*, *14*(1), 107. https://doi.org/10.1186/s13068-021-01957-3

Liu, L., You, Y., Deng, H., Guo, Y., & Meng, Y. (2019). Promoting hydrolysis of apple pomace by pectinase and cellulase to produce microbial oils using engineered *Yarrowia lipolytica*. *Biomass and Bioenergy*, *126*, 62–69. https://doi.org/10.1016/j.biombioe.2019.04.025

Maki, M., Leung, K. T., & Qin, W. (2009). The prospects of cellulase-producing bacteria for the bioconversion of lignocellulosic biomass. *International Journal of Biological Sciences*, 500–516. https://doi.org/10.7150/ijbs.5.500

Modenbach, A. A., & Nokes, S. E. (2012a). The use of high-solids loadings in biomass pretreatment-a review. *Biotechnology and Bioengineering*, *109*(6), 1430–1442. https://doi.org/10.1002/bit.24464

Modenbach, A. A., & Nokes, S. E. (2012b). The use of high-solids loadings in biomass pretreatment-a review. *Biotechnology and Bioengineering*, *109*(6), 1430–1442. https://doi.org/10.1002/bit.24464

Motta, F. L., Andrade, C. C. P., & Santana, M. H. A. (2013). A review of xylanase production by the fermentation of xylan: Classification, characterization and applications. *Sustainable Degradation of Lignocellulosic Biomass-Techniques, Applications and Commercialization*, *1*, 251–276.

Ostadjoo, S., Hammerer, F., Dietrich, K., Dumont, M. J., Friščić, T., & Auclair, K. (2019a). Efficient enzymatic hydrolysis of biomass hemicellulose in the absence of bulk water. *Molecules*, *24*(23), 4206. https://doi.org/10.3390/molecules24234206

Ostadjoo, S., Hammerer, F., Dietrich, K., Dumont, M. J., Friščić, T., & Auclair, K. (2019b). Efficient enzymatic hydrolysis of biomass hemicellulose in the absence of bulk water. *Molecules*, *24*(23), 4206. https://doi.org/10.3390/molecules24234206

Patel, A. K., Dixit, P., Pandey, A., & Singhania, R. R. (2020). Promising enzymes for biomass processing. In S. P. Singh, A. Pandey, R. R. Singhania, C. Larroche, & Z. Li (Eds.), *Biomass, biofuels, biochemicals* (pp. 245–271). Elsevier. ISBN 9780128198209. https://doi.org/10.1016/B978-0-12-819820-9.00013-2

Patel, A. K., Singhania, R. R., & Pandey, A. (2014). Biofuels from biomass. In *Novel combustion concepts for sustainable energy development* (pp. 25–44). Springer. https://doi.org/10.1007/978-81-322-2211-8_3

Peng, Z.-Q., Li, C., Lin, Y., Wu, S.-S., Gan, L.-H., Liu, J., Yang, S.-L., Zeng, X.-H., & Lin, L. (2021). Cellulase production and efficient saccharification of biomass by a new mutant *Trichoderma afroharzianum* MEA-12. *Biotechnology for Biofuels*, *14*(1), 219. https://doi.org/10.1186/s13068-021-02072-z

Ponnusamy, V. K., Nguyen, D. D., Dharmaraja, J., Shobana, S., Banu, J. R., Saratale, R. G., Chang, S. W., & Kumar, G. (2019). A review on lignin structure, pretreatments, fermentation reactions and biorefinery potential. *Bioresource Technology*, *271*, 462–472. https://doi.org/10.1016/j.biortech.2018.09.070

Ribeiro, B. D., Castro, A. M. de, Coelho, M. A. Z., & Freire, D. M. G. (2011). Production and use of lipases in bioenergy: A review from the feedstocks to biodiesel production. *Enzyme Research*, *2011*, 1–16. https://doi.org/10.4061/2011/615803

Saini, J. K., Singhania, R. R., Alok, S., Reetu, S., Ravi, G., Deepak, T., Anshu, M., & Mukund, A. (2016). Improvement of wheat straw hydrolysis by cellulolytic blends of two Penicillium spp. *Renewable Energy*, *98*, 43–50.

Singh, R. S., Singh, T., & Pandey, A. (2019). Microbial enzymes—An overview. In *Advances in enzyme technology* (pp. 1–40). Elsevier. https://doi.org/10.1016/B978-0-444-64114-4.00001-7

Singhania, R. R., Adsul, M., Pandey, A., & Patel, A. K. (2017). Cellulases. In *Current developments in biotechnology and bioengineering* (pp. 73–101). Elsevier. https://doi.org/10.1016/B978-0-444-63662-1.00004-X

Singhania, R. R., Dixit, P., Patel, A. K., Giri, B. S., Kuo, C. H., Chen, C. W., & Di Dong, C. (2021a). Role and significance of lytic polysaccharide monooxygenases (LPMOs) in lignocellulose deconstruction. *Bioresource Technology*, *335*, 125261.

Singhania, R. R., Ruiz, H. A., Awasthi, M. K., Dong, C. D., Chen, C. W., & Patel, A. K. (2021b). Challenges in cellulase bioprocess for biofuel applications. *Renewable and Sustainable Energy Reviews*, *151*(C), 111622.

Singhania, R. R., Patel, A. K., Raj, T., Chen, C. W., Ponnusamy, V. K., Tahir, N., Kim, S.-H., & Dong, C. D. (2022a). Lignin valorisation via enzymes: A sustainable approach. *Fuel*, *311*, 122608.

Singhania, R. R., Patel, A. K., Raj, T., Tsai, M.-L., Chen, C.-W., & Dong, C.-D. (2022b). Advances and challenges in biocatalysts application for high solid-loading of biomass for 2nd generation bio-ethanol production. *Catalysts*, *12*(6), 615. https://doi.org/10.3390/catal12060615

Tursi, A. (2019). A review on biomass: Importance, chemistry, classification, and conversion. *Biofuel Research Journal*, *6*(2), 962–979. https://doi.org/10.18331/BRJ2019.6.2.3

Wang, C., Yang, J., Wen, J., Bian, J., Li, M., Peng, F., & Sun, R. (2019). Structure and distribution changes of Eucalyptus hemicelluloses during hydrothermal and alkaline pretreatments. *International Journal of Biological Macromolecules*, *133*, 514–521. https://doi.org/10.1016/j.ijbiomac.2019.04.127

Weiss, N., Börjesson, J., Pedersen, L. S., & Meyer, A. S. (2013). Enzymatic lignocellulose hydrolysis: Improved cellulase productivity by insoluble solids recycling. *Biotechnology for Biofuels*, *6*(1), 5. https://doi.org/10.1186/1754-6834-6-5

Williams, C. L., Emerson, R. M., & Tumuluru, J. S. (2017). Biomass compositional analysis for conversion to renewable fuels and chemicals. In *Biomass Volume Estimation and Valorization for Energy* (pp. 251–270). DOI: 10.5772/65777.

Yokoyama, S., & Matsumura, Y. (2008). The Asian biomass handbook: A guide for biomass production and utilization. *The Japan Institute of Energy*, *1*, 61–62.

2 Fungal Cellulases: Production and Applications

Abraham Mathew, Meenu Mathew, and Reeta Rani Singhania

2.1 INTRODUCTION

Fungi are considered the vultures of the plant kingdom. The mycota kingdom includes unicellular to multicellular eukaryotic forms with a heterotrophic absorptive mode of nutrition. The number of fungal species identified remains over 150,000 and through high-throughput sequencing, the number is estimated to be 13.2 million species. Many are saprophytic forms that decompose dead remains thereby sustaining biogeochemical cycling of nutrients. Some are parasitic causing diseases to plants and animals, and yet others are mutualistic.

Fungi are pioneers in decomposing recalcitrant residues and thus play a pivotal role in sustaining life. The plant cell wall is composed of polysaccharides like cellulose, hemicellulose, pectin and lignin. Most fungi can decompose lignocellulosic biomass and they do so by using enzymatic or chemical machinery. The chemical method involves the generation of free radicals that randomly attack cellulose chains resulting in its cleavage. The enzymatic method is more pronounced whereby a cocktail of enzymes called cellulase results in the complete degradation of cellulose into glucose molecules. The enzymatic machinery in each fungus is so diverse to allow successful utilization of microhabitats thus reducing competition among species. Thus isozymes and allozymes are common, each varying in their molecular mass, amino acid composition, kinetic parameters, stability, optimum pH and temperature for activity, resistance to inhibition, resistance to protease attack, etc. Research in bioethanol production focuses on this wide variability and the effective cocktail development for faster bioconversion of lignocellulose into glucose.

2.2 CELLULOSE

Cellulose is a polymer of glucose residues joined by β-1,4 linkages. These straight-chain molecules get packed into crystalline forms and include polymorphs I, II and III. Form I is the native form found in the plant cell wall and is again a mixture of I α and I β forms. Interchain hydrogen bonds are absent and hence van der Waals force seems to play a great role in stability. Treatment with alkali or with ionic liquids like 1-butyl,3-methylimidazolium chloride can convert polymorph I into polymorph II. Mild treatments with agents like ammonia can convert type I and type II to type III. Type II and type III cellulose have interchain hydrogen bonding and staggered chains resulting in better digestibility by cellulase enzymes. Optimizing pretreatment methods not only exposes the cellulose chain by removing lignin and hemicellulose but also modifies the structure of cellulose making it more amenable to an enzyme attack. Cellulose contains amorphous and crystalline regions, and amorphous regions are thought to be generated by strain in the microfibrils. Thus methods that induce strain resulting in a more amorphous state can enhance cellulose digestibility. In plants, cellulose molecules are arranged as microfibrils, each with 36 chains. Recent studies indicate 24-chain and 18-chain models of cellulose arrangement in microfibrils. Microfibrils do not remain straight but show twisting when solvated in water. Figure 2.1 shows the cellulose microfibril structure of cellulose present in the plant cell wall.

DOI: 10.1201/9781003335313-3

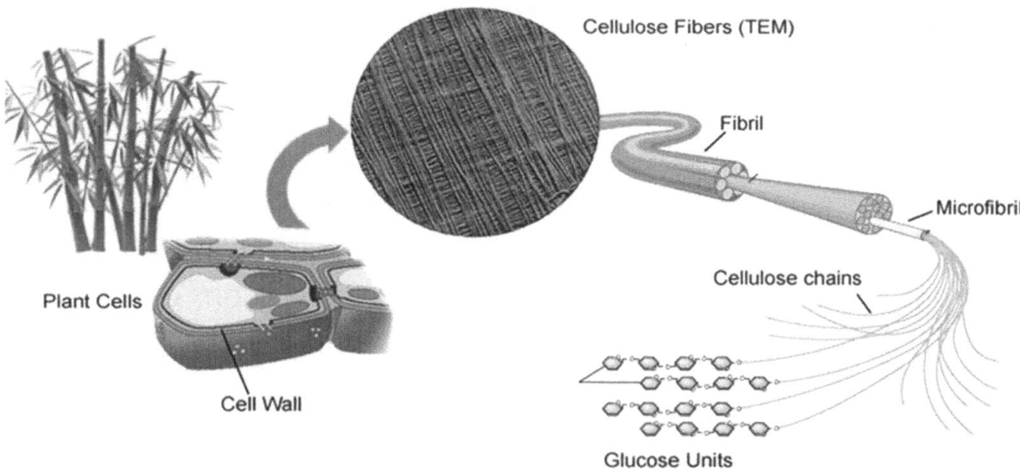

FIGURE 2.1 Organization of cellulose in plants.

2.3 CELLULASES

Cellulases are hydrolytic enzymes that cleave β-glycosidic bonds in the presence of water. Cellulases include three major types of enzymes: endoglucanases that cleave cellulose chains in the amorphous regions, exoglucanases that act on chain ends to produce cellobiose and β-glucosidases that act on cellobiose-producing glucose molecules. The synergistic action of the three results in the conversion of cellulose into glucose molecules. Figure 2.2 shows the synergistic behaviour of cellulase components.

Endoglucanase cleaves the internal β-glucosidic bonds resulting in shorter chains. Thus they depolymerize the cellulose chains. They can also act on cellodextrins. Fungal endoglucanases are found in eight glycosyl hydrolase (GH) families: 5, 6, 7, 9, 12, 45, 48 and 74. Exoglucanase or cellobiohydrolases hydrolyze cellulose chains from the reducing end (GH 7 and GH 48) or non-reducing end (GH 5, GH 6 and GH 48) resulting in the release of cellobiose. Exoglucanases are of

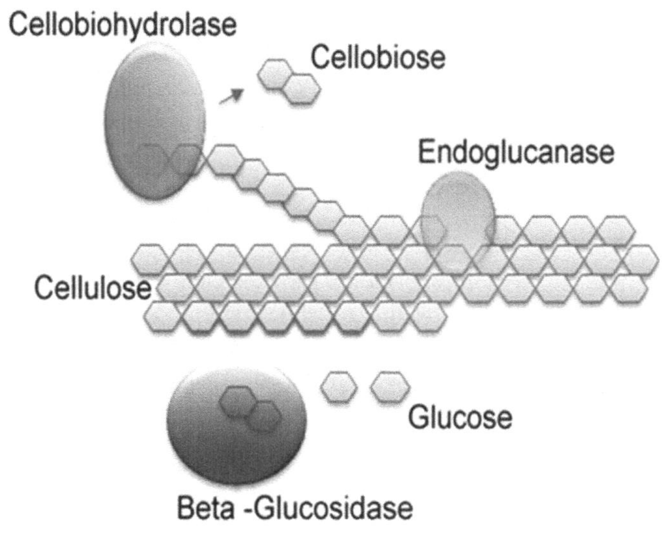

FIGURE 2.2 Mechanism of cellulase action.

prime importance to cellulolysis as they facilitate the production of cellobiose, the substrate upon which β-glucosidase acts and produces glucose. Their active sites are tunnel-like and hence accept only chain ends as substrates. Such a processive mechanism of catalysis is considered an effective system to overcome the high physical and chemical stability of cellulose (Sorensen *et al.*, 2015). β-glucosidases cleave β-glycosidic linkages in disaccharides and oligosaccharides. With an increase in chain length, their activity tends to decrease. Based on substrate specificity, they can be classified as cellobiase, aryl β-glucosidase and broad specificity β-glucosidase. Cellobiases hydrolyze cello oligosaccharides, especially cellobiose. Aryl β-glucosidase shows a preference for aryl glucoside and the broad specificity β-glucosidase shows activity to both cello oligosaccharides and aryl glycosidases. β-glucosidases are grouped into five glycosyl hydrolase families: GH1, GH2, GH3, GH5 and GH30.

Among the cellulase enzymes, β-glucosidase is of the most research importance because of its unusual features. In cellulose hydrolysis, often β-glucosidase is the rate-limiting cause. Even in hypercellulase-producing mutants of *Trichoderma reesei*, β-glucosidase production and secretion are comparatively low. Most of the β-glucosidases are subjected to product inhibition. When the concentration of glucose increases, the enzymes get inhibited. The level of inhibition varies with the taxa. The decreased β-glucosidase activity results in the accumulation of cellobiose, which in turn inhibits endoglucanase and exoglucanase. Also several reports point towards substrate-level inhibition of β-glucosidase. This type of inhibition is thought to arise due to the trans-glycosylation property of β-glucosidase.

Glucose-tolerant β-glucosidases are in great demand, especially in the saccharification process. β-glucosidases show wide variation in the nature of inhibition to the glucose. Some enzymes get inhibited even under low concentrations of glucose and the rate progressively increases with glucose concentration. Certain enzymes remain uninhibited until the glucose concentration reaches a threshold and thereafter gets inhibited. This threshold level varies with enzymes and hence β-glucosidase from different sources and isoforms even from a single strain shows different inhibition rates to glucose. The exact mechanism of inhibition is not yet known. Glucose at higher concentrations may prevent the entry of substrate or may prevent the release of hydrolyzed substrate. β-glucosidase from *Aspergillus oryzae* was found to have a high glucose tolerance with a *Ki* value of 1.36 M (Riou *et al.*, 1998); a different strain of *A. oryzae* showed a *Ki* value of 953 mM (Gunata & Vallier, 1999). The enzyme from *Candida peltata* was found to have a high tolerance to glucose with a *Ki* value of 1.4 M (Saha & Bothast, 1996). A single microbe can produce several isoforms of β-glucosidases and gene expression varies with the substrate. Four different isoforms were seen when *Aspergillus niger* was grown under SmF with lactose or cellulose as the carbon source, while the number was reduced to two when wheat bran or rice straw was used as the substrate (Singhania *et al.*, 2011). Both intracellular and secreted isoforms are also seen each having different optimum values of pH and temperature and varying thermal stabilities (Elyas *et al.*, 2010).

2.3.1 Sources of Cellulase

Mostly, microbes secrete enzymes to satisfy their growth. Thus most of the cellulolytic fungi are moderate producers of cellulase. Even strains that show high production of any one cellulase may show limited production of the others. Studies on fungal cellulase for saccharification started with the discovery *of Trichoderma reesei* QM6a from the Solomon Islands. Since then, the search for fungal strains from natural sources has been underway with several new strains being discovered with promising results, especially in the thermostability of cellulase and the glucose tolerance in the β-glucosidase enzyme. With the estimated number of 150,000 species, each with several strains and the yet-to-be-discovered millions of fungal taxa, screening from natural source offer great prospects in isolating potential fungal taxa. Studies on glucose-tolerant β-glucosidase by screening

608 fungal taxa isolated from the soil and water showed 17 strains with high titre of β-glucosidases. Of this, eight strains belonged to *Paecilomyces* sp. Production optimization via response surface methodology resulted in high titres of β-glucosidases, but further studies indicated it to be more of aryl β-glucosidase than cellobiase. Thus, screening technologies need to be modified to use cellobiose-conjugated fluorescence or analogues as substrates rather than para-nitrophenyl β-D-glucopyranoside or methyl umbelliferyl β-D-glucopyranoside.

An extensive literature survey on primary screening points out fungal taxa capable of secreting moderate levels of cellulase or enzymes with specific characteristics such as glucose tolerance and thermostability. Known hypercellulase producers that are used for the industrial production of cellulase include *Penicillium*, *Trichoderma*, *Neurospora*, *Aspergillus*, *Myeciliopthora* and *Humicola*. Selective screening of these taxa or to specific species within the taxa can be an effective methodology. The Cel 12A from *Humicola grisea* showed a T_m of 68.7°C, which was 14.3°C higher than the *T. reesei* enzyme. The thermal stability was due to the additional three cysteines found in the *H. grisea* Cel12A enzyme (Sandgren *et al.*, 2003). Cel 12A belonging to the GH 12 family from *T. reesei* was found to induce extension of heat-inactivated type I cell walls with changes in wall rheology similar to auxin-induced plant growth (Yuan *et al.*, 2001). This can expose cellulosic chains for enzymatic degradation. Thus, screening should not be restricted to cellulase production alone, but also other factors such as wall degrading enzymes, optimum conditions of activity, and kinetic properties that can enhance biomass saccharification. Based on such taxonomical data, standard culture collections from all over the world can be screened, and thus is a way to reduce the time and effort in identifying potential strains for cellulase production.

Another possibility is the stain improvement strategies whereby hypercellulase producers can be made. Conidial irradiation of *T. reesei* QM6a with a linear accelerator resulted in the production of strain QM9123; similar irradiation resulted in QM9414 (cellulase production of 30 U/L/h). Further mutation studies in the new strain using UV light and nitrosoguanidine resulted in the production of hypercellulolytic strains including MCB-77 (cellulase production of 90 U/L/h), RUT-C30 (cellulase protein 2.2 g/L) and RL-P37, L 27. Mutational studies thus form the base of the currently used strains of *T. reesei* RUT-C30 and RL-P37. The major change in the genome of RUT-C30 was the truncated *cre1* gene that resulted in the partial removal of catabolite repression by glucose. This resulted in the enhancement of extracellular protein titre by 20 g/L. Industrial strains are reported to have 100 g/L protein titre, but experimental data on strain development and fermentation process are unavailable. A hypercellulase-producing strain *Aspergillus* SU14 showed a 2.2-fold increase in cellulase production through mutations. The wild type showed a cellulase production of 9.65 U/g, which was enhanced after a series of mutations using Co^{60} γ rays. The mutants produced were irradiated with UV radiations and then were treated with NTG to develop a hypercellulase strain SU14-M15 secreting 19.2 U/g cellulase. Optimization in SSF mode resulted in cellulase yield 8.5 times more than the wild type (Vu *et al.*, 2011). UV mutagenesis of spores of *Aspergillus niger* enhanced cellulase yield twofold compared to the wild strain (Jafara *et al.*, 2017). The NTG-19 mutant form *Fusarium oxysporum* developed by UV irradiation followed by NTG mutation showed 80% more cellulase activity than the parent strain (Kuhad *et al.*, 1994). Compared to random mutations, site-directed mutations can specifically alter the desired gene. In *Gloeophyllum trabeum*, mutation of the 233rd amino acid affected the flexibility of loop 6 of GH5 cellulase resulting in changes in catalytic performance. Change of the asparagine amino acid to valine resulted in an optimum pH shift from 4.0 to 5.0, and all variants except aspartic acid showed a 10°C–20°C decrease in optimum temperature (Zheng *et al.*, 2018). Similar mutational studies in other strains of cellulase producers can be attempted that can enhance production or modify properties of cellulases for greater application in biomass hydrolysis.

Genetic recombination for homologous and heterologous expression of cellulase genes, inactivation and overexpression of genes can have profound effects on the quality and quantity of cellulase.

The first of this kind was in *T. reesei* with *cbh 1* gene inactivation followed by insertion of *egl 1* gene and its expression under CBH 1 promoter, resulting in overexpression of EG 1. Cellulase gene expression is adaptive in nature, and in *T. reesei* it is controlled by the action of at least five transcriptional activators (XYR1, ACE2, ACE3, BglR, HAP2/3/5 complex) and two repressors (ACE1 and the carbon catabolite repressor CRE1) (Castro *et al.*, 2016). Genetic changes like constitutive expression of XYR1, RNAi-mediated silencing of ACE1 could significantly enhance cellulase production and reduce catabolite repression by glucose (Wang *et al.*, 2013). A 26% overexpression of the *xyr1* gene in *Trichoderma harzianum* resulted in 36%, 37% and 66 % enhancement in β-glucosidase, xylanase and FPase activity, respectively. This resulted in a 25% hike in sugarcane bagasse saccharification in the initial period (da Silva Delabona *et al.*, 2017). Removal of genes encoding extracellular proteases like SLP1 and PEP1 was found to enhance protein secretion rates in *T. reesei* (Landowski *et al.*, 2015). Heterologous gene expression of the lacA gene from *Trametes* could enhance biomass saccharification by *T. reesei* (Zhang *et al.*, 2012). β-glucosidase gene from *Penicillium decumbens* ligated with CBH 1 promoter of *T. reesei* when inserted into *T. reesei* RUT-C30 resulted in 6- to 8-fold and 30-fold increase in β-glucosidase and filter paper activity, respectively (Ma *et al.*, 2011). Reconstruction of Expression Regulatory Network (REXRN) is a novel methodology for enhancing cellulase production. Overexpression of cellulose degradation regulator clrB in *Penicillium oxalicum* using promoter gpdA from *Aspergillus nidulans*, removal of the bgl2 gene that plays a negative role in cellulase and xylanase induction, and deletion of the Cre A gene that represses cellulase and hemicellulose resulted in trigenic recombinant RE 10. This strain showed increased cellulolytic ability and high cellulase expression and secretion when compared to the wild type and was also comparable to the industrial strain *P. oxalicum* mutant JU-A10-T (Yao *et al.*, 2015).

Use of the most modern technology, the CRISPER/Cas9 system, in the RUT-C30 strain resulted in an extracellular protein titre of 80.6 g/L indicating the potentiality of the technique in enhancing enzyme yield and heterologous gene expression. There was also a 72-fold increase in β-glucosidase and a 42-fold increase in xylanase production with saccharification efficiency similar to commercially available cellulase enzyme (Fonseca *et al.*, 2020). Protein engineering studies often aim to increase thermostability and enhance the half-life period of enzymes. Hence, comparisons of amino acid sequences and folding patterns among mesophilic and thermophilic organisms are to be carried out. Studies indicate a greater number of arginine, glutamic acid and methionine in thermostable endoglucanase, with quite a few changes in amino acids in protein folds and the catalytic site and changes in structural dynamics leading to the enhanced stability of enzymes (Nakabayashi *et al.*, 2019).

Heterologous expression of cellulase in plants is another area of research. Novel methodology involves the cloning and expression of microbial endoglucanase in plants. This causes internal hydrolysis of plant cell walls during plant growth, resulting in greater saccharification during enzyme hydrolysis. If the enzyme is thermostable, then it would be able to withstand pretreatment conditions, which can add to enhanced saccharification with the minimum addition of external endoglucanases. Nowadays, studies on strains that can perform simultaneous saccharification and fermentation are occurring. The work on this aspect started as early as 1983 with the *cbh 1* gene of *T. reesei* being successfully cloned into *E. coli* cells. Genetic recombination and heterologous gene expression resulting in strains of *Saccharomyces cerevisiae* with *T. reesei* genes for cellobiohydrolase supplemented with β-glucosidase is being effectively used in simultaneous saccharification and fermentation of microcrystalline cellulose (Ilmen *et al.*, 2011).

2.3.2 Production of Cellulase

Enzymes alone cost almost one-third or more for bioethanol produced, hence technological development has continued to focus on reducing the production cost of cellulases. The global focus has been on factors such as high titres of cellulases production and secretion using cheap and easily available substrates, production of glucose-tolerant cellobiase, ease of downstream processing, and suitability of creating a cocktail for effective saccharification. Cellulase production can either be by submerged fermentation (SmF) or by solid-state fermentation (SSF). Industrial production of cellulase is mostly done through the SmF method because of ease in automation, scale-up and containment of microbes. However, the process is cost-intensive due to low productivity and difficulty in downstream processing, and it also faces contamination risk. Stirred tank fermenters are generally used, but the shear force can break the mycelium, thereby they are less suitable for filamentous fungi. Cellulase production by SSF offers several advantages. It creates a natural environment for fungal growth and hence provides optimum conditions for growth and enzyme secretion. The enzyme titres are usually high compared to submerged fermentation, and contamination risk is minimal. Cheap substrate can be used for fermentation, thereby reducing the cost of production. The product can be extracted in concentrated form and hence downstream processing becomes cost effective. However, SSF has its disadvantages. Since there exists a gas phase between solid particles, heat transfer is difficult and heat accumulation can cause problems in growth and enzyme production. Containment is also difficult as fungi tend to sporulate in SSF, and a huge load of spores can be a real problem. Automation and scale-up are difficult, and the present fermenters designed for SSF can be effectively used only in limited processes.

SSF bioreactors can be grouped into four major types: Group I fermenters, where the substrate bed is static or mixed occasionally (e.g., tray fermenters); Group II fermenters in which the substrate bed is static but aerated forcefully (e.g., packed bed fermenters), Group III fermenters where the substrate bed is continuously or intermittently mixed, and air is circulated around the substrate bed but not vehemently through the bed (e.g., Rotating drum fermenters); and Group IV reactors where the bed is mixed and also forcefully aerated, like the gas–solid fluidized bed reactors (Mitchel *et al.*, 2006). There is a need to evolve fermenters for SSF that can be used at an industrial scale, and studies on new methodologies of mixing media within the fermenter, better heat and oxygen transfer, and aseptic incubation conditions are to be undertaken.

Suitable media that offer maximum enzyme production are to be selected. Cheap and easily available media are to be used to reduce the cost of production. Common substrates include wheat bran, rice bran, sugarcane bagasse, wheat straw, rice straw, sugarcane tops and corn stover. However, the presence of toxins due to the growth of microbes during storage and the presence of heavy metals in the media can inhibit the growth of inoculated fungi and the secretion of cellulases. Though crude media favour maximal growth, the addition of metabolite regulators such as precursors, inducers or inhibitors may be required to induce the production of cellulases. Sophorose, cellobiose, xylobiose, β-cellobiono-1,5-lactone, D-xylose, lactose and galactose are reported to induce cellulase and hemicellulase production in *T. reesei.* Since the occurrence of isoforms that differ in kinetic parameters and tolerance towards product inhibition is common in cellulolytic fungi, induction of the required isoform with specific chemicals is to be attempted. Extracellular proteases pose a great threat to secreted cellulase. Thus media compositions that favour minimum production of proteases or media that contain protease inhibitors are preferred for cellulase production.

Optimization of the media composition and process parameters enhances cellulase production. One factor at a time (OFAT) or the statistical method of media optimization can be attempted. The OFAT method is simple to perform and useful when a few parameters are to be optimized. The statistical method is based on the data obtained from initial studies or the data of previous works and the selected organism. The full factorial method employs all the possible combinations and gives insight into significant factors that contribute to the response. However, when the number of factors as well as the level of factors are enhanced, it becomes a tedious process. The fractional factorial method is the most common statistical method employed in optimization studies. It is a carefully selected subset of full factorial design and hence the number of runs needed for the optimization becomes less. The Plackett–Burman design can reveal significant factors and their contribution

TABLE 2.1
Commercial Cellulases from Fungal Sources

Commercial Name	Supplier	Unit Activity
Accellerase Trio	Genencor	2000–2600 CMC U/g >3000 ABX U/g >2000 pNPG U/g
Accellerase 1500	Genencor	2200–2800 CMC U/g 450–775 pNPG U/g
Accellerase 1000	Genencor	67.3 FPU/ml 84.2 CBU/ml
Accellerase XC	Genencor	1000–1400 CMC U/g
GC 220	Genencor	116 FPU/ml 215 CBU/ml
Spezyme CP	Genencor	55.2 FPU/ml 15.4 CBU/ml
Spenzyme CE	Genencor	2500 CMC IU/ML
Cellulase TRL	Solvay Enzymes	140 FPU/ml 10000 CBU/g
Cellulase 4000	Solvay Enzymes	4000 U/g
Celluclast	Novozymes	700 EGU/g
Cellubrix	Novozymes	56 FPU/ml 136 CBU/ml
Cellic Ctec2	Novozymes	223 FPU/ml
Novozyme 188	Novozymes	626.4 CBU/ml
Viscozyme	Novozymes	100 FBG/g
Palkonol MBW	Novozymes	12,000 MCU/ml
Cellulase AP30 K	Amano Enzymes	60 CBU/ml
Biocellulase A	Quest Intl	32 CBU/ml
Biocellulase TRI	Quest Intl	68 FPU/ml 200 CBU/ml
E-CELTR	Megazyme	85 EG U/ mg
E-CELTM	Megazyme	245 EG U/ mg

towards the response (cellulase production). Further optimization includes response surface methods such as Box–Benhken and central composite design that can be used to study the interaction of factors and to predict response with various combinations of factors. Slurry state fermentation using wheat bran as the carbon source resulted in a 2.52% increase in β-glucosidase production by *Byssochlamys fulva* (Rasheed *et al.*, 2015). Optimization of incubation conditions and moisture resulted in a 6.2-fold increase in cellulase production by *T. reesei* using wheat bran as substrate (Singhania *et al.*, 2007). Thus a tenfold to several hundredfold increase in production can be seen with optimization studies.

Hypercellulase producers are capable of secreting high titres of cellulase. However, they are usually deficient in glucose-tolerant β-glucosidases. Thus industrial cellulases are often a cocktail of various types of cellulase. The strategy to be employed is to study the mechanism of cellulose degradation including the mechanism of cell wall loosening during plant growth and to recreate these using engineered enzymes or proteins for saccharification. Industrial production of each such component and the development of suitable cocktails can significantly enhance saccharification and thus reduce the cost of bioethanol production (see Table 2.1 and Table 2.2).

TABLE 2.2
Lesser Known Fungal Taxa Producing Cellulase in SSF

Fungus	Substrate	Yield of Enzyme (IU/gds)			Reference
		FP	CMC	BGL	
Penicillium janthinellum	Wheat bran	20	—	25	Singhania *et al.*, 2015
Aspergillus fumigatus	Oil palm trunk	3.3	54.2	4.5	Ang *et al.*, 2013
Aspergillus fumigatus	Wheat bran, rice straw	102.5	826.2	255.1	Das *et al.*, 2013
Aspergillus fumigatus	Rice straw, wheat bran	ND	14.71	8.51	Sherief *et al.*, 2010
Aspergillus fumigatus	Rice straw	3.1	64.2		Sarkar and Aikat, 2012
Aspergillus sp.	Wheat bran	4.6	29		Gupta *et al.*, 2015
Aspergillus heteromorphus	Rice straw	225	14	186	Singh and Bishnoi, 2012
Aspergillus ellipticus	Wheat straw	117.2	725.1	29.6	Agrawal and Matkar, 2016
Aspergillus niger	Rice straw	19.5	129	100	Kang *et al.*, 2004
Emericella variecolor	Orange peel	39	236	197	Srivastava *et al.*, 2022
Myceliophthora thermophila	Sugarcane bagasse, wheat bran		357.5	45.4	Pereira *et al.*, 2015
Neurospora crassa	Wheat straw	97	13.3	5.8	Romero *et al.*, 1999
Penicillium janthinellum	Sugarcane bagasse	215	5.5	23.1	Adsul *et al.*, 2004
Phanerochaete chrysosporium	Soy hull	29.1	74.8	—	Jha *et al.*, 1995
Rhizopus oryzae	Rice husk, rice bran	2.3	5.1	ND	Kupski *et al.*, 2014
Thermoascus auranticus	Wheat straw	987	4.7	48.8	Kalogeris *et al.*, 1999
Trichoderma asperellum	Wheat bran	2.2	13.2	9.2	Raghuwanshi *et al.*, 2014
Trichoderma asperellum	Saw dust	0.95	2.06	—	Zapata *et al.*, 2018
Trichoderma asperellum	Paper waste	5.59	4.11	—	Zapata *et al.*, 2018
Trichoderma viridae	Sugarcane bagasse	10.1	331		Nathan *et al.*, 2014

2.4 BIOMASS HYDROLYSIS

Success in biomass hydrolysis depends on the type and quality of biomass, the enzyme cocktail used and the conditions employed for saccharification. Biomass-to-sugar conversion involves three sequential steps: reduction in the size of biomass, pretreatment and hydrolysis of biomass. Size reduction is done mechanically and this increases the bulk density, flowability and digestibility of biomass (Miao *et al.*, 2011). Pretreatment of lignocellulosic biomass aims at the removal of lignin and hemicellulose to expose cellulosic chains for hydrolysis. Pretreatment can be chemical using dilute acids, alkali, organic solvents or ammonia; physical, using hot water or steam explosion or microwave; biological, using lignin-degrading fungi or enzyme cocktails or combinations, such as acid-catalyzed steam explosion or ammonia fibre explosions. Alkali-based methods are more reliable compared to acid hydrolysis for lignin solubilization, with much of the hemicellulose left over as an insoluble polymeric form. This reduces the development of furan derivatives and phenolics, which are powerful inhibitors of subsequent alcohol fermentation (Boonmee, 2012; Hamelinck *et al.*, 2005; Saha *et al.*, 2005). Use of enzymes from a single source normally cannot offer satisfactory biomass hydrolysis. This is because microbes produce enzymes for their needs and any excess production accounts to the loss of energy that was reserved for growth and metabolite production. Biomass is a complex structure with carbohydrates, lipids, proteins, nucleic acids, and other primary and secondary metabolites. Thus, for the efficient hydrolysis of biomass, a combination of different enzymes is needed. Cellulases from *T. reesei* are supplemented with enzymes from other fungal sources. The addition of other hydrolytic enzymes like pectinase, ligninase and hemicellulase along with *T. reesei* cellulase is found to enhance saccharification. Use of 50 FPU cellulase from *T. reesei* and 10 U BGL from *A. niger* resulted in 26.3 g/l reducing sugar from pretreated rice straw (Sukumaran *et al.*, 2009). Enzyme cocktail development is not an easy task as several enzymes have different optimum conditions for activity. Thus a highly active β-glucosidase with a high *Ki* value to glucose may show minimal activity in saccharification, as its optimal condition can vary with the conditions employed for biomass hydrolysis. When the biomass is poor in quality and already contaminated with heavy metals and toxins, saccharification efficiency will be minimal even when the best enzyme cocktail is used.

Consolidated bioprocessing (CBP) is a process that combines cellulase production and hydrolysis in a single step. It can be implemented using recombinant bacterial strains of *Clostridium thermocellum*, which shows a saccharification level of 89.3% with pretreated wheat straw as the substrate (Liu *et al.*, 2019). Unicellular fungi showing a fast rate of multiplication with genes for cellulase synthesis and secretion could be as effective as bacterial counterparts. Thus strains that preferably degrade hemicellulose and lignin, use pentoses for growth and with high titres of cellulase secretion can be effectively used in CBP.

Simultaneous saccharification and fermentation is yet another process that combines the saccharification of biomass and fermentation of sugar. The concept was developed in the 1970s by Gauss and co-workers by combining the cellulase enzyme of *T. reesei*, plant biomass and fermentative microbes in a single vessel resulting in bioethanol production. The process can enhance bioethanol production, as glucose and cellobiose formed from enzyme hydrolysis are immediately used up for ethanol production, thereby minimizing glucose-mediated inhibition of cellulase. The capital investment cost is also low as both saccharification and fermentation can be carried out in a single vessel. Here also strains that preferably degrade hemicellulose and lignin, use pentoses for growth and with high titres of cellulase secretion and the ability to modulate metabolism from aerobic to anaerobic with ethanol as the end product of anaerobic fermentation could be the best candidate. The major hurdle for this technology is the varied optimal conditions for saccharification and fermentation. Saccharification is usually done at higher temperatures than fermentation. This can be overcome using thermophilic yeast strains that can work at higher temperatures of saccharification. Bacteria like *Zymomonas mobilis* can be used as an alternative to *Saccharomyces cerevisiae* because of their lower biomass yield and higher ethanol production. However, they cannot ferment pentoses and are less tolerant to lignocellulose-derived inhibitors (Olofsson *et al.*, 2008). *Byssochlamys fulva* AM 130 that produces glucose-tolerant β-glucosidase could also ferment glucose and xylose with an ethanol yield of 11.84 g/L in glucose-containing media (Krishnamoorthy *et al.*, 2021). Thus filamentous fungi also can be used in ethanol production rather than with unicellular yeast forms. The added advantage of this is on reutilization of fungal mycelium for subsequent batches of fermentation, provided that the fungal growth is not diffuse but more of a pelleted type. Extensive screening of alcohol-synthesizing fungi with an optimum growth pH and temperature similar to saccharification conditions and a genetic engineering approach to enhance ethanol production could yield promising results.

2.5 CONCLUSION AND PERSPECTIVES

Lignocellulose is the most abundant natural biomass and any process that can economically convert this into glucose could revolutionize the biofuel industry. Cellulases are excellent biocatalysts that can convert the cellulosic fraction of lignocellulosic biomass into sugars. Evolution in cellulase producers created several isoforms, each having different kinetic parameters and optimum reaction conditions to fine-tune the metabolism. Filamentous fungi are known for extracellular cellulase production. This ensures no excess production of glucose and thus offers a competitive advantage in the microbial world. Hypercellulase producers are often deficient in β-glucosidases and prone to product inhibition resulting in the halt of cellulase action. Most of the glucose-tolerant β-glucosidases are aryl-glucosidases and hence do not contribute much to cellulose hydrolysis.

Bioprocess development for hypercellulase production via fungi may pave the way for feasible technology for biomass to biofuel technology as cellulase itself is regarded as the crux of the whole technology. Better bioreactors with superior design may increase the productivity that is being worked on by various researchers and industrialists worldwide. Extensive screening for novel fungi, mutation studies on existing hypercellulase producers and novel technologies of genetic engineering are to be used for developing effective strains of fungi capable of secreting enzymes that can economically degrade cellulose. The development of strains that can perform consolidated bio-saccharification processes or simultaneous saccharification and fermentation processes can further reduce the cost of bioethanol. Even though the commercialization of biomass conversion technology has taken place globally, there is room for improvement, which is the reason for extensive work on cellulase. Collaboration between laboratories and industries may lead to game-changing technology for the sustainable conversion of biomass via cellulases in a cost-effective manner.

REFERENCES

Adsul MG, Ghule JE, Singh R, Shaikh H, Bastawdea KB, Gokhale DV, Varma AJ. 2004. Polysaccharides from bagasse: Applications in cellulase and xylanase production. *Carbohydrate Polymers.* 57(1):67–72.

Agrawal D, Matkar K. 2016. Enhancement of cellulase production by substrate manipulation in three *Aspergillus* sp. *Indian Journal of Applied Research.* 6(3):219–221.

Ang TN, Ngoh GC, Chua ASM. 2013. Development of a novel inoculum preparation method for solid-state fermentation: Cellophane film culture (CFC) technique. *Industrial Crops and Products.* 43:774–777.

Boonmee A. 2012. Hydrolysis of various Thai agricultural biomasses using the crude enzyme from *Aspergillus aculeatus* Izuka FR60 isolated from soil. *Brazilian Journal of Microbiology.* 43(2):456–466.

Castro DSL, de Paula RG, Antonieto AC, Persinoti GF, Silva-Rocha R, Silva RN. 2016. Understanding the role of the master regulator XYR1 in *Trichoderma reesei* by Global Transcriptional Analysis. *Frontiers in Microbiology.* 7:175.

da Silva Delabona P, Rodrigues GN, Zubieta MP, Ramoni J, Codima CA, Lima DJ, Farinas CS, da Cruz Pradella JG, Seiboth B. 2017. The relation between xyrl overexpression in *Trichoderma harzianum* and sugarcane bagasse saccharification performance. *Journal of Biotechnology.* 246:24–32.

Das A, Paul T, Halder SK, Jana A, Maity C, Mohapatra PKD, Pati BR, Mondal KC. 2013. Production of cellulolytic enzymes by *Aspergillus fumigatus* ABK9 in wheat bran-rice straw mixed substrate and use of cocktail enzymes for deinking of waste office paper pulp. *Bioresource Technology.* 128:290–296.

Elyas KK, Mathew A, Sukumaran RK, Ali PP, Sapna K, Sukumaran RK, Mol KR. 2010. Production optimization and properties of beta glucosidases from a marine fungus Aspergillus-SA 58. *New Biotechnology.* 27(4):347–351.

Fonseca LM, Parreiras LS, Murakami MT. 2020. Rational engineering of the *Trichoderma reesei* RUT-C30 strain into an industrially relevant platform for cellulase production. *Biotechnology for Biofuels.* 13:93.

Gunata Z, Vallier MJ. 1999. Production of a highly glucose-tolerant extracellular β-glucosidase by three *Aspergillus* strains. *Biotechnology Letters.* 21(3):219–223.

Gupta C, Jain P, Kumar D, Dixit AK, Jain RK. 2015. Production of cellulase enzyme from isolated fungus and its application as efficient refining aid for production of security paper. *International Journal of Applied Microbiology and Biotechnology Research.* 3:11–19.

Hamelinck CN, Hooijdonk GV, Faaij APC. 2005. Ethanol from lignocellulosic biomass: Techno-economic performance in short, middle and long term. *Biomass and Bioenergy.* 28(4):384–410.

Ilmen M, den Haan R, Brevnova E, Mc Bride J, Wiswall E, Froehlich A, Koivula A, Voutilainen SP, Siika-aho M, Grange DC, Thorngren N, Ahlgren S, Mellon M, Deleault K, Rajarhia V, van Zyl WH, Penttila M. 2011. High level secretion of cellobiohydrolases by *Saccharomyces cerevisiae. Biotechnology for Biofuels.* 4:30.

Jafara NM, Jafarizadeh-Malmiri H, Hameh-Mivehroud M, Adibpour M. 2017. Optimization of UV irradiation mutation conditions for cellulase production by mutant strains of *Aspergillus niger* through solid state fermentations. *Green Processing and Synthesis.* 6(3):333–340.

Jha K, Khare SK, Gandhi AP. 1995. Solid-state fermentation of soyhull for the production of cellulase. *Bioresource Technology.* 54(3):321–322.

Kalogeris E, Fountoukides G, Kekos D, Macris BJ. 1999. Design of a solid-state bioreactor for thermophilic microorganisms. *Bioresource Technology*. 67(3):313–315.

Kang SW, Park YS, Lee JS, Hong SI, Kim SW. 2004. Production of cellulases and hemicellulases by *Aspergillus niger* KK2 from lignocellulosic biomass. *Bioresource Technology*. 91(2):153–156.

Krishnamoorthy J, Mathew A, Kooloth-Valappil P, Adarsh VP, Puthiyamadam A, Pandey A, Sukumaren RK. 2021. Ethanol production by a filamentous fungal strain *Byssochlamys fulva* AM 130 under alternating aerobic and oxygen-limited conditions. *Systems Microbiology and Biomanufacturing*. 1(1):111–121.

Kuhad RC, Kumar M, Singh A. 1994. A hypercellulolytic mutant of *Fusarium oxysporum*. *Letters in Applied Microbiology*. 19(5):397–400.

Kupski L, Pagnussatt FA, Buffon JG, Furlong EB. 2014. Endoglucanase and total cellulase from newly isolated *Rhizopus oryzae* and *Trichoderma reesei*: Production, characterization, and thermal stability. *Applied Biochemistry and Biotechnology*. 172(1):458–468.

Landowski CP, Huuskonen A, Wahl R, Westerholm-Parvinen A, Kanerva A, Hanninen AL, Salovuori N, Penttilä M, Natunen J, Ostermeier C, Helk B, Saarinen J, Saloheimo M. 2015. Enabling low cost biopharmaceuticals: A systematic approach to delete proteases from a well-known protein production host *Trichoderma reesei*. *PLOS ONE*. 10(8):e0134723.

Liu S, Liu YJ, Feng Y, Li B, Cui Q. 2019. Construction of consolidated bio-saccharification biocatalyst and process optimization for highly efficient lignocellulose solubilisation. *Biotechnology for Biofuels*. 12:35.

Ma L, Zhang J, Zou G, Wang C, Zhou Z. 2011. Improvement of cellulase activity in *Trichoderma reesei* by heterologous gene expression of a beta-glucosidase gene from *Penicillium decumbens*. *Enzyme Micro. Technol*. 49(4):366–371.

Miao Z, Grift TE, Hansen AC, Ting KC. 2011. Energy requirement for comminution of biomass in relation to particle physical properties. *Industrial Crops and Products*. 33(2):504–513.

Mitchel DA, Krieger N, Berovic M. 2006. *Solid State Fermentation Bioreactors: Fundamentals of Design and Operation* (Berovic M, Ed.). Springer–Verlag, Berlin, pp. 13–439.

Nakabayashi M, Kamachi S, Malle D, Yanamoto T, Kishishita S, Fujii T, Inoue H, Ishikawa K. 2019. Construction of thermostable cellobiohydrolase I from the fungus *Talaromyces cellulolyticus* by protein engineering. *Protein Engineering, Design and Selection*. 32(1):33–40.

Nathan VK, Rani ME, Rathinasamy G, Dhiraviam KN, Jayavel S. 2014. Process optimization and production kinetics for cellulase production by *Trichoderma viride* VKF3. *SpringerPlus*. 3:92.

Olofsson K, Bertilsson M, Liden G. 2008. A short review on SSF – An interesting process option for ethanol production from lignocellulosic feedstocks. *Biotechnology for Biofuels*. 1(1):7.

Pereira JDC, Marques NP, Rodrigues A, Oliveira TBD, Boscolo M, Silva RD, Gomes E, Martins DAB. 2015. Thermophilic fungi as new sources for production of cellulases and xylanases with potential use in sugarcane bagasse saccharification. *Journal of Applied Microbiology*. 118(4):928–939.

Raghuwanshi S, Deswal D, Karp M, Kuhad RC. 2014. Bioprocessing of enhanced cellulase production from a mutant of *Trichoderma asperellum* RCK2011 and its application in hydrolysis of cellulose. *Fuel*. 124:183–189.

Rasheed T, Mathew A, Mathew M, Sukumaran RK, Elyas KK. 2015. Optimization of process parameters for β-glucosidase production by *Byssochlamys fulva* in slurry state fermentation. *Biotechnological Research*. 1(1):101–107.

Riou C, Salmon JM, Vallier MJ, Gunata Z, Barre P. 1998. Purification, characterization and substrate specificity of a novel highly glucose-tolerant β-glucosidase from *Aspergillus oryzae*. *Applied and Environmental Microbiology*. 64(10):3607–3614.

Romero MD, Aguado J, Gonzalez L, Ladero M. 1999. Cellulase production by *Neurospora crassa* on wheat straw. *Enzyme and Microbial Technology*. 25(3–5):244–250.

Saha BC, Bothast RJ. 1996. Production, purification and characterization of a highly glucose-tolerant novel β-glucosidase from *Candida peltata*. *Applied and Environmental Microbiology*. 62(9):3165–3170.

Saha BC, Iten LB, Cotta MA, Wu YV. 2005. Dilute acid pretreatment, enzymatic saccharification and fermentation of wheat straw to ethanol. *Process Biochemistry*. 40(12):3693–3700.

Sandgren M, Gualfetti PJ, Paech C, Paech S, Shaw A, Gross LS, Saldajeno M, Berglund GI, Jones TA, Mitchinson C. 2003. The *Humicola grisea* Cel12A enzyme structure at 1.2 A resolution and the impact of its free cysteine residues on thermal stability. *Protein Science*. 12(12):2782–2793.

Sarkar N, Aikat K. 2012. Cellulase and xylanase production from rice straw by a locally isolated fungus *Aspergillus fumigatus* NITDGPKA3 under solid state fermentation – statistical optimization by response surface methodology. *Journal of Technology Innovations in Renewable Energy*. 1:54–62.

Sherief AA, El-Tanash AB, Atia N. 2010. Cellulase production by *Aspergillus fumigatus* grown on mixed substrate of rice straw and wheat bran. *Research Journal of Microbiology*. 5(3):199–211.

Singh A, Bishnoi NR. 2012. Optimization of ethanol production from microwave alkali pretreated rice straw using statistical experimental designs by Saccharomyces cerevisiae. *Industrial Crops and Products.* 37(1):334–341.

Singhania RR, Reetu S, Mukund A, Saini JK, Anshu M, Deepak T. 2015. An integrative process for bio-ethanol production employing SSF produced cellulase without extraction. *Biochemical Engineering Journal.* 102:45–48.

Singhania RR, Sukumaran RK, Pandey A. 2007. Improved cellulase production by Trichoderma reesei RUT C30 under SSF through process optimization. *Applied Biochemistry and Biotechnology.* 42(1):60–70.

Singhania RR, Sukumaran RK, Rajasree KP, Mathew A, Gottumukkala L, Pandey A. 2011. Properties of a major β-glucosidase-BGL1 from *Aspergillus niger* NII-08121 expressed differentially in response to carbon sources. *Process Biochemistry.* 46(7):1521–1524.

Sorensen TH, Cruys-Bagger N, Windahl MS, Badino SF, Borch K, Westh P. 2015. Temperature effects on kinetic parameters and substrate affinity of Cel7A cellobiohydrolases. *Journal of Biological Chemistry.* 290(36):22193–22202.

Srivastava N, Mohammad A, Pal DB, Srivastava M, Alshahrani Y, Ahmad I, Singh R, Mishra PK, Yoon T, Gupta VK. 2022. Enhancement of fungal cellulase production using pretreated orange peel waste and its application in improved bioconversion of rice husk under the influence of nickel cobaltite nanoparticles. *Biomass Conversion and Biorefinery.* 2022. https://doi.org/10.1007/s13399-022-03070-3

Sukumaran RK, Singhania RR, Mathew GM, Pandey A. 2009. Cellulase production using biomass feed stock and its application in lignocellulose saccharification for bio-ethanol production. *Renewable Energy.* 34(2):421–424.

Vu VH, Pham TA, Kim K. 2011. Improvement of fungal cellulase production by mutation and optimization of solid state fermentation. *Mycobiology.* 39(1):20–25.

Wang S, Liu G, Wang J, Yu J, Huang B, Xing M. 2013. Enhancing cellulase production in *Trichoderma reesei* RUT C30 through combined manipulation of activating and repressing genes. *Journal of Industrial Microbiology and Biotechnology.* 40(6):633–641.

Yao G, Li Z, Gao L, Wu R, Kan Q, Liu G, Qu Y. 2015. Redesigning the regulatory pathway to enhance cellulase production in Penicillium oxalicum. *Biotechnology for Biofuels.* 8:71. doi: 10.1186/s13068-015-0253-8. PMID: 25949521; PMCID: PMC4422585.

Yuan S, Wu Y, Cosgrove DJ. 2001. A fungal endoglucanase with plant cell wall extension activity. *Plant Physiology.* 127(1):324–333.

Zapata Y-M, Galviz-Quezada A, Osorio-Echeverri V-M. 2018. Cellulases production on paper and sawdust using native *Trichoderma asperellum. Universitas Scientiarum.* 23(3):5–18.

Zhang J, Qu Y, Xiao P, Wang X, Wang T, He F. 2012. Improved biomass saccharification by *Trichoderma reesei* through heterologous expression of lacA gene from *Trametes* sp AH28-2. *Journal of Bioscience and Bioengineering.* 113(6):697–703.

Zheng F, Tu T, Wang X, Wang Y, Ma R, Su X, Xie X, Yao B, Luo H. 2018. Enhancing the catalytic activity of a novel GH5 cellulase GTCel5 from *Gloeophyllum trabeum* CBS 900.73 by site directed mutagenesis on loop 6. *Biotechnology for Biofuels.* 11:76.

3 Bacterial Cellulosomes
Production and Applications

Archana Bhaskar, Siddhant Dubey, and Reeta Rani Singhania

3.1 INTRODUCTION

Cellulosome is a multienzyme complex fixed within the cell surface that effectively degrades the polysaccharides/lignocellulosic substrates of the plant cell wall in a synergistic manner generated by cellulolytic anaerobic microbes. These macromolecular structures have a modular design and are made up of two primary components: cohesin-bearing scaffoldins, which are non-catalytic structural proteins, and various dockerin-bearing enzymes that tenaciously bind to scaffoldins. Cellulosomes attach to the cell walls of plants via scaffoldin CBM3a that binds to a variety of polysaccharide arrangements, including non-crystalline and crystalline cellulose, as well as xyloglucan. More than 70 CAZymes, including endo-1,4-xylanase, exo1,4-glucanase, endo-1,4-glucanase acetyl xylan esterase, lichenase, and chitinase, are encoded by the *Clostridium thermocellum* genome (Chukwuma et al., 2021).

Organic carbon synthesis is a significant biological activity and a fundamental energy source. Plants use solar energy for photosynthesis and convert it into organic carbon that is used by heterotrophic animals. The primary source of carbon and energy is the cell wall polysaccharides of plants, which are mainly constituted of cellulose and hemicellulose (Champreda et al., 2019). However, there is a limited number of bacteria available to degrade these polysaccharides. Bayer, Lamed and colleagues in 1980 while working on *C. thermocellum* found a large protein aggregation of a cellulose system that can degrade cellulose. This cellulose system is termed a "cellulosome". *C. thermocellum* is a gram-positive thermophilic anaerobic soil bacterium. It has many enzymes including cellulases, pectinases, hemicellulases, carbohydrate esterases, polysaccharides and glycoside hydrolases. All these microbes that degrade the cell wall of plants have a significant association with extracellular enzymes that work simultaneously to degrade the composite arrangements, amorphous substrates and even resistant crystalline. Cellulases and hemicellulases are extremely complicated enzymes with molecular complex structures in that the catalytic modules are attached to non-catalytic modules that perform vital roles in carbohydrates: protein binding and interactions between proteins and proteins (dockerin) (Sharma et al., 2022).

The cellulosome consists of two subunits, i.e., non-catalytic scaffoldin and active catalytic elements. There is a high-affinity recognition between scaffoldin subunit cohesin modules and enzyme-borne dockerin modules that facilitate the formation of cellulosomes. Several cohesin modules are commonly found in the scaffoldin, allowing a variety of enzymes to be organized into the cellulosome unit. In addition, several scaffoldins have been discovered in certain species, adding to the complexity of the cellulosome assembly. Theoretically, approximately a hundred components can be assembled within a single cellulolyticus cellulosome. The cellulose-specific carbohydrate-binding module (CBM), which serves as the primary binding factor for the specific identification of polysaccharide substrates, is another crucial cellulosomal element. In this chapter, bacterial cellulosome and its advances in production and application for biomass hydrolysis are discussed (Gilmore et al., 2015).

 DOI: 10.1201/9781003335313-4

3.2 LIGNOCELLULOSIC BIOMASS

The increasing world population and industrialization have increased the demand for fossil fuel, affecting society in many ways, such as the energy crisis, climatic change and global warming. The first generation of fossil fuel derived from crops has been reduced in recent years due to land unavailability. Therefore, it is necessary to develop an alternative natural or renewable resource that can fulfil human needs without affecting the environment. The biomass obtained from the waste of plants was considered to be the most suitable raw material for the second generation of fossil fuels, as it is environmentally friendly, more economical and available copiously throughout the year (Alves et al., 2021).

Waste that comes from plants, wood, animals and municipalities is known as biomass. Thus, biomass is defined as renewable organic matter that produces fuel that can be used as a second-generation source of energy. The primary component of plant biomass is lignocellulose. Lignocellulose is obtained from various agriculture, forest and crop waste, such as rice and wheat straw, sugarcane bagasse, sawdust and grasses. Generally, 90% of biomass consists of cellulose, hemicellulose and lignin, and the remaining 10% consists of proteins and ash. These biomass components wrapped around each other forming non-degradable complex polymers. Therefore, the conversion of biomass to energy requires a wide variety of enzymes, which increases the cost, is time-consuming and not feasible.

Cellulose is made up of linear β-(1,4)-linked D-glucopyranose chains attached by hydrogen bonding. It accounts for 30%–40% of the lignocellulose. These are the polymers of glucose arranged in bundles.

Hemicellulose primarily consists of pentose (C5) sugars and a small amount of OH groups and uronic acids (Champreda et al., 2019). The major component of hardwood hemicellulose is xylan and the major component of softwood hemicellulose is mannan. It accounts for 20%–30% of the lignocellulose. It is a short polymer chain of different sugars that stick to bundles of cellulose.

Depending on the type of biomass, lignin consists of aromatic p-hydroxyphenyl, coniferyl alcohol, sinapyl alcohol and guaiacyl. It accounts for 20%–25% of the lignocellulose. It is a propylphenol polymer implanted in hemicellulose. It provides firmness to the lignocellulose structure.

Lignocellulosic biomass is considered to be the most suitable alternative for the production of second-generation biofuels. With advances in biotechnology, it has attracted the research interests of academic groups, governments, and companies (Haldar et al., 2016).

3.3 CONVERSION OF LIGNOCELLULOSE INTO BIOFUEL

In the past few decades, the conversion of biomass to biofuel has been considered an alternative method to generate energy to replace the rising demand for fossil fuels. Lignocellulose obtained from crop residues and other plant waste seems to be the most promising renewable raw materials. Generally, the process of biomass conversion includes mechanical pretreatment to reduce the particle size, followed by physicochemical pretreatment for the deconstruction to separate the cellulose, hemicellulose and lignin, followed by hydrolysis (acidic or enzymatic) (Ichikawa et al., 2019). Then, the sugars obtained are fermented for the production of bioethanol. The enzymes are produced by both aerobic and anaerobic bacteria and fungi (Ding et al., 2008). Alternatively, the biomass can be processed through a thermochemical route (pyrolysis, gasification, combustion) to obtain fuels and chemicals

3.3.1 PRETREATMENT OF BIOMASS

Due to the complexity of the biomass structure, the enzymatic hydrolysis method is the most promising, but it requires a pretreatment process to increase the accessibility of enzymes by three- to tenfold. Pretreatment removes the barriers of the microfibril network and breaks the compact structure of lignocellulose to enhance the hydrolysis rate. Various types of pretreatment methods are used to increase the digestibility and finally to improve the yield of biofuels. When choosing the pretreatment method, it should be kept in mind that apart from breaking the components of lignocellulose, it must be compatible with the overall process.

The kinds of pretreatment methods available are physical (mechanical extrusion), chemical (acid, alkali, organic solvents), physiochemical (steam explosion, liquid hot water, wet oxidation, microwave) and biological (microorganisms secreting enzymes). The method selection (individual or hybrid) should aim to lower the pretreatment technology cost, including recovery of lignin, the digestibility of enzymes (so a minimum amount is required), and the power and thermal requirements (Qin et al., 2017).

3.3.2 Hydrolysis of Biomass

The conventional method of hydrolysis (e.g., acid hydrolysis) is not considered green and sustainable due to the environmental effects and high cost. In recent years, the enzymatic hydrolysis process has gained wide acceptance due to lower use of chemical consumptions, ease of operation, high efficiency, and decrease in the denaturation of cellulose and hemicellulose structure. However, the crystalline structure of cellulose poses a challenge to the accessibility of enzymes since they are unable to disrupt hydrogen bonds, leading to poor efficiency. There are other technical challenges such as particle, hemicellulose and lignin hindrances; no constant ratio of cell wall components; and end-product inhibition by glucose (Jiang et al., 2021). There are several aerobic or anaerobic bacteria and fungi possessing hemicellulases, laccases and cellulases that work synergistically and degrade the network of cellulose, hemicellulose and lignin (Figure 3.1). The degradation strategies

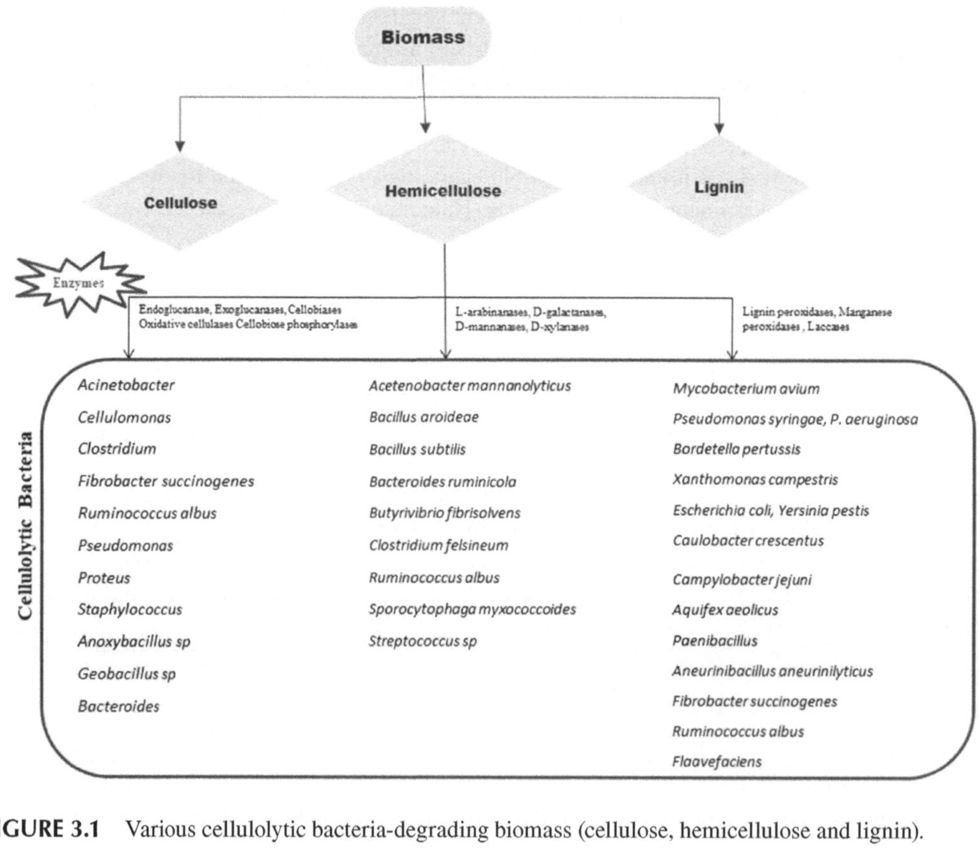

FIGURE 3.1 Various cellulolytic bacteria-degrading biomass (cellulose, hemicellulose and lignin).

of aerobic and anaerobic are different. The aerobic microorganism secretes enzymes that are free to diffuse directly into their substrate for degradation, whereas, in anaerobic microorganisms, the enzyme complexes are arranged sequentially and located on the cellular surface and known as cellulosomes (Lin & Lu, 2021).

3.3.3 Cellulosomes

Cellulosomes are secreted by both fungi and bacteria. The aerobic fungi that produce extracellular enzymes (hemicellulases, cellulases and laccases) that degrades lignocellulose belong mainly to the class basidiomycetes (*Phanerochaete chrysosporium, Fomitopsis palustris*) and ascomycetes (*Trichoderma harzianum, Aspergillus niger*). Some anaerobic fungi of class chytridiomycete (*Neocallimastix frontalis, Caecomyces communis*) found in gastrointestinal tract ruminants are also able to secrete multi-enzyme complexes. The cellulosomes secreted by bacteria are considered highly beneficial because of their efficiency and specificity in degrading the biomass naturally (Artzi et al., 2017).

3.3.3.1 Bacterial Cellulosomes

The extracellular enzyme complex produced by the bacteria, known as cellulosome, contains cellulases and hemicellulases that degrade the lignocellulose (cellulose, hemicellulose, pectin). Cellulosome is secreted by both aerobic and anaerobic bacteria. An aerobic bacterium occurs in water, decaying organic materials, soils and agriculture waste, and produces a complex of enzymes able to degrade cellulose. Examples of aerobic bacteria are *Streptomyces lividans* and *Pseudomonas fluorescens*. Generally, anaerobic bacteria are found in different environmental niches such as the rumen of ruminant animals, polluted water, sewage, organic waste of plants or compost, and also in symbiotic relation with some termites such as Lepidoptera (Mayer, 2019). Bacterial cellulosomes have many advantages, as they have a short generation time and so can be easily cultured for further use; can resist various environmental stresses such as changes in pH, temperature, light, oxygen, etc.; can speed up and elevate the production rate of products; and can degrade not only crystalline cellulose but also the whole cell wall structure of the plant.

Bacterial cellulolytic bacteria are classified based on the following physiological groups:

1) Aerobic bacteria, which is further divided into gram-positive bacteria (*Thermobifida* and *Cellulomona*) and gliding bacteria (*Cytophaga* and *Sporocytophaga*). Other examples are *Acidothermus cellulolyticus, Cellulomonas biazotea, Acinetobacter junii, Cellvibrio gilvus* and *Eubacterium cellulosolvens.*
2) Anaerobic gram-positive bacteria (*Clostridium* and *Ruminococcus*) and some gram-negative bacteria (*Acetivibrio* and *Butyvivrio*) that are fermentable. Examples of anaerobic bacteria are listed in Table 3.1.

3.3.3.2 Application of Bacterial Cellulosomes

With the advancement of biotechnology and nanotechnology, different types of multienzymatic systems, or bacterial cellulosomes, have been developed to enhance the degradation of biomass. The interaction of various components in bacterial cellulosomes has biocompatibility properties that make them highly efficient for hydrolysis and the ability to manipulate or engineer the microorganism for the conversion of lignocellulose to sugars and subsequently to biofuels (Mayer, 2019). Therefore, it has a wide range of applications. Bacterial cellulosomes are used in diverse fields such as industries, agriculture, biomedicine, environment, biorefineries, bioadsorption, bioremediation, production of useful chemicals, and biosensors (cf. Table 3.2 and Figure 3.2).

3.3.3.2.1 Cellulosome Components

As mentioned earlier, cellulosomes are made up of multiple types of proteins that are usually organized in a modular fashion, with every module performing a unique function within the

TABLE 3.1
Anaerobic Bacteria-Producing Cellulosomes

Family	Bacteria	Source of Isolation	Cellulosome Production	Growth Temperature
Lachnospiraceae	*Butyrivibrio fibrisolvens*	Rumen	Yes	Mesophilic
	Ruminococcus flavefaciens	Rumen	Yes	Mesophilic
	Ruminococcus succinogenes	Rumen	Yes	Mesophilic
	Ruminococcus albus	Rumen	Yes	Mesophilic
Syntrophomonodaceae	*Caldocellulosiruptor saccharolyticu*	Hot spring	No	Thermophilic
	Caldocellulosiruptor lactoaceticus	Not defined	No	Thermophilic
	Caldocellulosiruptor kristjanssonii	Hot spring	No	Thermophilic
	Anaerocellum thermophilum	Not defined	No	Thermophilic
Eubacteriaceae	*Eubacterium cellulolyticum*	Rumen	Not defined	Mesophilic
Clostridiaceae	*Clostridium acetobutylicum*	Soil	Yes	Mesophilic
	Clostridium chartatabidum	Rumen	Yes	Mesophilic
	Clostridium cellulovorans	Wood fermenter	Yes	Mesophilic
	Clostridium herbivorans	Pig intestine	Not defined	Mesophilic
	Clostridium cellulosi	Manure	Not defined	Thermophilic
	Clostridium cellobioparum	Rumen	Yes	Mesophilic
	Clostridium papyrosolvens	Paper mill	Yes	Mesophilic
	Clostridium josui	Compost	Yes	Thermophilic
	Clostridium cellulolyticum	Compost	Yes	Mesophilic
	Clostridium aldrichii	Wood fermenter	Not defined	Mesophilic
	Clostridium stercorarium	Compost	No	Thermophilic
	Clostridium thermocellum	Sewage soil	Yes	Thermophilic
	Clostridium cellulofermentans	Manure	Not defined	Mesophilic
	Clostridium celerescens	Manure	Not defined	Mesophilic
	Clostridium thermopapyrolyticum	Mud	Not defined	Thermophilic
	Clostridium thermocopriae	Hot spring	Not defined	Thermophilic
	Clostridium sp. C7	Mud	Yes	Mesophilic
	Bacteroides sp. P-1	Rotting biomass	Yes	Thermophilic
	Bacteroides cellulosolvens	Sewage	Yes	Mesophilic
	Acetivibrio cellulolyticus	Sewage	Yes	Mesophilic
	Acetivibrio cellulosolvens	Sewage	Not defined	Mesophilic

macromolecular cellulolytic complex. Various functions of cellulosomes are cellular tethering, assembly, mechanical support, enzymatic activity and substrate binding. The assembly parts set the cellulosomal system apart from other enzymatic systems. The cellulosome consists of large non-enzymatic scaffold proteins known as scaffoldins (CbpA, CipA, CipC). Scaffoldins have a number of cohesin domains (Coh), cellulose-binding domains (CBDs), dockerin II domains, hydrophilic domains and enzymatic coding domains. The cohesin and dockerin modules form strong and specialized protein–protein interactions, allowing the various cellulosomal components to assemble into a highly effective cellulolytic nanomachine (Alves et al., 2021).

3.3.3.2.1.1 Scaffoldin Cellulosome's enzymatic subunits are combined with a specific scaffoldin component that organizes them into the complex. In some circumstances, the cell surface appears to be intimately related to the majority of cellulolytic activity. Cellulolytic bacteria are commonly defined by protuberance-like surface structures. The cell surface protuberances must be stabilized in order to be seen in an electron microscope; for this, either particular labelling methods

TABLE 3.2
Bacterial Cellulosome Strains Used for Biomass Hydrolysis and Its Applications

Cellulosome-Producing Strains	Biomass (Carbon Source)	Yield (%)	End Product	Application	Reference
Acetivibrio cellulolyticus Cip V	Sewage		Methane	Enzyme immobilization	Jensen et al., 2021
Achromobacter sp.	Rice straw, wheat straw, sugarcane bagasse	73.10 and 91.13	Biosurfactant	Antimicrobial	Joi and Sharma, 2022
Actinobacteria	Olive pomace, molasses		Laccase, xylanase, organic acids, other compounds	Bioremediation	Detman et al., 2018; Alvarez et al., 2017
Aeromonas hydrophila	Wheat, rice, sugarcane, pea straw	100	Strawase, cellulase	Wastewater treatment	Busi et al., 2016
Bacillus sp.	Sawdust, algae, paper mill wastewater	40–100	Cellulase, CMCase and xylanase, methane, lipid, biodiesel	Paper, food, feed, textile, several other industries	Chin et al., 2020
Bacteroides cellulosolvens ClpBc	Sewage		Cellulases	Biorefineries	Mohanram et al., 2013
Burkholderia	Beechwood, rice straw, bran	28.5	Xylanase, cellulase	Bioethanol	Al-Battashi et al., 2019
Butyrivibrio fibrisolvens	Rumen		Lactic acid	Probiotic	Maia et al., 2016
Caldicellulosiruptor bescii	Switch grass	80	Acetate	Biofuels and biorefineries	Straub et al., 2019
Citrobacter sp.	Sugarcane bagasse		Glucose	Heavy metal removal	Jones et al., 2018
Clostridiales	Wheat straw	90	Methane		Xing et al., 2020
Clostridium acetobutylicum	Soil		Methane	Acetone, butanol, ethanol	Shao et al., 2007
Clostridium cellulolyticum CiPC	Compost		Methane	Biofuels	Desvaux, 2005
Clostridium cellulovorans CbpA	Wood fermenter		Methane	Biofuels	Tamaru et al., 2010
Clostridium josui CipC	Compost		Methane	Biofuels	Desvaux, 2005
Clostridium papyrosolvens	Paper mill		Methane	Biofuels	Desvaux, 2005
Clostridium thermocellum	Sewage soil, wheat, cattail	80	Ethanol, cellulases	Biorefineries	Froese et al., 2020
Firmicutes	Wheat, rice, sugarcane, molasses, corn stover		Cellulase, xylanase	Starch, paper, food, feed, textile, several other industries	Kanokratana et al., 2018
Klebsiella oxytoca	Wheat, rice, sugarcane, pea straw	100	Strawase, cellulase	Production of 2,3-butanediol	Guragain et al., 2016
Ochrobactrum sp.	Palm oil mill effluent, wheat straw	71	CMCase, xylanase	Depolymerize lignin	Tsegaye et al., 2018

(Continued)

TABLE 3.2 (CONTINUED)
Bacterial Cellulosome Strains Used for Biomass Hydrolysis and Its Applications

Cellulosome-Producing Strains	Biomass (Carbon Source)	Yield (%)	End Product	Application	Reference
Paenibacillus	Sawdust, switch grass	40–100	Cellulase, biofuels	Softwood paper pulp bleaching, biofuels production, food industry	Ticona et al., 2021
Pantoea ananatisSd-1	Pesticide carbaryl, rice straw	46	Biofuels	Sustainable energy production	Ma et al., 2016
Proteobacteria	Corn stover, molasses, energy grass		Enzymes, biogas	Herbicides	Kanokratana et al., 2018
Pseudomonas sp	Corn stover, wheat, rice, sugarcane, pea straw	99.8	Strawase, cellulose, bioflocculant	Bioremediation	Guo et al., 2017,
Ruminococcus sps	Rumen	83	Methane	Biofuel	Ohara et al., 2000
Saccharomyces cerevisiae	Paper sludge	53	Ethanol	Food and beverage	Parapouli et al., 2020
Sphingobacterium sp.	Corn husk, peanut husk, rice bran, sugarcane bagasse, paddy straw, coffee cherry husk	60	Cellulase, xylanase, pectinase, mannanase, laccase	Textile dye decolorization	Tamboli et al., 2011
Streptomyces spp., *Streptomyces lividans*	Rice waste biomass, wood straw, local grass powder, sugar cane barboja, sugarcane bagasse	40–100	Cellulase, triacylglycerol, endoglucanase, exoglucanase, cellobiases, filter paperase, amylase, xylanase	Lignocellulose-based biorefinery	Saratale et al., 2017
Tepidimicrobium spp.	Napier grass, swine manure	36.6	Acetate/ethanol	Biofuel and biogas	Wongwilaiwalin et al., 2018
Thermoanaerobacter	Wheat and cattail biomass	80	Ethanol	Biofuel	Froese et al., 2020

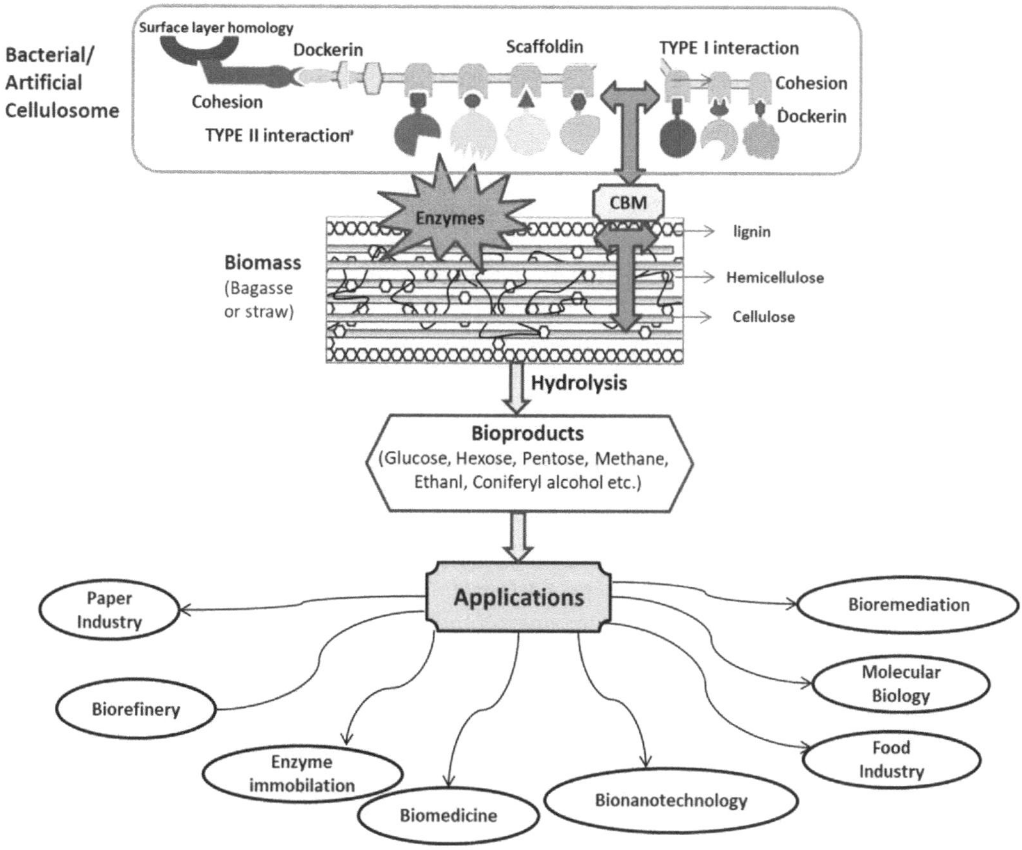

FIGURE 3.2 Bacterial cellulosome hydrolyzing biomass and its application in different fields.

(e.g., antibody or lectin-based labelling) or more general labelling methods (e.g., anionic site tagging with cationized ferritin) can be used. Scaffoldins are divided into four categories based on their function: main, anchoring, adapter and cell-free. The fundamental structural homology of cohesins and dockerins is used to classify them into different types (I, II, III and R). Type I interactions are mostly connected with cellulosome synthesis, whereas type II interactions are linked to cell wall anchoring. As a result, the C-terminus among the major scaffoldins contains a type II dockerin that interacts with equivalent type II cohesin modules in cell-bound anchoring scaffoldins. Many scaffoldins include the X-module, an Ig-like stabilizing module located close to the type II dockerin's N-terminal end (Alves et al., 2021).

Complex cellulosomes have scaffoldins that mediate the cellulosome's interaction with the bacterial cell surface. The importance of this cell-bound configuration compared to the alternative cell-free cellulosome is reflected in the clustering of cellulosomal genes, with main and anchoring scaffoldin genes typically clustered together. Through an SLH domain or a recognition motif, anchoring scaffoldins are either non-covalently or covalently connected to the organism's surface. On more complex and highly organized cellulosomes, another form of scaffoldin, the adaptor scaffoldin, has been discovered. Polyvalent adaptor scaffoldins with several cohesin modules can boost the quantity of CAZymes in the cellulosome. On the other hand, monovalent adaptor scaffoldins employ a single cohesin module to facilitate interaction between the enzymes with divergent dockerins and the principal scaffoldin. They can also act as a repertoire modification by changing the specificity of CAZymes that fill the major scaffoldin in response to substrate availability. Cohesin modules are the only components of cell-free scaffoldins, which are linked together by linker sections. They produce untethered cellulosomes, such as those produced by *Clostridium clarifavum* or *Hungateiclostridium thermocellum*, which can coexist with cell-bound cellulosomes in the extracellular space (Alves et al., 2021).

3.3.3.2.2 Dockerin Dockerin modules in the cellulosomal catalytic factors and cohesin modules on scaffoldin form a high-affinity protein–protein interaction (>109 M1) that allows cellulosomes to assemble. Dockers, which have been found to have a non-catalytic function in a number of cellulosomal enzymes, are made up of 70 amino acids divided into two segments, each being around 22 residues long. Dockers are found at the C-terminus of cellulosomal enzymes in a single copy. So every duplicated segment's first 12 residues are similar to the calcium-binding loop of EF-hand motifs, where the calcium-binding residues, aspartate or asparagine, are conserved. More recently, structural studies revealed that similarity to the EF-hand motif is limited to the calcium-binding loop and the F helix, implying that the F-hand motif exists. Dockerin stability and function have been demonstrated to be dependent on calcium; dockerins seem unable to bind with cohesins in the presence of EDTA that chelates calcium (Fontes and Gilbert, 2010).

3.3.3.2.3 Cohesin Cohesins are 150-residue modules that are found in scaffoldins as tandem repetitions. Beguin and colleagues discovered that scaffoldins' internal repetitions (cohesins) preferentially bind to non-catalytic dockerin modules found in cellulosomal cellulases and hemicellulases. As a result, scaffoldins, which are cohesin-containing proteins, play an essential role in cellulosome assembly and have been found in nearly all cellulosome-producing bacteria. Scaffoldins can have anywhere from 1 to 11 cohesin modules, with most having more than 4. The dockerins present in diverse cellulosomal enzymes appear to be unable to be distinguished by the cohesin domains of certain scaffoldins. Both cohesins and dockerins are quite similar within a species, and the acids immediately involved in protein–protein recognition are largely conserved. Proteomic study of individual cellulosomes from *C. thermocellum*, *C. cellulolyticum* and *C. cellulovorans* revealed that each cellulosome has a unique set of enzymes. There is a unique and specific interaction between the cohesin and dockerin domains of cellulosomal enzymes. This was known when dockerins present in *C. cellulolyticum* enzymes do not assemble with CipA of *C. thermocellum*. When organisms are developed in different carbon sources, changes in cellulosome composition are also visible, but the makeup of individual cellulosome molecules remains heterogeneous. The interaction of the cohesin–dockerin is helpful in the assembly of cellulosomes. This interaction is valuable for the progress of biosensors. The cohesin markers are merged with fluorescent, a substance that increases the sensitivity and specificity of developed biosensors (Stahl et al., 2012).

In biotechnology and nanotechnology processes, various types of enzymes are used. The recycling of these enzymes is an important issue and different methods are used to overcome this problem. The CBM and cohesin of *C. cellulovorans* were fused and immobilized onto a cellulose matrix that allows the complementary requisite of dockerin-tagged recombinant protein, which can be overturned by using EDTA. In another method, one CBM family from *C. thermocellum* and three types of cohesin–dockerin tags were used for the production of immobilized enzymes in *E. coli*, enhancing the enzymatic reaction as well as enzyme recycling. Nowadays, magnetic cellulose nanoparticles are applied for enzyme removal or enzyme recycling (Liu et al., 2022).

3.3.3.3 Cellulosome Catalytic Components

In addition to dockerin domains, cellulosomal hydrolase plant cell wall has one or two catalytic modules and one or more CBMs. The discovery of multiple proteins that are constituents of these enzyme complexes is possible by gene cloning of cellulosome-containing bacteria. The structural and chemical elaboration of the cellulosome substrate, the cell wall plant, may be reflected in the exceptional diversity of enzymes, which is intrinsically complicated due to their multimodular organization. For example, the cellulosomal catalytic machinery, which comprises cellulases–glucanases, mannanases, xylanases, galactanases, xyloglucanases, arabinanases, xylan, arabinofuranosidases, rhamnogalacturonan lyases, pectic esterases, pectate lyases, and proteinases, and complex polysaccharide-degrading enzymes, like their substrates, are extraordinarily varied and specialized.

3.3.3.4 Carbohydrate-Binding Module (CBM)

CBMs are variable and divided into families based on amino acid sequence. Generally, CBMs are found attached to scaffoldins, which helps in tightly binding with cellulose substrates such as crystalline form and also similar structures like chitin. They also serve as domains of non-cellulosomal cellulases and various cellulosomal.

Glycoside hydrolases, carbohydrate esterases, and polysaccharide lyases are among the enzymes that are classified into various categories based on primary sequence homology and are stored in the CAZy database (cazy.org), which is constantly growing. CBMs are grouped under families in the CAZy database, similar to CAZyme catalytic modules, based on their main structural homology. There are now 86 distinct families of CBM recorded in the CAZy database. CBMs are categorized into three groups based on their function: type A CBMs bind to crystalline polysaccharides (mainly cellulose), type B modules interact with interior sections of single glycan chains and type C CBMs identify tiny saccharides in the context of a complex carbohydrate. The breakdown of complex polysaccharides frequently necessitates the cooperation of many CAZymes. Synergy may occur as a result of novel mechanisms of action towards a particular substrate, such as when an endoglucanase hydrolyzes the substrate, resulting in extra strand ends that can be split by an exoglucanase. This can emerge from the hydrolysis of two closely related substrates, in which one enzyme's activity makes the hidden substrate available to the second enzyme's action (Shoseyov et al., 2006).

Cel48A, formerly known as CelS, and Cel8A, known initially as CelA, are the primary exo- and endocellulases discovered in *C. thermocellum* cellulosomes, according to several proteomic studies. Cel48A has a (α/α)6 barrel fold with a tunnel-shaped substrate-binding region, according to its structure. The enzyme targets the reducing end of cellulose chains. The polymer weaves its way down the tunnel, cleaving the glycosidic link between the second and third glucose residues to release cellobiose (Liu et al., 2022).

A vast number of dockerin-bearing enzymes have been discovered due to genome sequencing of various cellulosome-producing bacteria. Most known cellulosome-producing bacteria produce substantial quantities of a single-family 48 glycoside hydrolase (GH48) exoglucanase, which is a vital component of these structures' very effective cellulolytic function. However, these bacteria, release a large number of glycoside hydrolases from the family 9, which function in tandem with GH48 since GH48 has minimal action on cellulose on its own. Many GH9 enzymes have been identified from various cellulosomic systems, each with unique activity, binding capacities to cellulosic substrates and synergy with the main GH48 exoglucanase. Most of the GH9 enzyme is linked to a CBM3c that does not effectively attach to crystalline cellulose. Instead, CBM3c collaborates with the neighbouring catalytic domain to transiently bind the incoming cellulose chain before it is passed into the GH9 active site for hydrolysis. The deletion of CBM3c modules from GH9 enzymes reduces their activity and causes them to shift to an endo route of attack. As a result of CBM3c's modulator impact, GH9 enzymes work as processive cellulases in an exo-mode way (Liu et al., 2022). The products of GH9–CBM3c hydrolysis may vary. For example, CelR, a prominent cellulosomal GH9 endoglucanase, generates cellotetraose as its principal hydrolysis byproduct. Whether *C. thermocellum* needs such a large number of GH5 and GH9 enzymes to conduct plant cell wall hydrolysis is unknown.

CBM has many unique properties including small size, stability, elasticity, manipulability, robust identification, and the ability to bind with enzymes/microorganisms to expand their function. So, it can be used for various purposes, e.g., in the paper industry, bioremediation, and biomaterials and biofilms. Today, industrialization, urbanization, and the use of chemicals and pesticides have led to the accumulation of heavy metals in the environment that have a major effect on living organisms including humans. Some pollutants or toxic substances are persistent in the soil or water for a longer time and reach the groundwater. Hence, the removal or detoxification of these harmful substances is essential. CBM of bacterial cellulosome was engineered with decontaminating enzymes that made it immobilized on the cellulose cell surface and can be used for the production of various chemicals that degrade the pollutants. For example, the commonly used pesticide atrazine can be destroyed by immobilizing CBM with atrazine chlorohydrolase (Liu et al., 2022). The immobilized cellulosomes act as sorbents for the removal of various heavy metals. Based on specific cohesin–dockerin interactions it can be used to create biosensors that are used as biomarkers to detect the pathogen and environmental pollutants that degrade them. The fusion of CBM with protein–ligand was developed for mammalian cellular applications including cell separation technique, locomotion of cells, and stem cell factors.

One of the major applications of CBM is in the paper industry. Fibers are the basis of paper products. To increase their bonding capacity and strengthen the quality of paper, the fibers are refined and beaten, and sometimes chemicals or enzymes are used. However, the enzymes used cellulase and laccase, have many disadvantages including the swelling and foliation of paper. Recently, CBMs were used alone or along with other negatively charged reagents such as anionic polyacrylamide to increase the bonding with negatively charged fibers and thus the paper strength (Shi et al., 2014). For example, the use of CBM–polyethylene glycol improved the drainability of *Pinus sylvestris* and *Eucalyptus globulus* pulps without affecting the physical properties of the paper.

CBMs can also be tools for substrate-specific or enzyme-substrate recognition as they interact with polysaccharide to form probes and exists in pure form. In the medical field, CBM is able to generate new recognition sites as it binds specifically to insoluble substrates. Thus, CBM helps deliver drugs to their target location (Liu et al., 2022). CBM is environmentally friendly and can be applied in different fields, but preservation, production and transportation are still critical prerequisites for use in industrial applications.

3.4 BACTERIAL CELLULOSOME PRODUCTION

Mesophilic bacteria are known as the bacteria that create basic cellulosome systems; these bacteria release a tiny complex as a single scaffoldin that can include approximately nine enzymatic subunits, depending on the number of cohesin modules on the scaffoldin. *Clostridium cellulolyticum*, *Clostridium cellulovorans*, *Clostridium josui*, *Clostridium papyrosolvens*, *Ruminococcus albus* and *Ruminococcus bromii* are examples of these bacteria. *R. albus* can be obtained from the rumen of cows and includes a large number of dockerin-containing proteins, but just one cohesin sequence was discovered in its genome (Shoseyov et al., 2006). The lack of scaffoldins under this species suggests that these cellulolytic bacteria do not have a true cellulosome, and why this bacterium makes dockerin modules is unknown. *R. bromii*, which can be isolated from a human gut, has the capacity to break down starch but not cellulose. The *Acetivibrio cellulolyticus* genome has scaA, scaB, scaC and scaD scaffoldin genes. The scaffoldin ScaA of *Acetivibrio cellulolyticus* and *Clostridium thermocellum* has similar traits. Dockerin modules found in any of its starch-degrading enzymes have been shown to interact with cohesins found in the genome. In that condition, cellulosomes are also referred to as amylosomes (Doi, 2008).

3.4.1 Artificial Bacterial Cellulosomes

Although the production of cellulosomes from anaerobic microorganisms has done many wonders in the field of biotechnology, they are not able to be used on an industrial scale. The cultivation of native bacteria-producing cellulosome requires an anaerobic condition that is difficult and costly to maintain, further restricting the use of this cellulosome on a large scale. Cellulosome has inspired various inventive biotechnological solutions because of its constructing nature. One of the most prominent and effective projects in the area of cellulosomics is the development of artificial cellulosome chimeras such as designer cellulosomes, and mini-cellulosomes that can easily depolymerize the cellulosic substrates for the production of biofuels by converting them into simple sugars or monomer units (Dorival et al., 2021). Today, the physiology, structure and mechanism of cellulosome gene regulation are well understood. Thus recombinant microorganism is used to construct this artificial cellulosomal complex. Different types of artificial cellulosomes and their applications are illustrated in Table 3.3.

3.4.1.1 *In vitro* Synthesis of Designer Cellulosomes

Designer cellulosomes have been considered as a method for a better understanding of cellulosome function and biotechnological waste management applications. For the effective breakdown of specific substrates, small artificial constructed complexes of enzyme cellulosomes have been created. The designer cellulosome idea is based on the development of chimeric scaffoldins with divergent cohesins and corresponding dockerin-bearing enzymes that have very high affinity and specificity for the degradation of polymeric cellulose. The notion of designer cellulosome was first suggested by Bayer et al. in 1994 but it was constructed by Fierobe et al. (2001).

The production *in vivo* and *in vitro* of the derivatives of cellulosome has been employed for numerous biotechnological reasons, including studying cellulosomal function and attempting to increase the efficiency of native cellulosomes. The nature of their cohesins is the fundamental difference between designer cellulosomes and mini-cellulosomes. The first consists of native scaffoldins that have been shortened and contain cohesins that are similar to those present in the original complexes (Gonclaves et al., 2014). Designer cellulosomes, on the other hand, have a chimeric scaffoldin composed of cohesins of various unique features that can originate from several species, enabling exact control over the position and number of copies of certain enzymes.

Various tactics have been used for the manufacturing of designer cellulosomes such as shortened scaffoldin that has a specific cohesin domain or chimeric scaffoldin that has different cohesin and dockerin moieties. In addition to cellulosomal enzymes, different microorganisms such as *Neocallimastix patriciarum* were also oppressed to get free enzymes for the construction of artificial cellulosomes.

3.4.1.2 Chimeric Scaffoldin

In this type of designer cellulosome, linker segments were used to maintain the distance between various cohesin subunits. The linkers have 5 to 8 residues of amino acids in some microorganisms (*C. acetobutylicum*, *C. cellulolyticum*) and about 20 to 40 residues in other microorganisms (*C. thermocellum*, *R. flavefaciens*). The linkers provide flexibility and the appropriate conformation of designer scaffoldins and cohesins that allow for spatial positioning which helps in appropriate association with substrate. This allows the enzymes to work simultaneously and synergistically.

The cohesin domains were made from many types of species that work with the desired enzymes of dockerin that work efficiently in the hydrolysis of complex substrate molecules.

3.4.1.3 Surface Enzymes

Surface enzymes are another important point to be considered when constructing designer cellulosomes as they protect them from thermal degradation and proteases present in host cells. They help in hydrolyzing cellulose to their monosaccharides by working in a synergistic manner and thus prevent the enzyme desorption irremediable from its substrate, which in turn increases the field of biofuels. A designer cellulosome, protein A cellulosome complex, was prepared by using protein A of *Staphylococcus* bacteria in which the Ig-binding domain acts as a cohesin module and engineered affibody as dockerin modules. This complex along with CBMs bind to the surface of cellulose for

TABLE 3.3
Artificial Cellulosomes Used for Biomass Hydrolysis and Its Application

Artificial DesignedCellulosome	Scaffoldin	Enzyme Source	Substrate	Saccharification or Yield	Application	Reference
CBM3-CgIT Chimeric	*Clostridiumthermocellum*	*Thermoanaerobacter brockii b*-glucosidase	Rice straw	70%	Biofuel	Waeonukul et al., 2013
Mini-cellulosome	Recombinant miniCbpA (CBM-Coh1-Coh2)	Endoglucanase (BsCel5), processive endoglucanase (CtCel9), cellobiohydrolase (CpCel48) from *B. subtilis*, *C. thermocellum*, *C. phytofermentans*	Avicel	Not determined	Biorefinery, bionanotechnology	Hyeon et al., 2016
Nanocluster	CdSe-ZnS core-shell quantum dots	Clostridium thermocellum endoglucanase CelA, Clostridium cellulolyticum exoglucanase CelE	Not determined	Not determined	Biofuel production, bioremediation, drug design	Tsai et al., 2013
Hybrid mini-cellulosomes *Clostridium thermocellum* and *Clostridium cellulolyticum* cohesin	Parental Scaf4	Cel48F, Cel9G	Crystalline cellulose	2-fold	Immobilization, bioethanol production, paper, pulp, pharmaceutical industry	Hyeon et al., 2016
Rosettasome	Group II chaperonins fromthe hyperthermoacidophilic archaeon *S. shibata*	Cel9B, Cel9k, Cel9R, Cel48S (*C. thermocellum*)	Switchgrass	>50 %	Industrial biotechnology	Gonçalves et al., 2014; Chundawat et al., 2016
Cellulases, helical consensus ankyrin protein fusions	Ankyrin	CelA *Clostridium thermocellum*, Cel12A *Thermotoga marítima*	Soluble and insoluble substrates	6.2-fold	Cellular degradation, biofuel production	Gonçalves et al., 2014
Yeast ScafII	Anchoring/adaptor scaffoldin	*Clostridium. thermocellum*, *R. flavefaciens*, *Clostridium cellulolyticum*	Steam-exploded corn stover	28.22% yield	Ethanol production	Tian et al., 2019

(*Continued*)

TABLE 3.3 (CONTINUED)
Artificial Cellulosomes Used for Biomass Hydrolysis and Its Application

Artificial DesignedCellulosome	Scaffoldin	Enzyme Source	Substrate	Saccharification or Yield	Application	Reference
Mini-cellulosome (CMB; celCCA; celCCE; Ccel 2454)	Chimeric mini-scaffoldin protein	Endoglucanase (celCCA), cellobiohydrolase (celCCE), β-glucosidase (Ccel_2454) from *Clostridium cellulolyticum*	Carboxymethyl cellulose, phosphoric acid-swollen cellulose, or Avicel	45% yield	Ethanol production	Hu & Zu, 2019
Recombinant cellulosome *Lactobacillus plantarum*	Chimeric scaffoldins	CBP (*Saccharomyces cerevisiae*) GH5 and GH9 (*C. papyrosolvens* cellulases), Cel9G and Cel5A endoglucanase (*C. cellulolyticum*)	Wheat straw	13% yield	Bioethanol, probiotics	Stern et al., 2018
Mini-cellulosomes in Pichia pastoris X33	Scaffoldin CipA	Cellulose-binding module of *C. acetobutylicum* and Nasutitermes takasagoensis endoglucanase (NtEG)	Carboxymethyl cellulose, microcrystal cellulose, and filter paper	3.2-fold yield	Ethanol production	Hu & Zhu, 2019
Mini CipA:CelA:CBHII:BGLI cellulosome	Mini CipA Scaffoldin	Endoglucanase (*C. thermocellum* CelA), exoglucanase (*Trichoderma reesei* CBHII), β-glucosidase BGLI (*C. thermocellum*, or *Aspergillus aculeatus*)	Birchwood xylan	1.80 g/l yield	Biorefinery	Hu & Zhu, 2019

hydrolysis. Apart from the aforementioned methods, streptavidin, DNA and polystyrene nanoparticles are used in the construction of artificial cellulosomes (Liu et al., 2022).

One of the most actively investigated uses of designer cellulosomes is in the transformation of lignocellulosic biomass into fermentable sugar, which is still a major obstacle in biofuel production. Cellulosomes are used to develop, construct and test unique catalytic multiprotein complexes mixing fungal and bacterial enzymes from non-aggregating systems that show promise in biomass saccharification. Lytic polysaccharide monooxygenases, expansins and laccases are just a few enzymes that are not usually present in cellulosomes but have effectively contributed to increasing the efficiency of cellulosomes with the addition of novel complementing catalytic activity. There have been many attempts to develop genetically altered microorganisms that can produce and display mini-cellulosomes or designer cellulosomes on their cell surface (Hu & Zhu, 2019). The goal is to develop bacteria that can be employed in consolidated bioprocessing (CBP), which entails fermenting the cell walls of plant-producing fuels in a single bioreactor without the need for enzymes. In biorefineries, the designer cellulosome enzymes along with recombinant microorganisms are used for the hydrolysis of complex biomass into biofuels that make the environment clean and green. To increase the thermostability and activity of scaffold in artificial cellulosomes, ankyrin along with endocellulase catalytic domains are used.

3.4.1.4 Thermostable Designer Cellulosomes

Industrial uses of thermophilic microorganisms and enzymatic processes are extremely appealing. Enhanced stability over long periods, improved reaction rates, process flexibility, quicker diffusion rates, improved substrate solubility and reduced contamination problems are just a few of their advantages. Furthermore, the increased growing temperature may make it easier to extract important products like ethanol.

Studies of designer cellulosomes' long-term stability *in vitro* at higher temperatures (>50°C) have found that thermophilic cellulosomes are of great importance in hydrolytic conversion of biomass, enzyme recycling and many industries. *C. thermocellum* and *C. clariflavum* are two microorganisms for thermophilic CBP since they are inherently ethanologenic and may be modified to handle reasonably high ethanol concentrations (Hu & Zhu, 2019). Engineering the metabolic pathways of *C. thermocellum* for the synthesis of ethanol and isobutanol has also been suggested. Simultaneously, the coculture of *C. thermocellum* with microorganisms that are solventogenic in nature is being investigated intensively, since it may reduce the cost of converting lignocellulose into biofuels. Complementary growth with a suitable thermophile, such as *Thermoanaerobacter* spp., can compensate for *C. thermocellum*'s lack of pentose metabolism. *C. clariflavum* has still not been utilized to make solvents since it is a very new cellulosomal species. On the other hand, *C. clariflavum* strain 4-2a is a very appealing candidate for CBP due to its capacity to use pentoses. However, other difficulties, such as enzymatic hydrolysis using solvents and alcohol, and the expense of growing supporting anaerobic conditions remain obstacles to the successful implementation of CBP (Vazana et al., 2012)

Although enzymes are used in various industries, due to their thermosensitive properties they can't be recycled thus increasing the cost. The constructed thermostable designer cellulosome enzymes come from thermophilic microbes like *C. straminosolvens*, *C. thermocellum*, *C. clariflavum*, *A. fulgidus* and their close species. These designer cellulosomes along with scaffoldin can be recycled and are applied in many industries such as clearing clogged pipes, wastewater treatment and biomining. However, due to the narrow stability of scaffoldin components and interactions between cohesin–dockerin in designer cellulosomes, their application has been limited. The ideal technique for the creation of thermophilic CBP is to design a cellulosome-producing thermophilic microbe as the manufacture of biofuels while creating a solventogenic thermophile to hydrolyze lignocellulose needs the introduction of additional enzymes to accomplish total biomass breakdown (Gilmore et al., 2015). However, more advances in thermophilic designer cellulosomes have yet to be discovered for spontaneously converting plant biomass into biofuels.

3.4.2 *In vivo* Synthesis of Mini-Cellulosome

Generally, bacteria are used for *in vivo* synthesis of mini-cellulosomes as it is easy to alter their genes, have fast growth capacity and are able to secrete proteins extracellularly (Wang et al., 2021). There are many examples of *in vivo* synthesis of mini-cellulosomes. Some of them are as follows:

- *Bacillus subtilus* is used because it has many scaffoldin units, endoglucanase enzymatic activity and other important structures required for the conversion of complex biomass into useful products.
- *Clostridium acetobutylicum* has scaffoldins from *C. thermocellum* and *C. cellulolyticum* that bind with complementary dockerins. It can convert polysaccharides into solvents and acids.
- Cloning of *C. acetobutylicum* and *C. cellulolyticum* has been done for the synthesis of mannanase and scaffoldin complex that degrades the complex network of galactomannan.
- Apart from bacteria, modified yeast is also used for the synthesis of mini-cellulosomes. On cell surfaces of *Saccharomyces cerevisiae*, manipulation can be done by embedding *Aspergillus aculeatus* beta-glucosidase and endoglucanase, cellobiohydrolase, and the alpha-agglutinin surface display system from *T. reesei* that help in the production of second-generation biofuel by the fermentation process. The simultaneous use of the fermentation process and enzymatic hydrolysis has increased the yield of products as it removes the inhibiting hydrolytic end products.

Thus, the construction of cellulosome chimeras or mini-cellulosomes has revolutionized the world of nanotechnology and modern chemistry. The construction of designer cellulosomes is not an easy task but it proved to be an attractive alternative for the emerging bioenergy field, in particular for the production of bioethanol from lignocellulosic biomass. However, apart from the innumerable benefits, there are some drawbacks. Designer cellulosomes have species-specific cohesin–dockerin interactions that narrow their utilization and they are also less active than their natural counterparts. Studies are underway to construct multiple scaffoldin, broad substrate-range mini-cellulosomes and engineered microorganisms or engineered cellulosomal protein complexes to increase the efficacy at industrial scale for the production of biofuels.

3.5 FACTORS AFFECTING CELLULOSOMES

Cellulosomes are the protein complex from *C. thermocellum* that degrade the biomass lignocellulose. These cellulosomes were responsible for degrading the cellulose substrates. The construction of multienzyme complexes and many advances in cellulosomes have accelerated enzyme reactions and avoided enzyme-to-enzyme interactions (Wang et al., 2019). This self-assembly cascades of cellulosomes has been used in various field. But still, some factors affect the yield of products. They are explained in the following sections.

3.5.1 Substrate-Related Factors

This includes lignin assay, surface properties like hydrogen binding, hydrophobicity, electrostatic interaction, accessibility to the inner pore and pretreatment methods. Lignocellulosic biomass has a low exploitable interfacial area with merely an outer surface that is not suitable for microbial digesting. Commercial exploitation of lignocellulosic biomass into value-added bioproducts relies heavily on preprocessing. To initiate biocatalysis, direct physical interaction between cellulosome producers and lignocellulosic surfaces is required. The lignocellulosic biomass consists of external and internal surfaces, of which the external surface area is determined by the material's size and the internal surface area is determined by the pore size and distribution. Several enzymes on cellulosomes can compensate for cellulosomes' inability to assault the binding sites in the pores. As a result, the accessible surface area has a substantial impact on the digestibility of lignocellulose for cellulosomes, which will gradually rise with the enzymatic hydrolysis induced by the elimination of partial cellulose and hemicellulose. Aside from surface area, the cellulosomic hydrolysis rate is also affected by the hydrolysis stages. Despite the availability of more surface area, the rate of hydrolysis is generally brisk in the beginning and becomes significantly slower in the later stages. Higher crystallinity areas of the substrates, inactivation of the hydrolytic enzymes and rising lignin concentrations should all contribute to the slower hydrolysis rate (Wang et al., 2019).

Pretreatment, on the other hand, is a critical step in increasing the accessible surface area of the substrate for cellulosomes. Cellulosome actions are measured either immediately after mechanical size reduction or after an additional preparation with sodium hypochlorite to lower lignin content and increase enzymatic hydrolysis. The results showed that using both native and designer cellulosomes had no significant influence on the lignin concentration of the substrate when no chemical pretreatment was used (Wang et al., 2019). Despite the capacity of microbial enzymes to dissolve lignin and improve cellulase access to cellulose, cellulosomes scarcely reduce cellulose content and/or degrade hemicellulose without lignocellulosic substrate preparation (Van & Pletschke, 2012). However, chemical pretreatment of lignocellulose before enzymatic digestion was shown to produce lignocellulose-derived byproducts such as phenolic compounds, which inhibited enzymatic saccharification even more. Pretreatment activities modify multiple important elements at the same time, making it difficult to forecast their efficacy directly. Transesterification or simply the yields of target products via fermentation are commonly used to assess the success of pretreatment.

3.5.2 Carbon Source Factors

Extracellular carbon sources regulate the enzymes and structural components of the cellulosome, ensuring the presence of appropriate elements to break down the available carbs. Although cellulosome enzyme complexes can degrade about 90% of native or treated cellulosic substrates, there are indeed significant hurdles to the successful hydrolysis of various carbon sources by cellulosome enzyme complexes. Several prior investigations have shown that cellulosome compositions change when cellulosome producers are grown on different carbon sources. For example, when *C. cellulovorans* ATCC 35296 was grown on carbon substrates such as Avicel, xylan and polysaccharides, the cellulosome was found to be heterogeneous, although the scaffoldin protein CbpA, endoglucanase EngE and cellobiohydrolase ExgS were all rather stable. Cellulosomes containing CbpA, EngE/EngK, ExgS/EngH and EngL were found to promote cellulase activity in cells grown on a variety of carbon sources, whereas cellulosomes derived from cellulose, pectin and mixed carbon had higher amounts of XynB, XynA and unknown proteins (35–45 kDa), and had higher xylanase activity. These findings suggest that the autogenous regulatory system in *C. cellulovorans* regulates the proportion of cellulosomal subpopulations which compose the cellulosomal population (Van et al., 2012).

Various carbon sources have a considerable impact on the expression profiles of both cellulosomes and non-cellulosomal enzymes, but cellulosomal proteomes were more influenced than non-cellulosomal proteomes. It is well established that the buildup of carbon monomers like glucose, cellobiose and other hydrolysis end products inhibits cellulases and reduces glucose yields (Van et al., 2012). In addition to aerobic cellulase, cellulosome kinetics research is limited due to its complex structure, and catalytic mechanism describe the use of a recombinant *B. subtilis* anchoring cellulosome.

3.5.3 Cellulosome-Related Factors

The potential of cellulosomes to digest cellulose is greatly affected by changes in pH and temperature. The cellulosomal enzyme activities of *C. thermocellum* JYT01 were investigated with the availability of formate, acetate, and lactate to determine the effects of organic acid anions on cellulosome-induced cellulose hydrolysis. Surprisingly, while these anions hindered cell growth, they also acted as promoters of cellulosome activity at different concentrations, whereas negative effects were only realized at their critical concentrations.

3.5.4 Chemical Factors

Enhancing cellulosome function and/or developing a cohesiveness between cellulosomic saccharification and succeeding fermentation can potentially boost CBP efficiency. Cellular growth and fermentation efficiency are frequently inhibited by industrial chemicals created by the pretreatment process or derived from fermentation byproducts. Ethanol levels beyond 2% (v/v) decreased the cellulosome activity of the wild-type strain, whereas those evolved strains remained alive when ethanol concentrations were increased up to 8% (v/v). *C. thermocellum* cellulosomes were found to handle higher ethanol concentrations than commercial enzymes. Furthermore, furfural and phenols are often considered negative side products generated during the processing of lignocellulosic biomass because of their inhibition effect on cell growth. Cellulosomes from *C. thermocellum* showed tolerance to different concentrations of furfural and p-hydroxybenzoic acid. Furthermore, researchers created an *ex vivo* cellulosome-microbe complex on the *Bacillus subtilis* surface that included a mini-scaffoldin coupled to three dockerin-containing cellulase components, namely endoglucanase Cel5, processive endoglucanase Cel9 and cellobiohydrolase Cel48. β-glucosidase without a CBM normally acts in the bulk phase; the hydrolysis results show that a high concentration of cellodextrins in the boundary layer would impede cellulosome activity more effectively than short-chain products. As a result, cellulosomes that speed up the cellulose bioconversion rate can aid in the development of CBP bacteria that can hydrolyze refractory substrate efficiently at low secretory cellulase levels (Xu et al., 2010).

3.6 CONCLUSIONS AND PERSPECTIVES

Biomass, especially cellulose, hemicellulose and lignin, constitute a primary carbon and energy source. Only a limited number of natural bacterial cellulosomes have the ability to degrade the structural biomass. Different techniques have been used to develop sparingly viable and cost-effective cellulolytic enzymes from bacterial sources to hydrolyze biomass for the production of biofuels. This second-generation biofuel production has proven to be a boon to the environment as it decreases global warming, the energy crisis and environmental pollution.

Bacterial cellulosomes provide great potential for a range of future applications in biomass disintegration. The effective breakdown of plant cell walls, notably cellulose, is made possible by these complex protein combinations produced by certain bacteria. The known benefits of bacterial cellulosomes include their higher penetration in cellulose and higher catalytic rate; however, further study is required to maximize their production and utilization while understanding the way they work. Applications of bacterial cellulosomes are limited by their low productivity, which causes high financial input in the upstream process itself. Bacterial cellulosomes can improve the conversion of cellulose into fermentable sugars, which can subsequently be fermented into bioethanol or maybe butanol due to their capacity to break down complex plant cell walls. Cellulosome research may provide a spotlight on the composition and mechanism of cellulolytic enzymes. In order to develop more effective and reliable enzymes for industrial use, researchers may be able to improve how these enzymes interact inside the cellulosome complex. This could result in the creation of better enzyme mixtures for biomass degradation that have increased stability and catalytic activity.

Utilizing cellulosomes' characteristics is a major field of interest in biotechnology. Because they include certain cohesins, "mini-cellulosomes" may be produced that exclusively bind to particular enzymes. It is believed that mini-cellulosome constructs will eventually be employed to create synthetic metabolic pathways that enable the synthesis of any desired product. These constructs have been used with a great effect on cellulosome research in the study of the synergistic effects of cellulosomal enzymes and the cohesin–dockerin interaction.

Cellulosomes (designer or mini-cellulosomes) developed by the assembly of catalytic components (cohesin–dockerin interactions) of different microorganisms can be provided with specialized configuration through genomics, proteomics, molecular and structural biology, and nanotechnology. In order to make non-cellulose-degrading organisms become cellulose degraders, heterologous expression of cellulosome genes has attracted a lot of attention. Further advances in cellulosomes can be primed by the construction of recombinant cellulosomes, yeast-produced cellulosomes that can be used for cellulose hydrolysis, fermentation of sugars, enzyme recycling, wastewater treatment and production of biofuels. Thus, the future of cellulosomes is enormous and bright that can be precisely applied in various fields such as biosensors, bioadsorption, biorefineries, bioremediation and innumerable industries.

REFERENCES

Al-Battashi, H., Annamalai, N., Al-Kindi, S., Nair, A. S., Al-Bahry, S., Verma, J. P., & Sivakumar, N. (2019). Production of bioplastic (poly-3-hydroxybutyrate) using waste paper as a feedstock: Optimization of enzymatic hydrolysis and fermentation employing *Burkholderia sacchari. Journal of Cleaner Production, 214*, 236–247.

Alvarez, A., Saez, J. M., Costa, J. S. D., Colin, V. L., Fuentes, M. S., Cuozzo, S. A., & Amoroso, M. J. (2017). Actinobacteria: Current research and perspectives for bioremediation of pesticides and heavy metals. *Chemosphere, 166*, 41–62.

Alves, V. D., Fontes, C. M., & Bule, P. (2021). Cellulosomes: Highly efficient cellulolytic complexes. *Sub-Cellular Biochemistry*, 96, 323–354.

Artzi, L., Bayer, E. A., & Moraïs, S. (2017). Cellulosomes: Bacterial nanomachines for dismantling plant polysaccharides. *Nature Reviews in Microbiology, 15*(2), 83–95.

Busi, S., Chatterjee, R., Rajkumari, J., & Hnamte, S. (2016). Ecofriendly biosorption of dyes and metals by bacterial biomass of Aeromonas hydrophila RC1. *Journal of Environmental Biology, 37*(2), 267.

Champreda, V., Mhuantong, W., Lekakarn, H., Bunterngsook, B., Kanokratana, P., Zhao, X. Q. & Eurwilaichitr, L. (2019). Designing cellulolytic enzyme systems for biorefinery: From nature to application. *Journal of Bioscience Bioengineering*, 128(6):637–654. doi: 10.1016/j.jbiosc.2019.05.007.

Chin, K. L., Nurliyana, M. Y., H'ng, P. S., Lee, C. L., Go, W. Z., Khoo, P. S., Raja Nazrin, R. A., & Ashikin, S. N. (2020). Effects of bacterial bio-augmentation on the methane potential from facultative digestion of palm oil mill effluent and empty fruit bunch. *Waste and Biomass Valorization, 11*(7), 3407–3418.

Chukwuma, O. B., Rafatullah, M., Tajarudin, H. A., & Ismail, N. (2021). A review on bacterial contribution to lignocellulose breakdown into useful bio-products. *International Journal of Environmental Research and Public Health, 18*(11), Article 11. https://doi.org/10.3390/ijerph18116001

Chundawat, S. P., Paavola, C. D., Raman, B., Nouailler, M., Chan, S. L., Mielenz, J. R., & Dale, B. E. (2016). Saccharification of thermochemically pretreated cellulosic biomass using native and engineered cellulosomal enzyme systems. *Reaction Chemistry and Engineering, 1*(6), 616–628.

Desvaux, M. (2005). The cellulosome of Clostridium cellulolyticum. *Enzyme and Microbial Technology, 37*(4), 373–385.

Detman, A., Bucha, M., Simoneit, B. R. T., Mielecki, D., Piwowarczyk, C., Chojnacka, A., Błaszczyk, M. K., Jędrysek, M. O., Marynowski, L., & Sikora, A. (2018). Lignite biodegradation under conditions of acidic molasses fermentation. *International Journal of Coal Geology, 196*, 274–287.

Ding, S. Y., Xu, Q., Crowley, M., Zeng, Y., Nimlos, M., Lamed, R., & Himmel, M. E. (2008). A biophysical perspective on the cellulosome: New opportunities for biomass conversion. *Current Opinion in Biotechnology, 19*(3), 218–227.

Doi, R. H. (2008). Cellulases of mesophilic microorganisms: Cellulosome and noncellulosome producers. *Annals of the New York Academy of Sciences, 1125*(1), 267–279.

Dorival, J., Moraïs, S., Labourel, A., Rozycki, B., Cazade, P. A., Dabin, J., & Czjzek, M. (2021). Mapping the deformability of natural and designed cellulosomes in solution. *Biotechnology for Biofuels*, 15, 68 . https://doi.org/10.1186/s13068-022-02165-3

Fierobe, H. P., Mechaly, A., Tardif, C., Belaich, A., Lamed, R., Shoham, Y., Belaich, J. P., & Bayer, E. A. (2001). Design and production of active cellulosome chimeras: Selective incorporation of dockerin-containing enzymes into defined functional complexes. *Journal of Biological Chemistry, 276*(24), 21257–21261.

Fontes, C. M. G. A., & Gilbert, H. J. (2010). Cellulosomes: Highly efficient nanomachines designed to deconstruct plant cell wall complex carbohydrates. *Annual Review of Biochemistry, 79*(1), 655–681. https://doi.org/10.1146/annurev-biochem-091208-085603

Froese, A. G., Nguyen, T. N., Ayele, B. T., & Sparling, R. (2020). Digestibility of wheat and cattail biomass using a co-culture of thermophilic anaerobes for consolidated bioprocessing. *Bioenergy Research, 13*(1), 325–333.

Gilmore, S. P., Henske, J. K., & O'Malley, M. A. (2015). Driving biomass breakdown through engineered cellulosomes. *Bioengineered, 6*(4), 204–208.

Gonçalves, G. A., Mori, Y., & Kamiya, N. (2014). Biomolecular assembly strategies to develop potential artificial cellulosomes. *Sustainable Chemical Processes, 2*(1), 1–5.

Guo, H., Hong, C., Zheng, B., Lu, F., Jiang, D., & Qin, W. (2017). Bioflocculants' production in a biomass-degrading bacterium using untreated corn stover as carbon source and use of bioflocculants for microalgae harvest. *Biotechnology for Biofuels, 10*(1), 306. https://doi.org/10.1186/s13068-017-0987-6

Guragain, Y. N., Bastola, K. P., Madl, R. L., & Vadlani, P. V. (2016). Novel biomass pretreatment using alkaline organic solvents: A green approach for biomass fractionation and 2, 3-butanediol production. *Bioenergy Research, 9*(2), 643–655.

Haldar, D., Sen, D., & Gayen, K. (2016). A review on the production of fermentable sugars from lignocellulosic biomass through conventional and enzymatic route—A comparison. *International Journal of Green Energy, 13*(12), 1232–1253.

Hu, B. B., & Zhu, M. J. (2019). Reconstitution of cellulosome: Research progress and its application in biorefinery. *Biotechnology and Applied Biochemistry, 66*(5), 720–730.

Hyeon, J. E., Shin, S. K., & Han, S. O. (2016). Design of nanoscale enzyme complexes based on various scaffolding materials for biomass conversion and immobilization. *Biotechnology Journal, 11*(11), 1386–1396.

Ichikawa, S., Ogawa, S., Nishida, A., Kobayashi, Y., Kurosawa, T., & Karita, S. (2019). Cellulosomes localise on the surface of membrane vesicles from the cellulolytic bacterium Clostridium thermocellum. *FEMS Microbiology Letters, 366*(12), 1–9.

Jensen, M. B., De Jonge, N., Dolriis, M. D., Kragelund, C., Fischer, C. H., Eskesen, M. R., & Kofoed, M. V. W. (2021). Cellulolytic and xylanolytic microbial communities associated with lignocellulose-rich wheat straw degradation in anaerobic digestion. *Frontiers in Microbiology, 12*, 645174.

Jiang, Y., Zhang, X., Yuan, H., Huang, D., Wang, R., Liu, H., & Wang, T. (2021). Research progress and the biotechnological applications of multienzyme complex. *Applied Microbiology and Biotechnology, 105*(5), 1759–1777.

Joi, S., & Sharma, S. (2022). Antimicrobial applications of rhamnolipid biosurfactant produced from *Achromobacter* sp. (PS1) isolate using lignocellulosic hydrolysate. *Letters in Applied NanoBioscience,* 11(2), 4263–4271.

Jones, J. A. D., Kerr, R. G., Haltli, B. A., & Tinto, W. F. (2018). Temperature and pH effect on glucose production from pretreated bagasse by a novel species of Citrobacter and other bacteria. *Heliyon, 4*(6), e00657.

Kanokratana, P., Wongwilaiwalin, S., Mhuantong, W., Tangphatsornruang, S., Eurwilaichitr, L., & Champreda, V. (2018). Characterization of cellulolytic microbial consortium enriched on Napier grass using metagenomic approaches. *Journal of Bioscience and Bioengineering, 125*(4), 439–447.

Lin, C. Y., & Lu, C. (2021). Development perspectives of promising lignocellulose feedstocks for production of advanced generation biofuels: A review. *Renewable and Sustainable Energy Reviews, 136*, 110445.

Liu, Y., Wang, P., Tian, J., Seidi, F., Guo, J., Zhu, W., & Song, J. (2022). Carbohydrate-binding modules of potential resources: Occurrence in nature, function, and application in fiber recognition and treatment. *Polymers, 14*(9), 1806.

Ma, J., Zhang, K., Huang, M., Hector, S. B., Liu, B., Tong, C., Liu, Q., Zeng, J., Gao, Y., Xu, T., et al. (2016). Involvement of Fenton chemistry in rice straw degradation by the lignocellulolytic bacterium *Pantoea ananatis* Sd-1. *Biotechnology for Biofuels, 9*, 211

Maia, M. R., Marques, S., Cabrita, A. R., Wallace, R. J., Thompson, G., Fonseca, A. J., & Oliveira, H. M. (2016). Simple and versatile turbidimetric monitoring of bacterial growth in liquid cultures using a customized 3D printed culture tube holder and a miniaturized spectrophotometer: Application to facultative and strictly anaerobic bacteria. *Frontiers in Microbiology, 7*, 1381.

Mayer, F. (2019). The bacterial cellulosome-mini-review. *Laboratory Medicine, 1*(1), 102–105.

Mohanram, S., Amat, D., Choudhary, J., Arora, A., & Nain, L. (2013). Novel perspectives for evolving enzyme cocktails for lignocellulose hydrolysis in biorefineries. *Sustainable Chemical Processes, 1*(1), 15. https://doi.org/10.1186/2043-7129-1-15

Ohara, H., Karita, S., Kimura, T., Sakka, K., & Ohmiya, K. (2000). Characterization of the cellulolytic complex (cellulosome) from *Ruminococcus albus. Bioscience, Biotechnology and Biochemistry, 64*(2), 254–260.

Parapouli, M., Vasileiadis, A., Afendra, A. S., & Hatziloukas, E. (2020). Saccharomyces cerevisiae and its industrial applications. *AIMS Microbiology, 6*(1), 1.

Qin, L., Li, W. C., Zhu, J. Q., Li, B. Z., & Yuan, Y. J. (2017). Hydrolysis of lignocellulosic biomass to sugars. In *Production of Platform Chemicals from Sustainable Resources* (pp. 3–41). Singapore: Springer.

Saratale, G. D., Saratale, R. G., Ghodake, G. S., Jiang, Y. Y., Chang, J. S., Shin, H. S., & Kumar, G. (2017). Solid state fermentative lignocellulolytic enzymes production, characterization and its application in the saccharification of rice waste biomass for ethanol production: An integrated biotechnological approach. *Journal of the Taiwan Institute of Chemical Engineers, 76*, 51–58.

Shao, L., Hu, S., Yang, Y., Gu, Y., Chen, J., Yang, Y., & Yang, S. (2007). Targeted gene disruption by use of a group II intron (targetron) vector in *Clostridium acetobutylicum. Cell Research, 17*(11), 963–965.

Sharma, A., Balda, S., Capalash, N., & Sharma, P. (2022). Engineering multifunctional enzymes for agro-biomass utilization. *Bioresource Technology*, 347, 126706. doi: 10.1016/j.biortech.2022.126706.

Shi, P., Qiu, Z., Bai, Y., Yuan, T., Huang, H., Pan, X., Yang, P., Zhang, W., & Yao, B. (2012). A new xylanase from Streptomyces megasporus DSM 41476 with high yield of xylobiose. *World Journal of Microbiology and Biotechnology*, 28(2), 687–692. https://doi.org/10.1007/s11274-011-0863-7.

Shoseyov, O., Shani, Z., & Levy, I. (2006). Carbohydrate binding modules: Biochemical properties and novel applications. *Microbiology and Molecular Biology Reviews, 70*(2), 283–295.

Stahl, S. W., Nash, M. A., Fried, D. B., Slutzki, M., Barak, Y., Bayer, E. A., & Gaub, H. E. (2012). Single-molecule dissection of the high-affinity cohesin–dockerin complex. *Proceedings of the National Academy of Sciences, 109*(50), 20431–20436.

Stern, J., Moraïs, S., Ben-David, Y., Salama, R., Shamshoum, M., Lamed, R., Shoham, Y., Bayer, E. A., & Mizrahi, I. (2018). Assembly of synthetic functional cellulosomal structures onto the cell surface of Lactobacillus plantarum, a potent member of the gut microbiome. *Applied and Environmental Microbiology, 84*(8), e00282-18.

Straub, C. T., Khatibi, P. A., Otten, J. K., Adams, M. W. W., & Kelly, R. M. (2019). Lignocellulose solubilization and conversion by extremely thermophilic *Caldicellulosiruptor bescii* improves by maintaining metabolic activity. *Biotechnology and Bioengineering, 116*(8), 1901–1908.

Tamaru, Y., Miyake, H., Kuroda, K., Ueda, M., & Doi, R. H. (2010). Comparative genomics of the mesophilic cellulosome-producing *Clostridium cellulovorans* and its application to biofuel production via consolidated bioprocessing. *Environmental Technology, 31*(8–9), 889–903.

Tamboli, D. P., Telke, A. A., Dawkar, V. V., Jadhav, S. B., & Govindwar, S. P. (2011). Purification and characterization of bacterial aryl alcohol oxidase from *Sphingobacterium* sp. ATM and its uses in textile dye decolorization. *Biotechnology and Bioprocess Engineering, 16*(4), 661–668.

Tian, S., Du, J. L., Bai, Z. S., He, J., & Yang, X. S. (2019). Design and construction of synthetic cellulosome with three adaptor scaffoldins for cellulosic ethanol production from steam-exploded corn stover. *Cellulose, 26*(15), 8401–8415.

Ticona, A. R., Ullah, S. F., Hamann, P. R., Lopes, F. A., & Noronha, E. F. (2021). *Paenibacillus barengoltzii* A1_50L2 as a source of plant cell wall degrading enzymes and its use on lignocellulosic biomass hydrolysis. *Waste and Biomass Valorization, 12*(1), 393–405.

Tsai, S. L., Park, M., & Chen, W. (2013). Size-modulated synergy of cellulase clustering for enhanced cellulose hydrolysis. *Biotechnology Journal, 8*(2), 257–261.

Tsegaye, B., Balomajumder, C., & Roy, P. (2018). Biodegradation of wheat straw by *Ochrobactrum oryzae* BMP03 and Bacillus sp. BMP01 bacteria to enhance biofuel production by increasing total reducing sugars yield. *Environmental Science and Pollution Research, 25*(30), 30585–30596.

Van Dyk, J. S., & Pletschke, B. (2012). A review of lignocellulose bioconversion using enzymatic hydrolysis and synergistic cooperation between enzymes—Factors affecting enzymes, conversion and synergy. *Biotechnology Advances, 30*(6), 1458–1480.

Vazana, Y., Morai, S., Barak, Y., Lamed, R., & Bayer, E. A. (2012). Designer cellulosomes for enhanced hydrolysis of cellulosic substrates. *Methods in Enzymology, 510*, 429–452.

Waeonukul, R., Kosugi, A., Prawitwong, P., Deng, L., Tachaapaikoon, C., Pason, P., Ratanakhanokchai, K., Saito, M., & Mori, Y. (2013). Novel cellulase recycling method using a combination of *Clostridium thermocellum* cellulosomes and *Thermoanaerobacter brockii* β-glucosidase. *Bioresource Technology, 130*, 424–430.

Wang, H., Jiang, X., Qian, Y., & Yin, L. (2021). Constructing an efficient *Bacillus subtilis* spore display by using cohesin–dockerin interactions molecules, *26*(4), 1186.

Wongwilaiwalin, S., Mhuantong, W., Champreda, V., Tangphatsornruang, S., Panichnumsin, P., Ratanakhanokchai, K., & Tachaapaikoon, C. (2018). Structural and metabolic adaptation of cellulolytic microcosm in co-digested Napier grass-swine manure and its application in enhancing thermophilic biogas production. *RSC Advances*, *8*(52), 29806–29815.

Xing, B. S., Han, Y., Wang, X. C., Wen, J., Cao, S., Zhang, K.; Li, Q., & Yuan, H. (2020). Persistent action of cow rumen microorganisms in enhancing biodegradation of wheat straw by rumen fermentation. *Science of the Total Environment*, *715*, 136529.

Xu, C., Qin, Y., Li, Y., Ji, Y., Huang, J., Song, H., & Xu, J. (2010). Factors influencing cellulosome activity in consolidated bioprocessing of cellulosic ethanol. *Bioresource Technology*, *101*(24), 9560–9569. https://doi.org/10.1016/j.biortech.2010.07.065

4 Fungal β-Glucosidases
Production and Applications

Reshma M. Mathew, Meena Sankar, Athira Sreeja-Raju, Adarsh Velayudhanpillai Prasannakumari, Athulya Anoop Puthiyamadam, Meera Christopher, Jedy Jose, Valan Rebinro, and Rajeev K. Sukumaran

4.1 INTRODUCTION

Annually, plants convert about 100 billion tons of carbon dioxide into biomass. Most of this biomass is in the form of lignocellulose—a complex of lignin and the carbohydrate polymers cellulose and hemicellulose—that strengthens cell walls to enable vertical growth and provide protection against microbial attack. Presently, the earth is reeling from the unprecedented effects of climate change, which has prompted global efforts to mitigate the climate crisis and limit changes in temperature by switching to sustainable raw materials for the production of fuels and materials. Among others, one of the most promising sources of renewable energy is biomass, which can be processed into a number of clean-burning fuels, including bioethanol, biodiesel and biomethane. For developing countries that depend on crude oil imports to meet their domestic needs, biomass-based technologies ensure energy security, in addition to providing compliance with various treaties on climate change.

Lignocellulosic biomass, which encompasses agricultural, forestry, and industrial wastes and residues, can be upgraded to valuable chemicals and fuels via both thermochemical and biochemical processing. Thermochemical methods like pyrolysis and gasification use heat and pressure to mimic the geologic processes that result in the formation of crude oil and other fossil-based resources to produce bio-oil and syngas. Biochemical methods employ biological catalysts to release the sugars contained in lignocellulose, which are then fermented by microorganisms into usable fuels like ethanol and butanol, and other products like platform chemicals, solvents, organic acids and biopolymers. Compared to the thermochemical routes, biochemical methods offer advantages such as greener processing, mild operating conditions, higher productivities per unit quantity of feedstock and better energy efficiency of the overall process.

However, the lignocellulosic matrix has been evolutionarily engineered in such a way as to resist brutal environmental pressures. The composition of the cell wall differs among different types of plants, and softwood plants typically contain the highest percentage of lignin, while hardwood plants and grasses contain higher percentages of polysaccharides. Cellulose is a linear polymer of 500–15,000 glucose units that are linked by β-1,4 bonds. It makes up to 45% of the dry weight of biomass. Compact fibrils of cellulose are formed by hydrogen bonds that link chains of cellulose into fibrous strands. Hemicellulose is composed of several sugars including xylan, mannan, galactan and arabinan. Xylan is the major component of hemicellulose and is linked with β-1,4 glycosidic bonds. It makes up 25%–50% of the total dry weight in non-woody grasses, 15%–30 % in hardwood species, and 7%–10% in softwood species. The other five-carbon sugars associated with hemicellulose are connected to xylan by various types of linkages including α-1,2 and α-1,3 bonds. Hemicellulose is linked to cellulose in the plant cell wall by hydrogen bonding and linked to lignin by ionic interactions. Lignin is an amorphous highly branched polymer present in vascular plants and accounts for up to 30% of the dry weight of the plant biomass. The lignin polymer is composed

 DOI: 10.1201/9781003335313-5

of phenylpropane units made up of three types of monolignols: coniferyl alcohol with an aryl-OCH_3 group, sinapyl alcohol with two aryl-OCH_3 groups and p-coumaryl alcohol. These phenylpropane units are linked by β-aryl ether, di-aryl propane, biphenyl, diaryl ether, phenylcoumarin, spirodienone and pinoresinol bonds.

Further, cellulose, hemicellulose and lignin form structures called microfibrils, which are organized into macrofibrils that mediate the structural stability of the plant cell walls. Hydrogen bonds between the different layers of the polysaccharides further contribute to the resistance to microbial and oxidative degradation. All these levels of intra- and intermolecular bonding make it extremely challenging to extract fermentable sugars from the biomass, necessitating the use of complex multienzyme systems consisting of hydrolases and oxidases to break down the substrate. In nature, fungi, bacteria and protozoans secrete *cellulases* to synergistically catalyze the decomposition of cellulosic material. These enzymes sequentially cleave the 1,4-β-D-glycosidic linkages in the cellulose molecule to yield shorter polysaccharides, oligosaccharides and finally the monosaccharide component: D-glucose. According to the type of reaction catalyzed, cellulases are of three major types: endocellulases/endoglucanases (EC 3.2.1.4) randomly cleave internal bonds in the cellulose chain to create new chain ends; exocellulases/cellobiohydrolases (EC 3.2.1.91) processively cleave two to four glucose units from the ends of the exposed chains produced by endoglucanases, resulting in the generation of disaccharides (cellobiose) or oligo saccharides; and cellobiases/β-glucosidases (EC 3.2.1.21) hydrolyze the exocellulase product into individual monosaccharides.

β-glucosidases act on the β-(1,4) bond linking two glucose residues or glucose-substituted molecules, resulting in the release of glucose units. In nature, BGLs have a myriad of functions, such as biomass conversion in microorganisms, breakdown of glycolipids and process of lignification, defence against pests, phytohormones activation, catabolism of cell wall in plants, and mediation of both plant–microbes and plant–insect interactions. BGLs have a wide range of industrial applications including their use in biorefineries for glucose generation from biomass, in food and flavour processing, surfactant, cosmetic industry, and tobacco industry. In the hydrolysis of lignocellulose, the product of cellobiohydrolase (CBH) action, i.e., cellobiose, is an inhibitor of both CBHs and endoglucanases. Therefore, the rate of hydrolysis depends on the levels of BGL activity present, i.e., BGL is the rate-limiting enzyme in lignocellulose saccharification.

Among the natural producers of BGLs, filamentous fungi are generally regarded as cell factories for the production of these enzymes since they can attain rapid growth on low-cost cultivation media to secrete high titres of protein. Even though BGLs are produced by different genera and species of the fungal kingdom including Ascomycetes and Basidiomycetes, the ascomycete genera *Aspergillus* has been especially utilized for commercial BGL production, with *Aspergillus niger* being the industrial standard in BGL production. Considering the versatile nature of the enzyme and its indisputable importance for biorefinery applications, a lot of studies have been undertaken to elucidate the structure, functions, mechanism of action and dynamic properties of industrially useful BGLs. This chapter recapitulates our current knowledge of the roles of BGLs, especially fungal BGLs, in lignocellulose degradation, as well as the ongoing efforts to improve and expand the functional range of the enzymes by rational engineering.

4.2 β-GLUCOSIDASES: CLASSIFICATION AND MODE OF ACTION

β-Glucosidase (β-D-glucoside glucohydrolase, EC 3.2.1.21), which cleaves β-1,4 glucosidic linkages, is one of the widely studied significant enzymes that form the bottleneck in the degradation of lignocellulosic biomass and bioethanol production. β -glucosidases can be classified into different types based on their (a) substrate specificity and (b) nucleotide sequence identity (NSI) and folding similarities (Henrissat and Bairoch, 1996). Based on their substrate specificity, BGLs can be divided into three categories: (i) aryl-β-glucosidases, (ii) cellobiases and (iii) broad substrate specific enzymes. The CAZy database (www.cazy.org) shows the classification and distribution of glycosidases based on NSI into 176 glycoside hydrolase (GH) families, 214 subfamilies and 69 clans (last updated on 19 January 2023). β-glucosidases are generally grouped in GH1 and GH3 families of glycosyl hydrolases, but are also found in GH5, GH9 and GH30. GH1 includes β-glucosidases from archaebacteria, plants and mammals. They are members of the 4/7 superfamily with a common 8 fold β/α barrel domain (Jenkins et al., 1995). GH3 contains BGLs of bacterial, mold and yeast origin. Most of the fungal β-glucosidases belong to this group. They contain α/β TIM barrel domain at the N-terminal and a six-stranded β-sandwich domain at the C-terminal. The two domains are connected by a helix-like strand of 16 amino acid residues. The aglycone binding site is considered to be a narrow hydrophobic cleft (Bhatia et al., 2002).

4.2.1 Active Site

Studies have identified the significance of a pair of carboxylic acid residues in the catalytic activity of β-glucosidases and are generally conserved within different families of hydrolases. Two carboxylic acids are required as a general catalytic nucleophile/base and as a catalytic proton donor. For GH1, BGL glutamic acid acts as both a catalytic nucleophile and proton donor, whereas in GH3 BGL aspartic acid acts as a catalytic nucleophile and glutamic acid acts as a catalytic proton donor (www.cazy.org). The catalytic centre is present in the pocket at the interface of the two domains. Aspartic acid in the N-terminal domain acts as a catalytic nucleophile, whereas glutamic acid in the C-terminal domain acts as a proton donor (Singhania et al., 2013).

4.2.2 Mode of Action

The general mechanism of hydrolases can be divided into two types based on the stereochemical outcome of the product: (i) retaining (the anomeric configuration of the substrate is retained) and (b) inverting (the anomeric configuration of the product will be inverted). In inverting glycosidases, the carboxylic acid residues are located approximately 10 A° apart. They follow a single displacement mechanism. A water molecule attacks the anomeric centre of the bound substrate and the second carboxyl acts as a general acid to facilitate the departure of the leaving group through an oxocarbenium ion-like transition state (Figure 4.1).

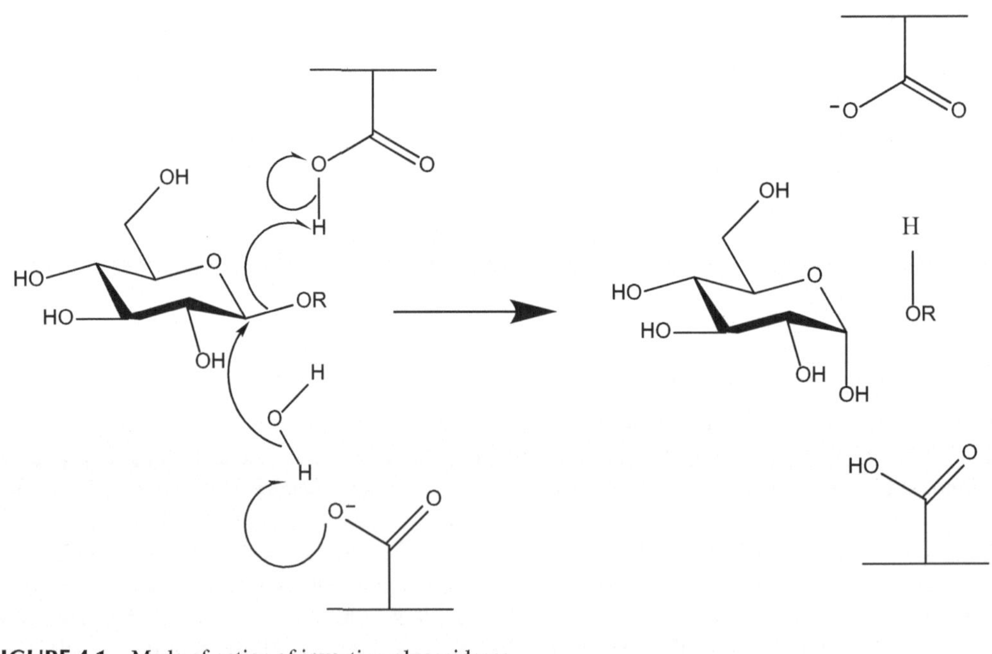

FIGURE 4.1 Mode of action of inverting glycosidases.

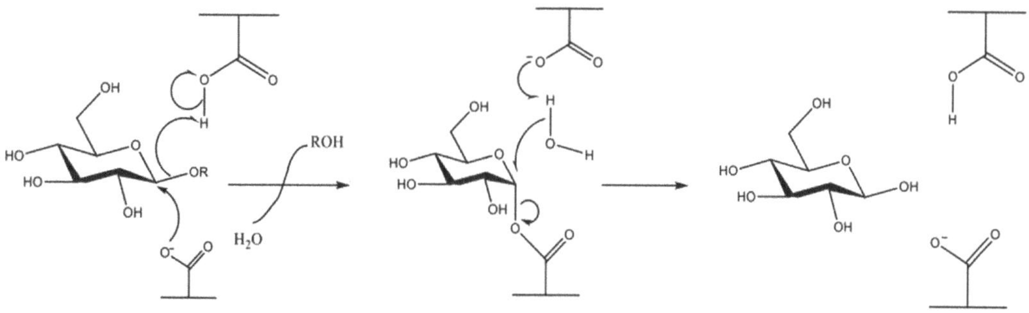

FIGURE 4.2 Mode of action of retaining glycosidases.

In retaining glycosidases, the two carboxylic acids are located approximately 5.5 A° apart and follow a double displacement mechanism. In this mechanism, one of the carboxyl groups protonates the glycosidic oxygen, whereas the other attacks the anomeric centre, thus forming a covalent glycosyl-enzyme intermediate in an oxocarbenium ion-like transition state. The second step involves the attack of water at the anomeric centre of the glycosyl-enzyme intermediate, catalyzed by the general base, which finally releases the product sugar with the same anomeric configuration as that of the substrate. This step also involves the formation of an oxocarbenium ion-like transition state (Figure 4.2). Most of the β-glucosidases in the GH3 group follow the retaining mechanism (Rye and Withers, 2000).

4.3 GLUCOSE-TOLERANT BGL: DEFINITION AND RELEVANCE IN BIOMASS HYDROLYSIS

β-Glucosidases, the enzymes carrying out the rate-limiting step in cellulose degradation, act by converting the final hydrolysis product cellobiose to glucose. However, the accumulation of glucose during the reaction leads to the inhibition of BGL, which leads to the accumulation of cellobiose. This cellobiose can inhibit other cellulase components, which in turn will significantly slow the rate of cellulose hydrolysis. This inhibitory effect will reduce the catalytic activity of enzymes, which is a major limitation in the biofuel industry. This indicates the importance of glucose-tolerant BGL (GT-BGL) to accelerate the cellulose breakdown process.

Numerous glucosidases have been discovered in a diverse range of organisms including bacteria, fungi, yeasts and metagenomic libraries. Glucose-tolerant β-glucosidase refers to an enzyme that can efficiently hydrolyze β-glycosidic bonds even in the presence of high concentrations of glucose. This is an important characteristic for the enzyme to have because, during biomass hydrolysis, glucose is produced as a by-product of the hydrolysis of cellulose and starch. Traditional β-glucosidases are not able to function efficiently in high glucose concentrations, which can lead to reduced yields and increased costs. GT-BGL, on the other hand, can function efficiently in high glucose concentrations, which results in higher yields and reduced costs.

Glucose-tolerant β-glucosidases are low molecular weight proteins synthesized in lower levels in comparison with high molecular weight highly expressed glucose sensitive β-glucosidase (Rajasree et al., 2013; Valappil et al., 2019). β-Glucosidases belonging to the GH1 family are ten- to hundred-fold more glucose tolerant in comparison to GH3 BGLs and a comparative structural study reveals a strong relationship between active-site accessibility and glucose tolerance. Matsuzawa et al. (2016) showed that in high glucose-tolerant GH1 β-glucosidase, Td2F2, an asparagine residue is located at position 223, which plays a key function in the glucose-tolerance trait.

Chan et al. (2016) attempted the heterologous expression of a GH1 β-glucosidase gene, GT-Bgl, from a thermophilic bacterium of *Anoxybacillus* sp. and found it to be more active and stable in a wide range of temperatures and pHs, and also have high resistance to glucose up to 5M, which further improves the enzymatic conversion of cellulosic biomass to glucose. Rajasree et al. (2013) discovered and characterized a β-glucosidase enzyme from the fungus *Aspergillus unguis* NII 08123, which is highly tolerant to glucose. A glucose-tolerant Bgl3 from a thermophilic fungus, *Malbranchea pulchella* (MpBgl3), has been characterized, which showed hypertolerance to glucose concentrations of up to 1 M. The comparative evaluation of the structure of MpBgl3 with GH1 BGL revealed that its deep active-site entrance had a similar shape and electrostatic characteristics to those of Bgl1 enzymes. This study also revealed that similar to GH1 enzymes, the depth, space and electrostatic properties of the catalytic site entry could influence MpBgl3's ability to tolerate glucose (Monteiro et al., 2020).

Kim et al. (2022) described the characterization of a β-glucosidase enzyme from the thermophilic bacterium *Thermoanaerobacterium saccharolyticum* capable of hydrolyzing hemicellulose. They purified the enzyme and characterized its biochemical properties, including its optimal temperature and pH for activity and determined the 3D structure. The enzyme had a unique structure compared to other β-glucosidases, with an extra domain not found in other enzymes of this class. This extra domain may contribute to the enzyme's glucose tolerance up to 0.3M and the 50% of activity reduced at 0.65M glucose concentration. The enzyme had a higher specific activity toward xylose, a sugar found in hemicellulose, than other known β-glucosidases. High conversion rates are necessary for biomass to be converted efficiently. Glucose accumulation will lower the hydrolysis rate by the inhibitory action of enzymes. For β-glucosidases, the end product is typically not removed during hydrolysis, so the actual reaction rate will differ more from the maximum reaction rate. Therefore, it is necessary that β-glucosidases have a high tolerance to glucose build-up. The hydrolysis must achieve higher glucose concentrations in order for the biomass conversion process to be profitable; as a result, the β-glucosidases must not be hindered by their product and must maintain high conversion rates at high glucose concentrations.

4.4 STRUCTURAL ASPECTS OF GT-BGL AND STRUCTURE–FUNCTION CORRELATION WITH GLUCOSE TOLERANCE

The β-glucosidase enzyme structure is composed of several domains, including a catalytic domain, a carbohydrate-binding domain and an oligosaccharide-binding domain. The catalytic domain contains the active site of the enzyme, where the hydrolysis of the β-glucosidic bond occurs. This domain typically contains several amino acid residues, such as Asp and Glu, that play important roles in the catalytic mechanism of the enzyme. The carbohydrate-binding domain is responsible for binding the substrate, or the complex carbohydrate, to the enzyme. This domain typically contains several amino acid residues, such as Trp and Tyr, that form specific interactions with the carbohydrate substrate. The oligosaccharide-binding domain is responsible for binding the product, or the oligosaccharide, to the enzyme. This domain typically contains several amino acid residues, such as Asn and Ser, that form specific interactions with the oligosaccharide product. The structure of β-glucosidase is a complex, multidomain enzyme that plays a vital role in the hydrolysis of complex carbohydrates.

Giuseppe et al. (2014) studied GH1 β-glucosidases from different organisms, including bacteria and fungi, using X-ray crystallography. They found that the enzymes have a unique structural feature called a "lid" that can open and close to allow the enzyme to bind to its substrate. The study also looked at how variations in the structure of the enzyme affected its ability to tolerate glucose. They found that certain changes in the enzyme's structure led to an increase in glucose tolerance, allowing the enzyme to continue to function even in the presence of high concentrations of glucose. Overall, the study provided new insights into the structural basis of glucose tolerance in GH1 β-glucosidases, and the implications for the development of new enzymes for use in industrial processes and biotechnology. The study also looked at how variations in the structure of the enzyme affected its ability to tolerate glucose. They found that certain changes in the enzyme's structure led to an increase in glucose tolerance, allowing the enzyme to continue to function even in the presence of high concentrations of glucose. Kao et al. (2019) reported *Chaetomella raphigera* β-glucosidase D2-BGL has "intriguing structural features" that make it an efficient supplement for hydrolyzing lignocellulosic biomass. Specifically, the enzyme has a high substrate affinity, allowing it to bind to and break down the materials in lignocellulosic biomass more effectively than other enzymes.

4.5 EFFECT OF TRANSGLYCOSYLATION ON BIOMASS HYDROLYSIS

Transglycosylation activity is generally observed at higher substrate concentrations above 100 mM cellobiose, thus playing a major role in removing the inhibitory levels of cellobiose and further activating β-glucosidases (Steinberg et al., 1977). However, β-glucosidases from *Magnaporthe grisea* and *Aspergillus oryzae* showed a transglycosylation activity at substrate concentrations where their hydrolytic activity was below their maximal rate. Transglycosylation and product (glucose) inhibition are the major reasons considered for the reduced hydrolytic rates at higher solid loadings in biomass conversions. However, the studies on the enzyme requirements for hydrolysis showed that product inhibition plays a major role in reduced hydrolytic activity than transglycosylation. In fact, the transglycosylated products in biomass conversions like cellotriose itself can act as the substrate for BGL and can play a major role in the induction of cellulases (Bohlin et al., 2013). Therefore, the enzyme required for the biomass conversion is highly susceptible to high substrate or product concentrations. These could affect the selection of β-glucosidases for biomass saccharification, where higher solid loadings are required, and these factors should be essentially considered along with their hydrolytic activity.

4.6 INDUSTRIAL PRODUCTION OF BGL

4.6.1 Organisms Used in Industrial Production of BGL and Their Characteristics

β-Glucosidase is one of the members of the biomass-degrading enzyme system and plays a vital role in enzymatic saccharification for biofuel production. Candidates with high activity and great stability over high temperatures and varied pHs are always preferred in industrial practice. BGL has a significant role in the biofuel industry, so improvements in its production are highly desirable for the industries.

β-glucosidase is an enzyme that is ubiquitous in nature and is produced by bacteria, fungi, plants and animals, including humans. Among them, both fungi and bacteria are well reported for BGL production, and members of the fungal genus *Aspergillus*, such as *Aspergillus phoenicus* and *Aspergillus niger* are reported to give a higher yield of the β-glucosidases enzyme. Bacterial strains such as *Bacillus subtilis* and *Acidothermus cellulolyticus* are also potential producers of β-glucosidase, yielding more thermostable BGLs compared to fungi, but these microorganisms are low producers of the enzyme (Li et al., 2018).

Aspergillus niger NII 08121 has shown the production of BGL isoforms when grown on different carbon sources indicating a difference in the induction of these isoforms. The production of BGL was observed in all carbon sources, but maximal production of moderately glucose-tolerant isoforms was observed with lactose induction (Singhania et al., 2011). Singhania et al. (2013) reported that the activity of β-glucosidases gets inhibited because of the feedback inhibition of the final product glucose. This limitation of BGLs while being considered as biomass hydrolyzing enzymes has led to an increased number of studies on glucose-tolerant and glucose-stimulated β-glucosidases in the past two decades (Salgado et al., 2018). Chen et al. (2017) reported that the production of the β-glucosidase enzyme with 45.44 U/mL activity was achieved when the BGL gene from *Bacillus licheniformis* was cloned into *Escherichia coli*. Tiwari et al. (2017) reported the production of the thermostable β-glucosidase using *Bacillus subtilis* RA10, which could efficiently convert cellulosic biomass into fermentable sugar.

Fungal β-glucosidases play a significant role in the biofuel industry because of their high catalytic efficiency on the hydrolysis of cellobiose and cello-oligomers. A combination of fungal β-glucosidases and *Trichoderma reesei* cellulases enzymes has the ability to hydrolyze lignocellulosic biomass efficiently (Kao et al., 2019). The major β-glucosidase (BGL1) upon purification and characterization showed interesting properties, including an optimum temperature of 70°C, optimum pH of 4.8 and extended stability at 50°C for more than 48 h. Normal enzymatic hydrolysis of biomass is performed at 50°C for 24–72 h and loss of enzyme activity is a major limitation in such processes. BGL1 with its extended stability is a very potent candidate to be used in enzyme cocktails for biomass hydrolysis. Liu et al. (2012) reported that *Aspergillus fumigatus Z5* gave a thermostable β-glucosidase active even at an elevated temperature, and also when added to the lignocellulosic biomass, it resulted in the removal of phenolic compounds, and hence it may be used for the degradation of polyphenols. Thermophilic β-glucosidase was isolated from *Thermotoga naphthophila* RUK-10 and it was used along with cellulase for the hydrolysis of untreated corn straw, leading to a 93.5% increase in the conversion rate of cellulose to glucose (Zang et al., 2018). A recombinant β-glucosidase from *Neocallimastix patriciarum*, a buffalo rumen fungus, expressed *Pichia pastoris* that could achieve saccharification efficiencies better than the commercially available Novozyme 188 BGL (Chen et al., 2012).

4.7 EFFECTS OF PH AND TEMPERATURE ON BGL ACTIVITY

Evaluation of the effects of pH and temperature on BGL activity is important for the selection of process conditions. The active sites on enzymes are frequently composed of ionizable groups which must be in the proper ionic form to maintain the conformation of the active site, bind the substrate or catalyze the reaction. In terms of the temperature effect on enzyme activity, a faster velocity is expected as the temperature rises, as an increase in temperature imparts more kinetic energy to the molecules resulting in more productive collisions per unit of time. However, if the molecule absorbs too much energy, the enzyme can be denatured and lose its catalytic activity. The statistical design of experiments and response surface methodology analysis were used to evaluate the effect of pH and temperature effects on a partially purified BGL (Baraldo Junior et al., 2014). The statistical analysis revealed that the optimum values of temperature for partially purified BGL were found to be between 40°C and 65°C, and the optimum pH range was between 4 and 5.5. Higher experimental values of enzymatic activity were found at the condition of the central point, which is at pH 4.5 and 55°C (Baraldo Junior et al., 2014) (Table 4.1). Several studies focusing on the characterization of BGL in terms of pH and temperature parameters have been made. The optimum range of temperature for BGL, between 40°C and 65°C, and the optimum pH range, between 4 and 5.5, are interesting characteristics for application in the hydrolysis of lignocellulosic materials, as this also

TABLE 4.1
Optimum pH and Temperature of Various BGLs

Microorganism Source	Optimum pH	Optimum Temperature (°C)	Reference
Penicillium pinophilum	5.0	60	(Joo et al., 2010)
Penicillium funiculosum	4.0 and 5.0	60	(Ramani et al., 2012)
Aspergillus glaucus	3.6	60	(Ma et al., 2011a)
Aspergillus terreus	5.0	60	(Elshafei et al., 2011)
Aspergillus niger	4.0	60	(Le Traon-Masson and Pellerin, 1998)
Aspergillus niger	4.5	60–70	(Dekker, 1986)
Aspergillus niger	4.5	55	(Watanabe et al., 1992)
Aspergillus niger	5.0	55	(Yan and Lin, 1997)
Aspergillus niger	4.0–5.5	40–65	(Baraldo Junior et al., 2014)

Source: Reproduced from Baraldo Junior et al., 2014.

constitutes the ideal range of these parameters for biomass hydrolyzing cellulases. For the simultaneous saccharification and fermentation process configuration, a compromise has to be found, since the optimal temperature for enzymatic hydrolysis is usually higher (around 55°C when using fungal enzymes) than that of fermentation by yeasts (around 30°C). Therefore, an enzyme with this relatively broad range of optimum temperatures would be of interest.

4.8 THERMOSTABILITY OF BGL

Evaluation of enzyme thermostability is very important toward defining the optimal range of process conditions in order to preserve the catalytic activity during hydrolysis. Thermostable enzymes can provide process advantages such as faster reaction rates, lower viscosity and increased solubility of the substrate. The usage of thermostable BGL in the hydrolysis of lignocellulosic biomass has several potential advantages including increased flexibility for the process configurations, longevity of the enzymes allowing extended hydrolysis times and higher specific activity (lower qualities of enzyme needed) Baraldo Junior et al. (2014) had studied the thermostability of partially purified BGL at 37°C and 50°C. These two temperatures were selected considering that the use of 37°C is an intermediate condition for a simultaneous saccharification and fermentation process configuration, since the yeast usually has an optimal temperature of around 30°C and fungal enzymes around 50°C. A higher stability at 37°C and 50°C was observed for BGL present in the crude extract in comparison to the partially purified enzyme, showing that the purification procedure negatively influenced BGL thermostability. This result was rather unexpected since the purification procedure usually favours enzyme stability due to the removal of proteases and other proteins that could form aggregates with the molecule of interest. A reduction in BGL stability after purification was reported and may be related to the removal of some stabilizing compounds originally present in the medium. In practice, different conditions such as the presence of the substrate and stabilizing agents can circumvent this characteristic

4.9 COMMERCIAL BGLS AND THEIR EFFICIENCY

The BGL produced by enzyme manufacturing companies are often not cost-effective for biorefinery applications due to the non-availability of a suitable microorganism, limitations due to glucose sensitivity and costly downstream processing. The first commercially reported β-glucosidase enzyme, Novozyme SP 188, was produced from *Aspergillus niger* and had poor glucose tolerance (Marie et al., 2010). Due to multiple applications of BGLs, various companies manufacture them under different brand names, as given in Table 4.2. Important patents on BGL are given in Table 4.3.

Accelerase 1500 produced by Genencor (Dupont) is a replacement for Accelerase 1000. The company claims that Accelerase 1500 has more BGL content and more substrate loading capacity than Accelerase 1000. CellicCtec3 produced by Novozymes has a reported efficiency of 90% and contains a high loading of BGL. The total revenue of biofuel enzymes in 2011 stood at 535.6 million USD. It is estimated that by 2024, the biofuel enzyme market will reach a compounded annual growth rate of 7.3%.

TABLE 4.2
Global Players in the Production of β-Glucosidase

Commercial Name	Company
Accelerase 1500	Genencor
BGL from almonds	Sigma-Aldrich
CellicCtec3	Novozyme
E-BGLUC	Megazyme

TABLE 4.3
Important Patents on BGL Production

Patent no	Country	Function	Organism	Reference
US 8394619B1	US	Novel β-glucosidase with high enzymatic activity	*Chaetomella raphigera*	https://patents.google.com/patent/US8394619B1/en
US 9175324B2	US	Recombinant C1 β-glucosidase		https://patents.google.com/patent/US9175324B2/en
US8486683B2	US	Modified BGL	*Trichoderma reesei*	https://pubchem.ncbi.nlm.nih.gov/patent/US8486683
11,162,087	US	Modified BGL	*Pseudo-trichonympha*	https://patents.justia.com/patent/11162087

4.10 GENETICALLY ENGINEERED BGL: IMPORTANCE, GENETIC MANIPULATIONS, STRUCTURAL CHARACTERISTICS AND ADVANTAGES

For the efficient and economic hydrolysis of lignocellulosic biomass, a cellulase system with an optimum cocktail of all the diverse cellulolytic enzymes such as cellulases, hemicellulases and accessory proteins at high titres is essential. *T. reesei*, being the industrial producer of cellulases, satisfies most of these conditions, except the fact that it is severely deficient in β-glucosidase activity, which results in incomplete hydrolysis (Bischof et al., 2016). Currently, this limitation is addressed by supplementing β-glucosidases from other sources in the cellulase enzyme preparation or through co-cultivation of cellulase-producing organisms with BGL-producing fungi, mainly from *Aspergillus* species (Fang et al., 2019). But, if the same organism can produce an optimum enzyme mixture through strain improvement, it would be more cost-effective. Earlier, the toolbox for the genetic engineering of fungi was limited, which made the strain improvement programmes for filamentous fungi notoriously difficult and resulted in the slow pace of fungal strain improvement programmes. With the advent of new genetic tools, the scenario has changed and genetic engineering has become common practice for cellulase improvement. Different strategies can be employed for the genetic engineering of cellulase producers, which include promoter alteration, the use of heterologous genes coding enzymes/proteins that are deficient in the cellulase producer and regulator manipulations that allow de-repression of cellulase genes or enhance their transcription (Figure 4.3). In the following section, different genetic engineering approaches used for the improvement of BGL enzyme production in *T. reesei*, the industrial workhorse for cellulase production, and other cellulolytic filamentous fungi are described.

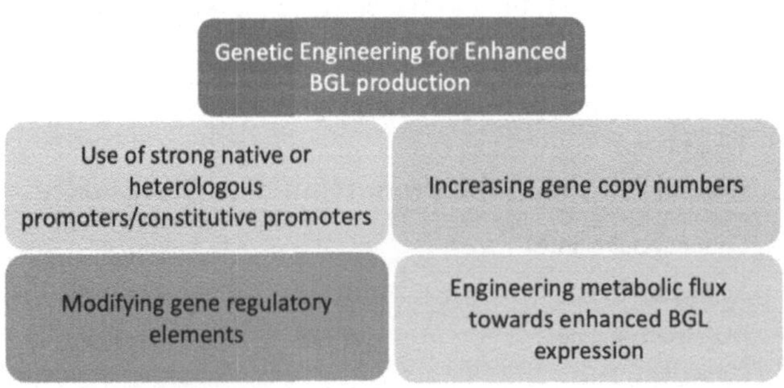

FIGURE 4.3 Genetic engineering approaches for improving BGL production in fungi.

4.10.1 Strain Improvement for Increased BGL Production in *Trichoderma reesei*

T. reesei strains with improved BGL production were constructed either by expression of the native *bgl* genes under a different promoter or by introducing heterologous *bgl* genes into the genome. In such an approach, a BGL-1 overproducing strain was constructed by placing an additional copy of *bgl1* under the control of the stronger homologous promoters *egl3* or the *xyn3*. The recombinant *T. reesei* strain had 4.0- and 7.5-fold higher β-glucosidase activity than the native strain (Rahman et al., 2009). Cellobiohydrolase1 (CBH1) accounts for 60% of the total secreted protein in *T. reesei* owing to its strong promoter, and the same has been used for the expression of many homologous and heterologous proteins (Li et al., 2012). Under the control of the modified four-copy *cbh1* promoter in *T. reesei*, the homologous *bgl1* gene was overexpressed, which resulted in significantly enhanced production of β-glucosidase (3.7-, 2.0- and 1.8-fold increase) and total cellulase (improvement up to 130% and 55%) in the BGLI-overexpressing strains than the parental strain *T. reesei* RUT-C30 (Zhang et al., 2010). In a hypercellulolytic variant SN1, characterized by high endoglucanase activity but low bgl activity, the β-glucosidase gene was overexpressed under the same promoter to optimize the cellulase system for the saccharification of corncob residues (Qian et al., 2016).

Heterologous *bgl* genes for overexpression in *T. reesei* are often selected from filamentous fungi belonging to genera *Aspergillus* and *Penicillium*. In such an approach, a β-glucosidase I coding sequence from *Penicillium decumbens* was ligated with the cellobiohydrolase I (cbh1) promoter of *T. reesei* and introduced into the genome of *T. reesei* strain RUT-C30. In comparison to that from the parent strain, the β-glucosidase activity of the enzyme complexes from two selected transformants increased six to eight times and their filter paper activity (FPAs) was enhanced by 30% on average. The transformant's saccharifying ability toward pretreated cornstalk was also significantly enhanced (Ma et al., 2011). Also, highly efficient ethanol production was reported from Japanese cedar by recombinant *T. reesei* expressing *A. aculeatus* β-glucosidase I. It was comparable to the commercial enzyme cocktail both in simultaneous saccharification and fermentation (SSF) or separate hydrolysis and fermentation (SHF) processes (Treebupachatsakul et al., 2015). In a study by Ellilä et al. (2017), a heterologous β-glucosidase (Cel3A) from *Talaromyces emersonii* was introduced into *T. reesei* along with overexpression of Xyr1 and invertase-expressing gene from *A. niger*. The manipulated strain has shown better hydrolytic productivity utilizing lignocellulosic biomass with a drop in processing cost.

4.10.2 Improved BGL Production in Other Filamentous Fungi

There are many strains which are considered potent alternatives to the *T. reesei* cellulase cocktail such as *Penicillium* strains. *Penicillium oxalicum* strains, which secrete a more balanced lignocellulose-degrading enzyme system and display higher β-glucosidase activity compared to *T. reesei*, were reported (Yao et al., 2016). Three β-glucosidases from *P. oxalicum*, which showed a promising enzymatic performance, were overexpressed either constitutively or inductively by means of two different types of promoters. A number of high-yield β-glucosidase producers and their β-glucosidase yields were elevated from 2-fold to 65-fold. Microcrystalline cellulose, filter paper and corncob residue were used as substrates to assess the hydrolysis efficiency of the enzyme complex (Yao et al., 2016). Wakai et al. (2019) constructed a genetically engineered *A. oryzae* strain that simultaneously produced cellobiohydrolase, endoglucanase and β-glucosidase by integrating multiple copies of the genes encoding these cellulases into fungal chromosomes. The resulting strain possessed 5–16 copies of each cellulase gene within the chromosome and showed approximately tenfold higher activity. In *Penicillium verruculosum*, cloning of *bgl1* from *A. niger* (AnBGL) along with an LPMO from *T. reesei* (TrLPMO) under the control of inducible *gla1* promoter led to the far better hydrolysis of lignocellulosic biomass in contrast with the wild-type strain (Bulakhov et al., 2017).

4.10.3 Improved BGL Production through Regulator Manipulation

Apart from directly engineering BGL, manipulation of cellulase regulators has also resulted in improved BGL activity. By the constitutive expression of xyr1 under the control of the strong *pki1* promoter from *T. reesei*, a 36% higher value of β-glucosidase activity was observed during cultivation in a stirred-tank bioreactor along with improvement in other cellulases (Silva Delabona et al., 2017). There are also reports in which the cellulase regulator manipulation resulted in improved cellulase activity, which included the β-glucosidase genes (Zhang et al., 2018). BglR (β-glucosidase regulator) was identified as a new transcription factor that upregulates the expression of specific genes encoding β-glucosidases (Nitta et al., 2012). This gene was proposed as a target for genetic manipulations since the regulator was specific for β-glucosidases.

4.11 ROLE OF BGL IN BIOMASS HYDROLYZING ENZYME BLENDS

β-Glucosidase is one of the critical enzymes in the biomass hydrolyzing enzyme cocktail, as the accumulation of its substrate, cellobiose, in the hydrolysate inhibits both endoglucanases and exoglucanases, thereby stalling the entire cellulase machinery and reducing the conversion efficiency. As a single organism is not capable of producing all the required enzymes at the desired level for efficient biomass saccharification, enzymes from different fungi are blended to form the cellulase cocktail that acts synergistically to degrade lignocellulose. Commercial cellulases contain high loadings of β-glucosidases to overcome the inhibition as well as loss arising due to substrate-related factors such as lignin and lignin-degraded products that irreversibly bind cellulase components and reduce the loadings of available enzymes (Li and Zheng, 2017). Commercial cellulase preparations are typically from *Trichoderma reesei,* which has low titres of BGL and are thus supplemented/blended with BGLs from other sources such as *Aspergillus.* Several reports have shown that the addition of BGL to commercial cellulase has improved the digestibility of different lignocellulosic materials. Agrawal et al. (2018) developed a synthetic enzyme mixture using commercial enzymes. They partially replaced cellulase (Celluclast) with commercial accessory enzymes like BGL, xylanase and laccase. A 75% hydrolytic improvement at 6 h of hydrolysis and a 30% hydrolytic improvement at 24 h of hydrolysis was obtained with Celluclast, BGL and xylanase in a protein ratio of 20.40:38.43:41.16, respectively. Additionally, the enzyme dosage requirement could be reduced by 25%. In addition to BGL supplementation, cellulase-producing fungi with low BGL titres are either grown in engineered media that produce cellulase with high BGL content or have been genetically modified to secrete more β-glucosidases to enhance the saccharification efficiency. As the industrial workhorse, *Trichoderma reesei* has low BGL titres and has been genetically modified by several groups to secrete BGL-enriched cellulase. According to the study by Qian et al. (2017), EG2 and BGL1 were overexpressed, which could significantly improve the cellulose conversion (up to 94.2%) for the delignified corncob residues after 48 h of enzymatic saccharification. Kao et al. (2019) expressed the heterologous fungal BGL from *Chaetomella raphigera* in *Pichia pastoris,* which produced a recombinant BGL with higher substrate affinity than the commercial β-glucosidase Novozyme 188. When this BGL was combined with cellulase from *Trichoderma* RUT-C30, it showed an improved cellulose conversion of 83% for acid-pretreated sugarcane bagasse and 63% for rice straw. The use of this recombinant BGL also halved the time required to produce maximal levels of ethanol by a semi-simultaneous saccharification and fermentation process.

In addition to increasing the BGL dosage in the cellulolytic mixture, glucose-tolerant BGLs have been used for maintaining the catalytic activity of BGL in the presence of high glucose concentration. Rajasree et al. (2013) have shown that GT-BGL retained an activity of nearly 60% in 1.5M glucose and improved the saccharification efficiency by 20% at 12 h of enzymatic saccharification. The addition of this purified BGL to a commercial cellulase preparation resulted in about 20% enhancement of the released reducing sugars.

4.12 CONCLUSIONS AND PERSPECTIVES

The wide distribution and characterization of beta-glucosidases and their applications in the bioethanol industry have been reviewed. While *Aspergillus niger* remains the major source of commercial BGLs, various other sources with the potential of being industrial sources are increasingly being reported. The physicochemical properties of BGLs determine their suitability in industrial processes and more BGLs are being studied for their properties that allow them to be optimally used in various applications. For biomass hydrolysis, glucose tolerance of BGL is a very important feature, and though comparatively rarer, there are multiple studies which describe GT-BGLs with exceptionally high glucose tolerance.

Some of the major hurdles in the commercialization of lignocellulosic ethanol and the biochemical production of industrially relevant chemicals and materials from biomass in biorefineries include the high cost of enzymes, slower reaction rates, feedback inhibition and enzyme inactivation due to lignin adsorption. β-glucosidases are known to be among the most critical enzymes for biomass conversion and are indispensable for efficient biomass saccharification within a limited time. Several studies have reported the importance of glucose-tolerant and thermotolerant BGLs and the development of enzyme cocktails having higher BGL titres either by genetic manipulation or by synergistic blending from different sources, but they are often limited to lab-scale studies with scaling-up challenges. Understanding the structure of GT-BGL has provided new insights to improve its functioning as well as tolerance to increased sugar concentrations. Scaling up the production of BGLs with improved features and incorporation of these BGLs in the enzyme cocktail can improve sugar yield and bring down the total enzyme dosage, which in turn can reduce the cost of biomass conversion.

REFERENCES

Agrawal, R., S. Semwal, R. Kumar, A. Mathur, R. P. Gupta, D. K. Tuli, and A. Satlewal. 2018. Synergistic Enzyme Cocktail to Enhance Hydrolysis of Steam Exploded Wheat Straw at Pilot Scale. *Frontiers in Energy Research* 6, 1–11. https://doi.org/10.3389/fenrg.2018.00122

Baraldo Junior, A., D. G. Borges, P. W. Tardioli, and C. S. Farinas. 2014. Characterization of β-Glucosidase Produced by *Aspergillus niger* under Solid-State Fermentation and Partially Purified Using MANAE-Agarose. *Biotechnology Research International* 2014, 1–8. https://doi.org/10.1155/2014/317092

Bhatia, Yukti, Saroj Mishra, and V. S. Bisaria. 2002. Microbial β-Glucosidases: Cloning, Properties, and Applications. *Critical Reviews in Biotechnology* 22(4), 375–407. https://doi.org/10.1080/07388550290789568

Bischof, Robert H., Jonas Ramoni, and Bernhard Seiboth. 2016. Cellulases and Beyond: The First 70 Years of the Enzyme Producer *Trichoderma REESEI. Microbial Cell Factories.* https://doi.org/10.1186/s12934-016-0507-6

Bohlin, Christina, Eigil Praestgaard, Martin J. Baumann, Kim Borch, Jens Praestgaard, Rune N. Monrad, and Peter Westh. 2013. A Comparative Study of Hydrolysis and Transglycosylation Activities of Fungal β-Glucosidases. *Applied Microbiology and Biotechnology* 97(1), 159–169. https://doi.org/10.1007/s00253-012-3875-9

Bulakhov, Alexander G., Pavel V. Volkov, Aleksandra M. Rozhkova, Alexander V. Gusakov, Vitaly A. Nemashkalov, Aidar D. Satrutdinov, and Arkady P. Sinitsyn. 2017. Using an Inducible Promoter of a Gene Encoding *Penicillium verruculosum* Glucoamylase for Production of Enzyme Preparations with Enhanced Cellulase Performance. *PLOS ONE* 12(1). https://doi.org/10.1371/journal.pone.0170404

Chan, Chia Sing, Lee Li Sin, Kok Gan Chan, Mohd Shahir Shamsir, Fazilah Abd Manan, Rajesh Kumar Sani, and Kian Mau Goh. 2016. Characterization of a Glucose-Tolerant β-Glucosidase from Anoxybacillus Sp. DT3-1. *Biotechnology for Biofuels* 9(1), 1–11. https://doi.org/10.1186/S13068-016-0587-X/FIGURES/6

Chen, H. L., Y. C. Chen, M. Y. J. Lu et al. 2012. A Highly Efficient β-Glucosidase from the Buffalo Rumen Fungus Neocallimastix Patriciarum W5. *Biotechnol Biofuels* 5, 24. https://doi.org/10.1186/1754-6834-5-24

Chen, Z., T. Meng, Z. Li, P. Liu, Y. Wang, N. He, and D. Liang. 2017. Characterization of a β-Glucosidase from Bacillus licheniformis and Its Effect on Bioflocculant Degradation. *AMB Express* 7(1). https://doi.org/10.1186/s13568-017-0501-3

Dekker, R. F. H. 1986. Kinetic, Inhibition, and Stability Properties of a Commercial β-d-Glucosidase (Cellobiase) Preparation from *Aspergillus niger* and Its Suitability in the Hydrolysis of Lignocellulose. *Biotechnology and Bioengineering* 28(9), 1438–1442. https://doi.org/10.1002/bit.260280918

Delabona, Priscila da Silva, Gisele Nunes Rodrigues, Mariane Paludetti Zubieta, Jonas Ramoni, Carla Aloia Codima, Deise Juliana Lima, Cristiane Sanchez Farinas, José Geraldo da Cruz Pradella, and Bernhard Seiboth. 2017. The Relation between Xyrl Overexpression in Trichoderma Harzianum and Sugarcane Bagasse Saccharification Performance. *Journal of Biotechnology* 246(March), 24–32. https://doi.org/10.1016/j.jbiotec.2017.02.002

Ellilä, Simo, Lucas Fonseca, Cristiane Uchima, Junio Cota, Gustavo Henrique Goldman, Markku Saloheimo, Vera Sacon, and Matti Siika-Aho. 2017. Development of a Low-Cost Cellulase Production Process Using *Trichoderma reesei* for Brazilian Biorefineries. *Biotechnology for Biofuels* 10(1). https://doi.org/10.1186/s13068-017-0717-0

Elshafei, A. M., M. M. Hassan, N. M. Morsi, and D. H. Elghonamy. 2011. Purification and Some Kinetic Properties of β-Glucosidase from *Aspergillus terreus* NRRL 265. *African Journal of Biotechnology* 10(84), 19556–19569. https://doi.org/10.5897/AJB10.2617

Fang, Hao, Runze Zhao, Chaofeng Li, and Chen Zhao. 2019. Simultaneous Enhancement of the B--Exo Synergism and Exo-Exo Synergism in *Trichoderma reesei* Cellulase to Increase the Cellulose Degrading Capability. *Microbial Cell Factories* 18(1). https://doi.org/10.1186/s12934-019-1060-x

Giuseppe, Priscila Oliveira De, Tatiana De Arruda Campos Brasil Souza, Flavio Henrique Moreira Souza, Leticia Maria Zanphorlin, Carla Botelho Machado, Richard John Ward, Joao Atilio Jorge, Rosa Dos Prazeres Melo Furriel, and Mario Tyago Murakami. 2014. Structural Basis for Glucose Tolerance in GH1 β-Glucosidases. *Acta Crystallographica Section D: Biological Crystallography* 70(6), 1631–1639. https://doi.org/10.1107/S1399004714006920

Henrissat, Bernard, and Amos Bairoch. 1996. Updating the Sequence-Based Classification of Glycosyl Hydrolases. *Biochemical Journal Letters* 315(2), 695–696. https://doi.org/10.1097/QAD.0000000000001318

Jenkins, John, Leila Lo Leggio, Gillian Harris, and Richard Pickersgill. 1995. β-Glucosidase, β-Galactosidase, Family A Cellulases, Family F Xylanases and Two Barley Glycanases Form a Superfamily of Enzymes Wit 8-Fold β/α Architecture and with Two Conserved Glutamates near the Carboxy-Terminal Ends of β-Strands Four and Seven. *FEBS Letters* 362(3), 281–285. https://doi.org/10.1016/0014-5793(95)00252-5

Joo, A. R., M. Jeya, K. M. Lee, K. M. Lee, H. J. Moon, Y. S. Kim, and J. K. Lee. 2010. Production and Characterization of β-1,4-Glucosidase from a Strain of Penicillium Pinophilum. *Process Biochemistry* 45(6), 851–858. https://doi.org/10.1016/j.procbio.2010.02.005

Kao, Mu Rong, Hsion Wen Kuo, Cheng Chung Lee, Kuan Ying Huang, Ting Yen Huang, Chen Wei Li, C. Will Chen, Andrew H. J. Wang, Su May Yu, and Tuan Hua David Ho. 2019. Chaetomella Raphigera β-Glucosidase D2-BGL Has Intriguing Structural Features and a High Substrate Affinity That Renders It an Efficient Cellulase Supplement for Lignocellulosic Biomass Hydrolysis. *Biotechnology for Biofuels* 12(1), 1–18. https://doi.org/10.1186/S13068-019-1599-0/FIGURES/12

Kim, In Jung, Uwe T. Bornscheuer, and Ki Hyun Nam. 2022. Biochemical and Structural Analysis of a Glucose-Tolerant β-Glucosidase from the Hemicellulose-Degrading Thermoanaerobacterium Saccharolyticum. *Molecules* 27(1). https://doi.org/10.3390/MOLECULES27010290

Le Traon-Masson, M. P., and P. Pellerin. 1998. Purification and Characterization of Two β-D-Glucosidases from an *Aspergillus niger* Enzyme Preparation: Affinity and Specificity toward Glucosylated Compounds Characteristic of the Processing of Fruits. *Enzyme and Microbial Technology* 22(5), 374–382. https://doi.org/10.1016/S0141-0229(97)00206-8

Li, Junxin, Juan Wang, Shaowen Wang, Miao Xing, Yu Shaowen, and Gang Liu. 2012. Achieving Efficient Protein Expression in Trichoderma reesei by Using Strong Constitutive Promoters. *Microbial Cell Factories* 11(1), 1. https://doi.org/10.1186/1475-2859-11-84

Li, X., and Y. Zheng. 2017. Lignin-Enzyme Interaction: Mechanism, Mitigation Approach, Modeling, and Research Prospects. *Biotechnology Advances* 35(4), 466–489. https://doi.org/10.1016/j.biotechadv.2017.03.010

Li, Y., M. Bu, P. Chen, X. Li, C. Chen, G. Gao, Y. Feng, W. Han, and Z. Zhang. 2018. Characterization of a Thermophilic Monosaccharide Stimulated β-Glucosidase from *Acidothermus cellulolyticus. Chemical Research in Chinese Universities* 34(2), 212–220. https://doi.org/10.1007/s40242-018-7408-7

Liu, D., R. Zhang, X. Yang, Z. Zhang, S. Song, Y. Miao, and Q. Shen. 2012. Characterization of a Thermostable β-Glucosidase from *Aspergillus fumigatus* Z5, and Its Functional Expression in Pichia pastoris X33. *Microbial Cell Factories* 11, 25. https://doi.org/10.1186/1475-2859-11-25

Ma, Liang, Jun Zhang, Gen Zou, Chengshu Wang, and Zhihua Zhou. 2011. Improvement of Cellulase Activity in *Trichoderma reesei* by Heterologous Expression of a B-Glucosidase Gene from *Penicillium decumbens*. *Enzyme and Microbial Technology* 49(4), 366–371. https://doi.org/10.1016/j.enzmictec.2011.06.013

Ma, S. J., B. Leng, X. Q. Xu, X. Z. Zhu, Y. Shi, Y. M. Tao, S. X. Chen, M. N. Long, and Q. X. Chen. 2011a. Purification and Characterization of β-1,4-Glucosidase from Aspergillus glaucus. *African Journal of Biotechnology* 10(84), 19607–19614. https://doi.org/10.5897/AJB11.2144

Marie, C., M. Hugues, H. Delphine, C. Dominique, M. Frédéric, and L. F. Nicolas. 2010. Comparative Kinetic Analysis of Two Fungal β-Glucosidases. *Biotechnology for Biofuels* 3, 1–8.

Matsuzawa, Tomohiko, Toshinori Jo, Taku Uchiyama, Jenny A. Manninen, Takatoshi Arakawa, Kentaro Miyazaki, Shinya Fushinobu, and Katsuro Yaoi. 2016. Crystal Structure and Identification of a Key Amino Acid for Glucose Tolerance, Substrate Specificity, and Transglycosylation Activity of Metagenomic β-Glucosidase Td2F2. *FEBS Journal*, 2340–2353. https://doi.org/10.1111/FEBS.13743

Monteiro, Lummy Maria Oliveira, Ana Claudia Vici, Matheus Pinto Pinheiro, Paulo Ricardo Heinen, Arthur Henrique Cavalcante de Oliveira, Richard John Ward, Rolf Alexander Prade, Marcos S. Buckeridge, and Maria de Lourdes Teixeira de Moraes Polizeli. 2020. A Highly Glucose Tolerant SS-Glucosidase from Malbranchea Pulchella (MpBg3) Enables Cellulose Saccharification. *Scientific Reports* 10(1), 1–12. https://doi.org/10.1038/s41598-020-63972-y

Nitta, Mikiko, Takanori Furukawa, Yosuke Shida, Kazuki Mori, Satoru Kuhara, Yasushi Morikawa, and Wataru Ogasawara. 2012. A New Zn(II) 2Cys 6-Type Transcription Factor BglR Regulates β-Glucosidase Expression in *Trichoderma reesei*. *Fungal Genetics and Biology: FG and B* 49(5), 388–397. https://doi.org/10.1016/j.fgb.2012.02.009

Qian, Y., L. Zhong, J. Gao, N. Sun, Y. Wang, G. Sun, Y. Qu, and Y. Zhong. 2017. Production of Highly Efficient Cellulase Mixtures by Genetically Exploiting the Potentials of *Trichoderma reesei* Endogenous Cellulases for Hydrolysis of Corncob Residues. *Microbial Cell Factories* 16(1), 1–16. https://doi.org/10.1186/s12934-017-0825-3

Qian, Yuanchao, Lixia Zhong, Yunhua Hou, Yinbo Qu, and Yaohua Zhong. 2016. Characterization and Strain Improvement of a Hypercellulytic Variant, *Trichoderma reesei* SN1, by Genetic Engineering for Optimized Cellulase Production in Biomass Conversion Improvement. *Frontiers in Microbiology* 7, 1–11. https://doi.org/10.3389/fmicb.2016.01349

Rahman, Zinnia, Yosuke Shida, Takanori Furukawa, Yota Suzuki, Hirofumi Okada, Wataru Ogasawara, and Yasushi Morikawa. 2009. Application of *Trichoderma reesei* Cellulase and Xylanase Promoters through Homologous Recombination for Enhanced Production of Extracellular β-Glucosidase I. *Bioscience, Biotechnology and Biochemistry* 73(5), 1083–1089. https://doi.org/10.1271/bbb.80852

Rajasree, Kuni Parambil, Gincy Marina Mathew, Ashok Pandey, and Rajeev Kumar Sukumaran. 2013. Highly Glucose Tolerant β-Glucosidase from Aspergillus Unguis: NII 08123 for Enhanced Hydrolysis of Biomass. *Journal of Industrial Microbiology and Biotechnology* 40(9), 967–975. https://doi.org/10.1007/S10295-013-1291-5

Ramani, G., B. Meera, C. Vanitha, M. Rao, and P. Gunasekaran. 2012. Production, Purification, and Characterization of a β-Glucosidase of *Penicillium funiculosum* NCL1. *Applied Biochemistry and Biotechnology* 167(5), 959–972. https://doi.org/10.1007/s12010-012-9645-4

Rye, Carl S., and Stephen G. Withers. 2000. Glycosidase Mechanisms. *Current Opinion in Structural Biology* 4(5), 573–580. https://doi.org/10.1016/j.sbi.2005.05.003

Salgado, J. C. S., L. P. Meleiro, S. Carli, and R. J. Ward. 2018. Glucose Tolerant and Glucose Stimulated B-Glucosidases – A Review. *Bioresource Technology* 267(July), 704–713. https://doi.org/10.1016/j.biortech.2018.07.137

Singhania, Reeta Rani, Anil Kumar Patel, Rajeev K. Sukumaran, Christian Larroche, and Ashok Pandey. 2013. 'Role and Significance of B-Glucosidases in the Hydrolysis of Cellulose for Bioethanol Production'. *Bioresource Technology* 127(January), 500–507. https://doi.org/10.1016/J.BIORTECH.2012.09.012

Singhania, R. R., R. K. Sukumaran, K. P. Rajasree, A. Mathew, L. Gottumukkala, and A. Pandey. 2011. Properties of a Major β-Glucosidase-BGL1 from *Aspergillus niger* NII-08121 Expressed Differentially in Response to Carbon Sources. *Process Biochemistry* 46(7), 1521–1524. https://doi.org/10.1016/j.procbio.2011.04.006

Steinberg, D., P. Vijayakumar, and E. T. Reese. 1977. β Glucosidase: Microbial Production and Effect on Enzymatic Hydrolysis of Cellulose. *Canadian Journal of Microbiology* 23(2), 139–147. https://doi.org/10.1139/m77-020

Tiwari, R., P. K. Singh, S. Singh, P. K. S. Nain, L. Nain, and P. Shukla. 2017. Bioprospecting of Novel Thermostable β-Glucosidase from *Bacillus subtilis* RA10 and Its Application in Biomass Hydrolysis. *Biotechnology for Biofuels* 10(1), 1–18. https://doi.org/10.1186/s13068-017-0932-8

Treebupachatsakul, Treesukon, Koki Shioya, Hikaru Nakazawa, Takashi Kawaguchi, Yasushi Morikawa, Yosuke Shida, Wataru Ogasawara, and Hirofumi Okada. 2015. Utilization of Recombinant Trichoderma reesei Expressing *Aspergillus aculeatus* β-Glucosidase I (JN11) for a More Economical Production of Ethanol from Lignocellulosic Biomass. *Journal of Bioscience and Bioengineering* 120(6), 657–665. https://doi.org/10.1016/j.jbiosc.2015.04.015

Valappil K. Prajeesh, Kuni Parambil Rajasree, Amith Abraham, Meera Christopher, and Rajeev K. Sukumaran. 2019. Characterization of a Glucose Tolerant β-Glucosidase from Aspergillus Unguis with High Potential as a Blend-In for Biomass Hydrolyzing Enzyme Cocktails. *Biotechnology Letters* 41(10), 1201–1211. https://doi.org/10.1007/S10529-019-02724-Z/METRICS

Wakai, Satoshi, Nanami Nakashima, Chiaki Ogino, Hiroko Tsutsumi, Yoji Hata, and Akihiko Kondo. 2019. Modified Expression of Multi-cellulases in a Filamentous Fungus *Aspergillus oryzae*. *Bioresource Technology* 276, 146–153. https://doi.org/10.1016/j.biortech.2018.12.117

Watanabe, T., T. Sato, S. Yoshioka, T. Koshijima, and M. Kuwahara. 1992. Purificication and Properties of Aspergillus niger β -Glucosidase. *European Journal of Biochemistry* 209(2), 651–659. https://doi.org/10.1111/j.1432-1033.1992.tb17332.x

Yao, Guangshan, Ruimei Wu, Qinbiao Kan, Liwei Gao, Meng Liu, Piao Yang, Jian Du, Zhonghai Li, and Yinbo Qu. 2016. Production of a High-Efficiency Cellulase Complex via B-Glucosidase Engineering in *Penicillium oxalicum*. *Biotechnology for Biofuels* 9(1), 1–11. https://doi.org/10.1186/s13068-016-0491-4

Yan, T. R., and C. L. Lin. 1997. Purification and Characterization of a Glucose-Tolerant β-Glucosidase from *Aspergillus niger* CCRC 31494. *Bioscience, Biotechnology and Biochemistry* 61(6), 965–970. https://doi.org/10.1271/bbb.61.965

Zhang Jiajia, Chuan Wu, Wei Wang, Wei Wang, and Dongzhi Wei. 2018. Construction of Enhanced Transcriptional Activators for Improving Cellulase Production in *Trichoderma reesei* RUT C30. *Bioresources and Bioprocessing* 5(1). https://doi.org/10.1186/s40643-018-0226-4

Zhang Jiwei, Yaohua Zhong, Xuena Zhao, and Tianhong Wang. 2010. Development of the Cellulolytic Fungus *Trichoderma reesei* Strain with Enhanced β-Glucosidase and Filter Paper Activity Using Strong Artificial Cellobiohydrolase 1 Promoter. *Bioresource Technology* 101(24), 9815–9818. https://doi.org/10.1016/j.biortech.2010.07.078

Zang Xiangyun, Meiting Liu, Yihong Fan, Jie Xu, Xiuhong Xu, and Hongtao Li. 2018. The Structural and Functional Contributions of β-Glucosidase-Producing Microbial Communities to Cellulose Degradation in Composting. *Biotechnology for Biofuels* 11(1), 1–13. https://doi.org/10.1186/s13068-018-1045-8

5 Xylanases
Production and Applications

Kim Kley Valladares-Diestra, Luciana Porto de Souza Vandenberghe, Sabrina Vieira, Ariane Fátima Murawski de Mello, Priscilla Zwiercheczewski de Oliveira, and Carlos Ricardo Soccol

5.1 INTRODUCTION

Plant cell wall polysaccharides are the most abundant organic compounds found in nature and are mainly composed of cellulose, hemicelluloses and lignin. Xylan is the major hemicellulose polymer found in cereals and hardwood, representing up to 30%–35% of the total dry weight. The enzymatic degradation of these components, for different applications, is performed by xylanases. Xylanases are glycosidases, which catalyze the hydrolysis of 1,4-ζ-D-xylosidic linkages in xylan (Vandenberghe et al., 2020). The heterogeneity and complexity of xylan have resulted in an assortment of xylanases with different specificities. Different isoforms act synergically to enhance the hydrolysis of xylan. Xylanases are usually grouped into the glycoside hydrolase (GH) families GH10 and GH11, based on amino acid sequence similarity (Valladares-Diestra et al., 2021).

Xylanases are widely produced by filamentous fungi of the genera *Aspergillus* and *Trichoderma* spp., which are induced by xylanolytic components of different materials such as agro-industrial by-products and/or residues. Thus, the reduction of enzyme production costs passes through the use of alternative substrates. In recent years, different agro-industrial by-products have been recycled, reused and introduced into biotechnological processes, thus creating new possibilities for the massive use of these residues in the development of value-added products such as xylanases and other biochemicals under the biorefinery concept (Valladares-Diestra et al., 2021). Applications are mainly turned to animal feed, pulp and paper, bioenergy, and biofuels. The use in stock improvement, biochemical production and other applications are also found (FMI, 2023). This chapter aims to present the production and advancements of xylanases in biomass-degrading applications.

5.2 XYLANASES: DESCRIPTION AND CLASSIFICATION

The importance of xylanases is linked to the increasing industrial demand for lignocellulosic biomass, where xylan represents up to 40% of the vegetal structure (Basit et al., 2018). The enzymatic hydrolysis of these materials is a complex challenge that requires the synergistic action of multiple enzymes – aimed at releasing fermentable sugar molecules – under specific conditions.

Characterized as depolymerizing enzymes, the two main enzymes that hydrolyze xylan are endoxylanases (endo-1,4-β-xylanase) and β-xylosidases (xylan-1,4-β-xylosidase) (Vandenberghe et al., 2020). Depolymerization occurs firstly by the action of endoxylanases through the random cleavage of the xylan structure into xylooligosaccharides. These xylooligosaccharides are then hydrolyzed by β-xylosidases, from the non-reducing end into xylose monomers. Side groups can be hydrolyzed by α-glucuronidase (α-glucosiduronase), α-arabinofuranosidase (α-l-arabinofuranosidase) and acetylxylan esterase (Basit et al., 2021).

DOI: 10.1201/9781003335313-6

TABLE 5.1
Functional and Structural Properties of Xylanases

GH Family	Clan	Xylanase Activity	Structure	Mechanism	Nucleophile–Proton Donor
GH5	GH-A	Endo-β-1,4-xylanase (EC 3.2.1.8), arabinoxylan-specific endo-β-1,4-xylanase (EC 3.2.1.-)	(β/α)8 barrel	Retaining	Glu/Glu
GH7	GH-B	Endo-β-1,4-glucanase (EC 3.2.1.4), Endo-β-1,3–1,4-glucanase (EC 3.2.1.73)	β-jelly roll	Retaining	Glu/Glu
GH8	GH-M	Endo-1,4-β-xylanase (EC 3.2.1.8)	(α/α)8 barrel	Inverting	Asp/Glu
GH10	GH-A	Endo-1,4-β-xylanase (EC 3.2.1.8), endo-1,3-β-xylanase (EC 3.2.1.32)	(β/α)8 barrel	Retaining	Glu/Glu
GH11	GH-C	Endo-1,4-β-xylanase (EC 3.2.1.8), exo-1,4-β-xylosidase (EC 3.2.1.-)	β-jelly roll	Retaining	Glu/Glu
GH43	GH-F	Xylanase (EC 3.2.1.8), β-xylosidase (EC 3.2.1.37)	5-fold β-propeller	Inverting	Asp/Glu

Source: CAZy, Carbohydrate-Active Enzymes Database, 2023 (http://www.cazy.org).

Initially, xylanases were classified based on three properties: (1) molecular mass and isoelectric point, (2) crystal structure and (3) catalytic or kinetic property. Considering the first property, xylanases were again classified into (a) high molecular weight (above 30 kDa) with a low isoelectric point (acidic) (HMWLI) and (b) low molecular weight (below 30 kDa) with a high isoelectric point (basic) (LMWHI). Considering that not all xylanases fit into these categories, several exceptions were observed according to properties 2 and 3, mainly because enzymes have different crystalline structures, catalytic domains and specific substrates. As a result, more accurate information is currently found, such as genomic aspects, sequencing, classifications into family groups, 3D crystal structures and biochemical functions. This advanced knowledge is available in the Carbohydrate-Active Enzymes Database (CAZy).

From CAZy (2023), it is observed that at least, 14 GH families are attributed to xylanases, based on homologies in structural elements and hydrophobic clusters. GH enzymes 5, 7, 8, 10, 11 and 43 have a single catalytic domain, while GH enzymes 16, 51 and 62 have two catalytic domains with bifunctional properties (Collins et al., 2005). GH enzymes 9, 12, 26, 30 and 44 have secondary xylanase activity. Hydrophobic clustering analysis of the catalytic domains and amino acid sequence similarities allowed the final classification of the xylanases into GH 10 and GH 11 (Verma and Satyanarayana, 2013). The catalytic properties of GH 10 and GH 11 are constantly being studied, while information about GH 5, 7, 8 and 43 is still limited (Bhardwaj et al., 2019). Table 5.1 lists details about xylanase activity, including mechanisms and structure. As it can be observed, there are two different mechanisms during xylanase action:

1) Retaining occurs when there is a double displacement, with the formation of intermediate residual configurations, and subsequent hydrolysis. This mechanism originates from the active site of the xylanase consisting of two carboxylic groups. Intermediate residues are formed from the protonation of the substrate by the first carboxylic acid group; the second group acts as an acid catalyst, carrying out a nucleophilic attack. The result of protonation is the inversion of the β configuration into α, due to the intermediate formation of the α-glycosyl enzyme. After the nucleophilic attack, there is an inversion of the α to β configuration through the transition state of oxo-carbonium ions (Bhardwaj et al., 2019).

2) Inverting occurs when there is a single acid-catalyzed displacement mechanism of a carboxylate ion. Usually, GH 8 and GH 43 act via inversion of the anomeric center with glutamate and aspartate as the main catalytic residue. This mechanism originates from the active site of the xylanase consisting of a single carboxylic group. An active nucleophilic water molecule attacks the anomeric carbon via enzymatic action to break glycosidic bonds and cause inversion of the anomeric carbon configuration (Bhardwaj et al., 2019).

Enzymes obtained from bacteria and fungi are the primary organisms hydrolyzing xylanases – endo-1,4-β-xylanase (EC 3.2.1.8) and β-xylosidase (EC 3.2.1.37) – defining characteristics and optimal environmental conditions for enzymatic activity, such as temperature and pH (Singh et al., 2019; Walia et al., 2017). Although each enzyme has specificities for enzymatic activity and biological function, in general, xylanases obtained from these microorganisms are thermostable with optimum temperatures ranging from 40°C to 80°C and pH ranging from 4.5 to 6.5 (Sharma et al., 2019; Singh et al., 2019). With the improvement of molecular identification tools, genetic sequencing allows the decoding of specific genes enabling optimized production assays, culminating in the discovery of several producing genera (Table 5.2).

TABLE 5.2
Identified Microorganisms Responsible for Xylanase Production

Group	Genus	Species	Optimal Temperature	Optimal pH
Bacteria	*Arthrobacter*	*Arthrobacter* sp.	40°C–90°C	4.5–10
	Bacillus	*B. firmus*, *B. arseniciselenatis*, *B. licheniformis*, *B. amyloliquefaciens*, *B. subtilis*		
	Clostridium	*C. thermocellum*, *C. papyrosolvens*		
	Cellulomonas	*C. fimi*, *C. flavigena*, *C. uda*		
	Pediococcus	*P. acidilactici*		
	Pseudomonas	*Pseudomonas* sp., *P. aeruginosa*, *P. fluorescens* subsp. *cellulosa*, *P. boreopolis*, *P. stutzeri*		
	Rhodothermus	*R. marinus*		
	Staphylococcus	*Staphylococcus* sp., *S. aureus*		
	Streptomyces	*Streptomyces* sp., *S. lividans*		
	Thermoactinomyces	*T. thalophilus*, *T. vulgaris*		
	Thermomonospora	*T. fusca*, *T. curvata*		
Fungus	*Aspergillus*	*A. niger*, *A. flavus*, *A. niveus*, *A. ochraceus*, *A. foetidus*, *A. fumigates*, *A. terreus*, *A. tamari*	48°C–75°C	4.0–8.0
	Fusarium	*Fusarium* sp., *F. solani*, *F. proliferatum*, *F. oxysporum*, *F. heterosporum*, *F. roseum*		
	Humicola	*Humicola* sp., *H. insolens*, *H. lanuginosa*, *H. grisea*		
	Rhizomucor	*R. pusillus*		
	Talaromyces	*T. byssochlamydoides*, *T. emersonii*		
	Thielavia	*Thielavia* sp., *T. terrestaris*		
	Trichoderma	*T. inhamatum*, *T. piluliferum*, *T. viride*, *T. longibrachiatum*, *T. asperellum*, *T. stromaticum*, *T. harzianum*, *T. reesei*		

Sources: Adapted from Bhardwaj et al. (2019), Singh et al. (2019), Sharma et al. (2019), Walia et al. (2017).

5.3 PRODUCTION OF XYLANASES

The production of xylanases is generally carried out by microorganisms such as bacteria, filamentous fungi and protozoa. Some higher organisms such as marine algae, insects, gastropods, arthropods and plants were also reported. However, filamentous fungi and bacteria are most often used for their industrial production (Table 5.2). Fungal xylanases of the genera *Aspergillus* and *Trichoderma* are the remarkable xylanases producers, including commercial xylanases, followed by the *Bacillus* genus representing the bacterial xylanases.

In most organisms, the production of xylanases is directly related to the presence of xylan, where this polysaccharide generates a positive induction in the gene expression of xylanases (Polizeli et al., 2005). As can be seen in Figure 5.1, the xylan induction starts through a cascade reaction involving the XlnR gene, which is one of the main genes involved in the expression of xylanases. This gene is activated by the presence of low concentrations of small xylooligosaccharides derived from xylan that cross the cell membrane (Moran-Aguilar et al., 2021). In turn, the xylanases released outside the cytoplasm increase the release of xylooligosaccharides through xylan hydrolysis, consequently, an increase in the induction and production of xylanases is observed. Moreover, the presence of xylooligosaccharides concomitantly induces the production of β-xylosidase enzymes that hydrolyze xylobiose to produce xylose. Xylose acts as a repressor in the gene expression of xylanases through the mechanism of carbon catabolite repression protein (Cre). This mechanism is the main factor for the repression of carbon catalytic enzymes (xylanases, cellulases and others), and it can be activated by relatively high concentrations of monosaccharides, such as xylose, glucose and fructose (Valladares-Diestra et al., 2022). It has been reported that concentrations above 70 mM of xylose and concentrations of free glucose in the culture medium remained stable throughout the

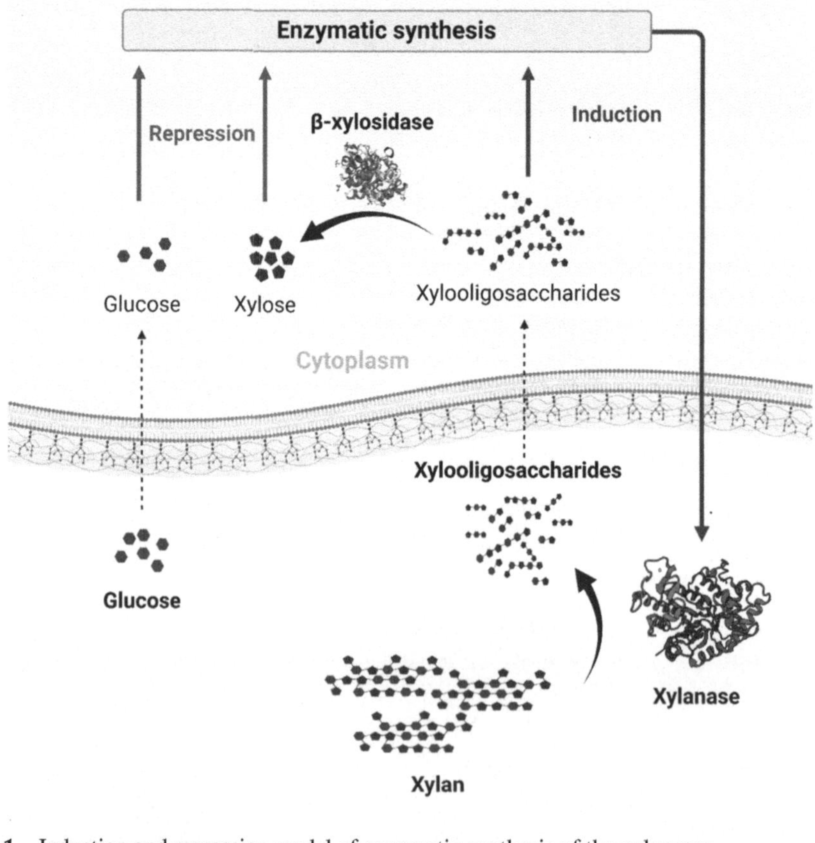

FIGURE 5.1 Induction and repression model of enzymatic synthesis of the xylanases.

fermentation process, with approximately 10 mM, leading to the activation of the Cre mechanism, which causes the repression of protein expression of xylanolytic enzymes (de Vries et al., 1999).

The heterogeneity and complexity of xylan have resulted in a wide range of xylanases with different specificities. Different isoforms cooperate to enhance the hydrolysis of xylan. The multiplicity in the production of xylanases can be further increased by post-translational modifications such as differential glycosylation, proteolysis or both. Therefore, each xylanase produced from different microorganisms under different types or culture conditions exhibits unique biochemical and catalytic properties.

Industrial production of microbial-derived xylanases can be performed in two configurations: discontinuous (batch) or continuous. Normally the discontinuous process is employed where an initial cultivation step is required for the target enzyme production. Fermentation parameters, such as aeration, agitation, temperature, pH and time, that must be optimized for each microorganism are commonly reported in the literature (Valladares-Diestra et al., 2021). The continuous process is mainly characterized by the immobilization of the biocatalyst particle (i.e., whole cell) onto inert materials (e.g., calcium alginate beads, porous glass, silica) or insoluble substrates (i.e., agro-industrial by-products), which minimizes pH and temperature fluctuations, allows the reuse of the biocatalyst particle and significantly improves the volumetric production of the bioreactor (Sánchez-Martínez et al., 2020). Besides the selection of appropriate carbon/nitrogen sources and micronutrients, operational conditions are also of the utmost importance for avoiding catabolite repression or low productivity (Kumar and Dubey, 2019).

The cost of the enzymes becomes a critical factor for the use of this technology at an industrial scale. Agro-industrial residues have also been investigated as potential carbon sources due to low cost, great availability and sustainability aspects. Agro-industrial residues are suitable supports for solid-state fermentation or submerged fermentation (Valladares-Diestra et al., 2021). Several authors have worked on the production of xylanases using agro-industrial residues. Di Marco et al. (2017) obtained 8.99 $U.mL^{-1}$ using *Paenibacillus* sp. and sugarcane bagasse as a substrate. Elegbede and Lateef (2017) used corncob as a carbon source for xylanase production with *Aspergillus fumigatus*, reaching 50.55 $U.mL^{-1}$. The production varies greatly according to the substrate used and the employed microorganism. These differences may be attributed to the medium composition and process conditions.

The type of bioreactors used in xylanase production also significantly affected the level of enzyme production. Valladares-Diestra et al. (2021) evaluated the xylanases production profile by *A. niger* using sugarcane bagasse and soybean meal in two types of bioreactors: stirred tank reactor (STR) and bubble column reactor (BCR). The highest xylanase activity in STR was 33.73 $U.mL^{-1}$ obtained after three days. In contrast, in the BCR, the maximum concentration of xylanase production was 60.45 $U.mL^{-1}$ at seven days. Thus, the BCR system exhibited better performance for xylanase production using agro-industrial residues in suspension. This shows that in the production of xylanases by filamentous fungi, the STR system presents a better homogenization but causes shear stress in the mycelia of the fungi caused by mechanical agitation. However, the agitation of the BCR is pneumatic, which is promoted by ascending bubbles, having a lower impact on filamentous fungi and increasing the production of xylanases (Libardi et al., 2019).

5.4 APPLICATION OF XYLANASES FOR BIOMASS DEGRADATION

In a biorefinery, enzymatic hydrolysis is the technology that converts lignocellulosic biomass into sugar monomers and value-added products. Due to the abundance and cost, the bioconversion of hemicellulose is a factor of interest in the industry and, consequently, the potential use of xylanases. These enzymes have applications in various industrial processes, such as in the food, paper, animal feed, biofuels and textile industries, with the synthesis of valuable products. As xylanase degrades the lignocellulosic biomass breaking the xylan polymer, biofuel production has been a focus of research, including the production of bioethanol from lignocellulosic agricultural residues exploited by the action of xylanases from various sources. In the pulp and paper industry, it is used to bleach wood pulp and improve paper quality. In the food industry, the application of xylanases consists of improving the quality and texture of the dough, acting on the sugars present in the flour that help increase the elasticity of the gluten network. The clarification of fruit juices must be free of excess polysaccharides to be effectively liquefied (Kaushal et al., 2021). Xylanases can hydrolyze the xylan polymer that is bound to cellulose and lignin in cellulosic fiber. Xylan rupture leads to a separation between these components, thereby increasing swelling in the fiber wall and improving lignin extraction from the pulp.

The specific characteristics of xylanases are industrially attractive, as each process requires enzymes under different conditions. In this way, it is necessary to determine the parameters that identify the potential that each enzyme presents, depending on the application. So, the specification of optimum pH and temperature, stability, and kinetic parameters is performed.

The overexpression of enzymatic components, as well as the improvement of properties such as thermostability, are improvements for lignocellulose bioprocessing. The production of a multienzymatic system of xylanases, in which each enzyme has a function, is a strategy for microorganisms to achieve effective xylan hydrolysis. Microorganisms also produce other proteins and metabolites along with the desired xylanase. Therefore, purification of the crude enzyme is necessary. The characterization of purified xylanase will help with knowledge of its stability and specificity for different substrates, and this facilitates the selection of the best use in the industrial process. With current biotechnological techniques, such as recombinant DNA technology, several attempts have been made to identify, isolate and clone the gene encoding for xylanase in an appropriate system. This approach helps in engineering efficient microorganisms for improved xylanase production with the desired properties (Bhardwaj et al., 2019).

Because it is used in various sectors, there is a need to improve the quality and catalytic efficiency of this enzyme so that it can be used in adverse environmental conditions that require high temperatures and extensive pH range. Strains isolated from different extreme environments can provide a solution to this problem since the enzymes of these microorganisms can withstand adverse conditions due to acclimatization to extreme environments. In addition, the xylanase enzyme can undergo technological advances such as recombinant DNA technology to produce an enzyme with the desired characteristics, modifying its amino acid distribution. One of the techniques that can be used for the immobilization of xylanases is the use of organic metal structures that have gained a lot of attention in the last decade. The possibility of forming networks for the support of enzymes based on suitable metal and organic ligands offers the opportunity to increase the catalytic efficiency of the enzyme (Kaushal et al., 2021). In the future, for the production of biofuels, applying molecular biology techniques and biotechnology, and xylanases, engineered with cellulolytic and ligninolytic activity, may be ideal for high yields (Chaudhary et al., 2021). There is a growing industrial interest in the xylanases of thermophilic organisms due to their ability to degrade xylan at high temperatures. Exploring the use of new enzymes and the combination of enzymes from different organisms, with different characteristics, can contribute to the development of a highly efficient lignocellulose conversion system.

5.5 MARKET AND COMMERCIAL PRODUCTS USING XYLANASES

Industrial enzymes are molecules that possess great market value, growth and development once they provide sustainable alternatives to conventional chemical processes and can enhance the quality of several products. Xylanases, along with pectinases and cellulases, take up to 20% of the industrial enzyme market share and can be applied in different contexts (Martínez-Pacheco et al., 2020). The rising awareness of the impacts of the usage of chemicals in several areas and the potential of substituting them with more sustainable processes using enzymes is the main driver of their market growth. This is the case of xylanase application in the pulp and paper industry, in which xylanases can substitute or reduce the amount of chemicals necessary for the bleaching process. One example of a commercial product with this objective is Xylio-Pre® from Novozymes. Xylanases can also be a potential substitute for chemical additives acting in dough conditioning in the baking industry, such as the commercial enzymes Pentopan® and Panzea® from Novozymes.

In lignocellulosic biomass processing, xylanases can be applied to obtain fermentable sugars that can be further applied for the production of added-value and sustainable molecules, such as the case of second-generation bioethanol – a biofuel that could potentially substitute petroleum-derived substances. Enzymatic hydrolysis usually occurs after the pretreatment step, and a set of different enzymes with specific activities need to be applied for the complete degradation of biomass, releasing C5 (such as arabinose and xylose) and C6 (glucose) sugars. Thereby, xylanases are usually combined with cellulolytic enzymes, and to make these bioprocesses profitable, enzymes need to be highly efficient, evoking the need for commercial products with these characteristics. The Danish multinational company Novozymes is one of the pioneers in this development and, in 2012, launched a plant for the research and production of enzymes for biofuels in the US (Novozymes, 2023). The products of the Cellic® brand stand out in this segment: Ctec mainly has cellulosic activity with xylanases acting as accessory enzymes and Htec presents xylanases as the main component. The constant development of new solutions by Novozymes recently led to Cellic Ctec 3, which is claimed to have an efficiency 1.5 times higher than its precursor, Cellic Ctec 2 (Novozymes, 2023). Another application of xylanases in conjunction with other enzymes is in the animal feeding field, in which these molecules help to increase the biodisponibility of nutrients and reduce the content of components that may hinder digestion. Enzymatic complexes may help to reduce feeding costs and usually include amylases for starch processing, cellulases and hemicellulases that act on the lignocellulosic fraction, and phytases to remove phytic acid. Enzyme-producing companies usually work together for the development of new solutions for animal feeding, for example, the Feed Alliance was founded in 1998 by Novozymes and DSM (DSM, 2023). Since its founding, the Feed Alliance has provided the development of enzymes with xylanase activity such as Ronozyme® Multigrain and VP (in which xylanase is an accessory compound) and WX2,000. A summary of the cited commercial products in this section and their characteristics is presented in Table 5.3.

TABLE 5.3
Commercial Xylanases and Their Characteristics

Product	Area of Application	Characteristics	Company*
Xylio-Pre®	Pulp and paper	High pH and temperature-tolerant xylanase that can be used either before or after the bleaching process to reduce chemical usage and process costs.	Novozymes (www.novozymes.com)
Pentopan® and Panzea®	Baking	Xylanases from different sources (fungal and bacterial, respectively) that act on dough conditioning, improving its quality and final product appearance.	Novozymes (www.novozymes.com)
Ronozyme® Multigrain and VP	Animal feeding	Enzymatic complexes with xylanase activity that enable the biodisponibility of nutrients of several feeding ingredients and promote feeding cost reduction. Both are available either in powder and liquid form.	Novozymes (www.novozymes.com) and DSM (www.dsm.com)
Ronozyme® WX2,000	Animal feeding	Mono-component xylanase responsible for degrading xylan in different cereals for poultry feeding. Available either in powder and liquid form.	Novozymes (www.novozymes.com)
Cellic® Ctec 2 and 3	Biomass processing	Enzymatic complexes with hemicellulase activity for lignocellulosic biomass degradation with the purpose of further producing, mainly second-generation ethanol.	Novozymes (www.novozymes.com)
Cellic® Htec 2	Biomass procesing	Endoxylanase with cellulase background useful for lignocellulosic biomass degradation. Usually combined with Cellic Ctec for higher conversion yields.	Novozymes (www.novozymes.com)

* Information about the commercial products and their characteristics were retrieved from the companies' websites.

5.6 CONCLUSIONS AND PERSPECTIVES

Xylanases are hydrolases that act on the complex structure of xylan, which is why there are different isoforms with different specificities that act synergically. These enzymes are usually grouped into glycoside hydrolase (GH) families based on amino acid sequence similarity. These enzymes are mainly produced from microbial sources where filamentous fungi, *Aspergillus* and *Trichoderma*, and bacteria, *Bacillus*, are most often employed for their industrial production using xylan-rich substrates. Xylanase enzymes undoubtedly have clear importance in the global enzyme market due to their great versatility, being applied in different industrial processes such as in the food, paper, animal feed, biofuels and textile industries. They are one of the main enzymatic catalysts used in the degradation of lignocellulosic biomass in large volumes for the production of new biomolecules and bioenergy. As an example, they can be used, concomitantly with cellulase complexes, in the enzymatic hydrolysis step to obtain sugar-rich hydrolysate to produce second-generation bioethanol in biorefineries in the context of a circular and sustainable economy. Several studies have demonstrated the ability of xylanases to produce xylooligosaccharides from agro-industrial residues. These prebiotics are highly employed, presenting important immunomodulatory, anticancer, growth-regulating, antioxidant, antimicrobial and other bioactive properties.

Recent developments show that the quality and catalytic efficiency of xylanases can be improved with the use of technological advances, such as recombinant DNA technology, to produce enzymes with desired characteristics. In this case, xylanases could be applied in adverse environmental conditions, such as at high temperatures and in extensive pH ranges. In addition, immobilization of these enzymes can occur with the use of different techniques. These techniques certainly enable the reuse of enzymes in different cycles, thus cheapening the processes in which they are applied.

REFERENCES

Basit, A., Jiang, W., Rahim, K., 2021. Xylanase and its industrial applications. In: Basso, T.P., Basso, T.O., Basso, L.C. (Eds.), *Biotechnological Applications of Biomass*. Intech open. https://doi.org/10.5772/intechopen.92156

Basit, A., Liu, J., Rahim, K., Jiang, W., Lou, H., 2018. Thermophilic xylanases: From bench to bottle. *Crit. Rev. Biotechnol. Biotechnol.* 38(7), 989–1002. https://doi.org/10.1080/07388551.2018.1425662

Bhardwaj, N., Kumar, B., Verma, P., 2019. A detailed overview of xylanases: An emerging biomolecule for current and future prospective. *Bioresour. Bioprocess.* 6(1), 40. https://doi.org/10.1186/s40643-019-0276-2

Chaudhary, R., Kuthiala, T., Singh, G., Rarotra, S., Kaur, A., Arya, S.K., Kumar, P., 2021. Current status of xylanase for biofuel production: A review on classification and characterization. *Biomass Convers. Biorefin.* https://doi.org/10.1007/s13399-021-01948-2

Collins, T., Gerday, C., Feller, G., 2005. Xylanases, xylanase families and extremophilic xylanases. *FEMS Microbiol. Rev.* 29(1), 3–23. https://doi.org/10.1016/j.femsre.2004.06.005

de Souza Vandenberghe, L.P., Karp, S.G., Pagnoncelli, M.G.B., von Linsingen Tavares, M., Junior, N.L., Diestra, K.V., Viesser, J.A., Soccol, C.R., 2020. Classification of enzymes and catalytic properties. In: Singh, S.P., Pandey, A., Singhania, R.R., Larroche, C., Li, Z. (Eds.), *Biomass, Biofuels, Biochemicals - Advances in Enzyme Catalysis and Technologies*. Elsevier, pp. 11–30.

de Vries, R.P., Visser, J., de Graaff, L.H., 1999. CreA modulates the XlnR-induced expression on xylose of Aspergillus niger genes involved in xylan degradation. *Res. Microbiol.* 150(4), 281–285. https://doi.org/10.1016/S0923-2508(99)80053-9

Di Marco, E., Soraire, P.M., Romero, C.M., Villegas, L.B., Martínez, M.A., 2017. Raw sugarcane bagasse as carbon source for xylanase production by *Paenibacillus* species: A potential degrader of agricultural wastes. *Environ. Sci. Pollut. Res. Int.* https://doi.org/10.1007/s11356-017-9494-3

DSM, 2023. DSM corporate [WWW Document]. https://www.dsm.com/corporate/home.html (accessed January 22, 2023).

Elegbede, J.A., Lateef, A., 2017. Valorization of Corn-Cob by fungal isolates for production of xylanase in submerged and solid state fermentation media and potential biotechnological applications. *Waste Biomass Valorization*, 1–15. https://doi.org/10.1007/s12649-017-9932-y

FMI, 2023. Xylanase market [WWW Document]. Future Mark Insights. https://www.futuremarketinsights.com/reports/xylanase-market (accessed January 28, 2023).

Kaushal, J., Khatri, M., Singh, G., Arya, S.K., 2021. A multifaceted enzyme conspicuous in fruit juice clarification: An elaborate review on xylanase. *Int. J. Biol. Macromol.* 193(B), 1350–1361.

Kumar, P., Dubey, K.K., 2019. Current perspectives and future strategies for fructooligosaccharides production through membrane bioreactor. In: Shukla, P. (Ed.), *Applied Microbiology and Bioengineering.* Academic Press, Boca Raton, pp. 185–202. https://doi.org/10.1016/b978-0-12-815407-6.00010-1

Libardi, N., Soccol, C.R., de Carvalho, J.C., de Souza Vandenberghe, L.P., 2019. Simultaneous cellulase production using domestic wastewater and bioprocess effluent treatment – A biorefinery approach. *Bioresour. Technol.* 276, 42–50. https://doi.org/10.1016/j.biortech.2018.12.088

Martínez-Pacheco, M.M., Flores-García, A., Zamudio-Jaramillo, M.A., Chávez-Parga, M.C., Alvarez-Navarrete, M., 2020. Optimization of production of xylanases with low cellulases in Fusarium solani by means of a solid state fermentation using statistical experimental design. *Rev. Argent. Microbiol.* 52(4), 328–338. https://doi.org/10.1016/j.ram.2019.12.003

Moran-Aguilar, M.G., Costa-Trigo, I., Calderón-Santoyo, M., Domínguez, J.M., Aguilar-Uscanga, M.G., 2021. Production of cellulases and xylanases in solid-state fermentation by different strains of *Aspergillus niger* using sugarcane bagasse and brewery spent grain. *Biochem. Eng. J.* 172, 108060. https://doi.org/10.1016/j.bej.2021.108060

Novozymes, 2023. Build your business with people [WWW Document]. Novozymes. https://biosolutions.novozymes.com (accessed 1.23.23).

Polizeli, M.L.T.M., Rizzatti, A.C.S., Monti, R., Terenzi, H.F., Jorge, J.A., Amorim, D.S., 2005. Xylanases from fungi: Properties and industrial applications. *Appl. Microbiol. Biotechnol.* 67(5), 577–591. https://doi.org/10.1007/s00253-005-1904-7

Sánchez-Martínez, M.J., Soto-Jover, S., Antolinos, V., Martínez-Hernández, G.B., López-Gómez, A., 2020. Manufacturing of short-chain fructooligosaccharides: from laboratory to industrial scale. *Food Eng. Rev.* 12(2), 149–172. https://doi.org/10.1007/s12393-020-09209-0

Sharma, A., Sharma, A., Singh, S., Kuhad, R.C., Nain, L., 2019. Thermophilic fungi and their enzymes for biorefineries. In: Tiquia-Arashiro, S., Grube, M. (Eds.), *Fungi in Extreme Environments: Ecological Role and Biotechnological Significance.* Springer Nature, pp. 479–502. https://doi.org/10.1007/978-3-030-19030-9_24

Singh, S., Sidhu, G.K., Kumar, V., Dhanjal, D.S., Datta, S., Singh, J., 2019. Fungal xylanases: Sources, types, and biotechnological applications. In: Yadav, A.N., Mishra, S., Singh, S., Gupta, A. (Eds.), *Recent Advancement in White Biotechnology Through Fungi, Fungal Biology.* Springer Nature, pp. 405–428. https://doi.org/10.1007/978-3-030-10480-1_12

Valladares-Diestra, K.K., Porto de Souza Vandenberghe, L., Soccol, C.R., 2021. A biorefinery approach for enzymatic complex production for the synthesis of xylooligosaccharides from sugarcane bagasse. *Bioresour. Technol.* 333, 125174. https://doi.org/10.1016/j.biortech.2021.125174

Valladares-Diestra, K.K., Porto de Souza Vandenberghe, L., Soccol, C.R., 2022. Integrated xylooligosaccharides production from imidazole-treated sugarcane bagasse with application of in house produced enzymes. *Bioresour. Technol.* 362, 127800.

Verma, D., Satyanarayana, T., 2013. Production of cellulase-free xylanase by the recombinant *Bacillus subtilis* and its applicability in paper pulp bleaching. *Biotechnol. Prog.* 29(6), 1441–1447. https://doi.org/10.1002/btpr.1826

Walia, A., Guleria, S., Mehta, P., Chauhan, A., Parkash, J., 2017. Microbial xylanases and their industrial application in pulp and paper biobleaching: A review. *3 Biotech* 7(1). https://doi.org/10.1007/s13205-016-0584-6

6 Lignin-Degrading Enzymes
Production and Applications

Amanjot Kaur, Himanshu, and Jitendra Kumar Saini

6.1 INTRODUCTION

Lignin is a complex organic polymer deposited in the cell wall of many plants and is a major component of lignocellulosic biomass along with cellulose and hemicellulose. It is highly heterogeneous and therefore resistant to physical, biological and chemical breakdown. However, some microorganisms are still capable of catalyzing lignin breakdown via lignin-degrading enzymes (LDEs), which catalyze lignin breakdown by oxidative mechanisms. LDEs are broadly categorized into two major groups: lignin-modifying enzymes (LMEs) and lignin-degrading auxiliary enzymes (LDAEs). LMEs can catalyze lignin breakdown on their own, whereas LDAEs are unable to solely catalyze such reactions and must act in concert with main enzymes, thereby, facilitating the degradation. Due to their diverse substrate range and superior oxidative properties, many LDEs are exploited in various industrial sectors, including textiles, food, cosmetics, bioenergy and pharmaceuticals. Many technological developments have taken place in the recent past with regard to production enhancements, improvements in catalytic efficiencies and newer/novel industrial applications. Therefore, the main aim of this chapter is to provide an up-to-date overview of basic concepts, microbial production, analysis and applications of lignin-degrading enzymes. A detailed emphasis on recent applications of ligninolytic enzymes in various sectors has been provided with an outlook for future needs and prospects of lignin-degrading enzymes.

6.2 STRUCTURE AND PROPERTIES OF LIGNIN

Lignin is a complex organic polymer deposited in the cell wall of many plants. It is a key structural component that provides mechanical strength to plants as well as resistance against various pathogens. Chemically, lignin is an intricate, three-dimensional, long-chain, heterogeneous polymer of phenylpropanes of C-3 attached to C-6 atom rings, interlinked by ether bonds. It possesses methoxy, a phenolic hydroxyl group and a few terminal aldehyde groups in their side chains. Heterogeneity of lignin arises due to the diversity and degree of cross-linking between various monolignol units. Its precursor includes p-coumaryl, coniferyl and sinapyl alcohol, which correspond to guaiacyl (G), syringyl (S) and p-hydroxyphenyl (H) units of lignin, respectively. Depending on the plant species, ratios of these precursors vary greatly, such as the abundance of G units in softwood, G–S units in hardwood and G–S–H units in herbaceous plants. Lignin makes about 10%–15% (wt.) of biomass of the primary cell wall of plants, which helps to build a rigid, impermeable and resistant structure against microbial attacks and oxidative stress (Ponnusamy et al., 2019). Lignin deposition is initiated from the middle lamella, followed by the primary cell wall and the S1 layer in the secondary wall. Eventually, lignin is located in the S2 and S3 layers of the secondary cell wall. Besides being the major structural component, lignin also allows the transportation of water and solutes through the vasculature system of plants. Naturally, lignin is white or colorless compound with high molecular weight. It has strong inter- and intramolecular hydrogen bonds due to which it is insoluble in any solvent. The antioxidant property of a polymer is due to aromatic rings of lignin present in the form

 DOI: 10.1201/9781003335313-7

of basic phenyl propane units. It possesses considerable antimicrobial and antifungal activities. As compared to the sugar component, lignin is more impervious to chemical/biological attacks and environmental stresses.

6.3 LIGNIN-DEGRADING ENZYMES: TYPES AND MECHANISMS OF ACTION

Though lignin is a very recalcitrant biological material, various microorganisms are capable of its breakdown to different extents. This degradation mechanism is distinctive in involving an oxidative mechanism and an array of ligninolytic enzymes. In nature, various organisms have the potential to degrade lignin, including bacteria (*Bacillus subtilis*, *Streptomyces cyaneus*, and *Azospirillum lipoferum*) and fungi (*Ganoderma lucidum, Trametes versicolor*, *Pycnoporus* sp., *Pleurotus ostreatus*, and *Cerrena maxima*), insects (*Manduca sexta*, *Nephotettix cincticeps*, and *Tribolium castaneum*), plants (*Rhus vernicifera*, *Mangifera indica*, and *Zea mays*). However, the most effective lignin degraders are found in ascomycetes, basidiomycetes and deuteromycetes, especially basidiomycetes which include white rot fungi (WRF). WRF possess the ability to utilize various combinations of oxidative lignin-modifying enzymes to release numerous nonspecific free radicals that begin the depolymerization of lignin by breaking carbon–carbon and various ether interunit linkages. These enzymes have been broadly classified into two major categories, namely, LMEs and LDAEs (Figure 6.1). LDAEs are additional enzymes involved in the complete lignin degradation and include cellobiose dehydrogenases, aryl alcohol oxidases, glyoxal oxidases, heme-thiolate haloperoxidases, pyranose 2-oxidases and glucose oxidases. LMEs are ligninolytic enzymes that are oxidative in nature and carry out oxidation-mediated lysis. These have been further subclassified into phenol oxidases and heme-peroxidases. As shown in Figure 6.1, enzymes under both subcategories belong to the oxidase superfamily and the major ones are discussed next.

According to the PeroxiBase, heme-peroxidases are mainly classified into two superfamilies: non-animal and animal peroxidases. The former includes four families: class I, II, III and other

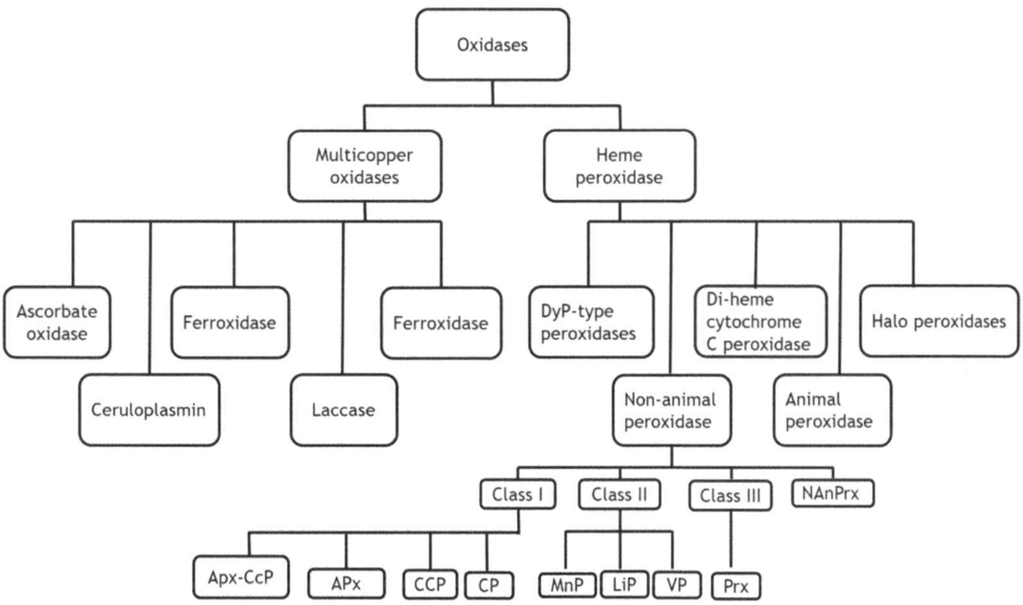

FIGURE 6.1 Major ligninolytic enzymes and their subcategories (Cagide & Castro-Sowinski, 2020). (APx, ascorbate peroxidase; CcP, cytochrome c peroxidase; CP, catalase peroxidase; MnP, manganese peroxidase; LiP, lignin peroxidase; VP, versatile peroxidase; NAnPrx, non-animal peroxidases).

non-animal peroxidases (Falade et al., 2017). Some of the heme-peroxidases are secreted to the milieu and possess lignin-degrading activity, such as manganese peroxidase (MnP), lignin peroxidase (LiP) and versatile peroxidase (VP), which are typically produced by fungi (Pollegioni et al., 2015).

6.3.1 Phenol Oxidases

Phenol oxidases are contained within multicopper oxidase (MCO) enzymes, which have one or more of the three different Cu-sites (type 1, 2 or 3). These enzymes facilitate the transfer of four electrons and reduce oxygen resulting in the formation of water molecules. This category of enzymes includes laccases, ascorbate oxidase, ferroxidase, nitrite reductase and ceruloplasmin.

6.3.1.1 Laccase

Laccase, also known as benzenediol oxygen reductases, belongs to the class oxidoreductase (EC 1.10.3.2). Due to its low specificity against the substrate, laccase has a high potential for degrading numerous inorganic or organic compounds that possess phenolic structure, including non-substituted or substituted phenolics or even compounds that are basically non-phenolic. Structurally, it has a primary structure of nearly 500 amino acid residues arranged in three continuous domains, with a beta-barrel topology. It is associated with MCO. It holds four copper molecules in its active sites that are engaged in the reduction of oxygen, resulting in the production of water molecules. The enzyme's four copper atoms are distributed in three types of places: T1, T2 and T3. T1 has elevated redox ability in the enzyme that plays a vital role in enzymatic catalysis by oxidation of substrate. T2 is a paramagnetic copper that is deprived of absorption in the visible range of the spectrum. This shows strong absorption under electron paramagnetic resonance. Core T2/T3 is also involved in the inhibition of enzymatic activity when interacting with certain anions. This enzyme also possesses various ligands such as mannose, NAG and glycerol. which bring considerable differences between various laccases.

Laccases are subdivided into high and low redox potential laccases depending upon their affinity to transfer electrons during catalysis, with the former ones synthesized by the fungi and the latter ones synthesized by a wide variety of organisms including insects, bacteria or plants.

Laccases play a variety of physiological roles, ranging from anabolic processes involved in plant morphogenesis to catabolic functions involved in cell wall lignin degradation. The activity of low redox potential laccases can be enhanced by supplementation of mediators, and sometimes laccases employ intermediate mechanisms to carry out the oxidation of some high redox potential lignin compounds of a non-phenolic nature. Naturally available mediator compounds which enhance laccase catalysis include vanillin; syringaldehyde; acetosyringone; and sinapic, coumaric and ferulic acids. Preferred mediator compounds of non-natural origin include azinobis-ethylbenzothiazoline-sulphonic acid, tetra-methyl-piperidinyloxy, hydroxybenztriazole, violuric acid, promazine and nitroso-naphthol-disulfonic acid.

The molecular mass of laccases ranges from 50 to 90 kDa; the optimal pH lies between 3.6 and 6.8. Laccases generally possess pH and thermal stability, which is one key feature associated with their large-scale industrial applications in different sectors.

6.3.2 Heme-Peroxidase

Inclusive of heme-peroxidases and other oxidoreductase enzymes, peroxidases facilitate the oxidation of various substrates and the reduction of hydrogen peroxide to generate water and oxidize varied types of substrates. Followed by laccases, lignin peroxidase and manganese peroxidase are the most crucial lignin-degrading enzymes. They belong to heme proteins as they have the protoporphyrin 9 as a prosthetic group. This group of enzymes catalyzes the reaction in three major steps. Initially, the oxidation of an enzyme results in the production of a modified enzyme and the release of water. Second, the modified enzyme facilitates the generation of free radicals along with the second modification of the enzyme by reduction of substrate (electron transfer). Finally, the second modified form of the enzyme reacts with the substrate to produce free radicals and water; meanwhile, the enzyme gets reduced to its native form. Some of the peroxidase enzymes are described in detail below:

6.3.2.1 Lignin Peroxidase (LiP)

LiP is an oxidoreductase enzyme which oxidizes a variety of aromatic compounds inclusive of lignin, veratryl alcohol and methoxy benzenes. LiPs are glycosylated monomeric enzymes with a molecular weight of ~40 KDa, consisting of amino acid residues, water molecules, heme group, carbohydrates and calcium ions. It has high redox potential ranging between 800 and 1500 mV. A specific property related to high redox potential is the presence of amino acid tryptophan (Trp171) on the enzyme surface. Its crystal structure indicates that the preserved Trp residue plays a crucial role in catalysis. This Trp171 residue mediates the electron transfer from substrate to the enzyme and can actively catalyze veratryl alcohol, which is responsible for the oxidation of aromatic compounds. Although this enzyme can actively oxidize many phenolic and non-phenolic aromatic compounds both in the presence and absence of redox mediators, its presence enhances their catalytic performance. Some research supports that veratryl alcohol safeguards the enzyme from the action of hydrogen peroxide, and acts as a redox mediator between enzyme and substrate. LiP has been characterized by a variety of microorganisms, however, *Phanerochaete chrysosporium* has been the model organism. Various isoforms of LiP are known, differing on the basis of level of glycosylation and isoelectric points. The type of isoform depends mostly upon medium composition. Catechol, guaiacol, syringic acid, vanillyl alcohol and acetosyringone are some of the phenolic compounds acted upon by LiP enzymes.

6.3.2.2 Manganese Peroxide (MnP)

The MnP class of lignin-degrading enzyme is secreted by white and brown rot basidiomycetous fungi and plays a vital role in waste decomposition. Fungi are the main producers of MnP, but many bacteria and algae are also reported to synthesize these enzymes, although at a slower pace. MnPs are glycoproteins that use manganese ions as mediators, with an average molecular weight of around 50 KDa and possess nearly 350 amino acid residues with a similarity of 43% compared to LiPs. Structurally, they have two domains with a heme group in the center, ten major helices, one minor helix and five disulfide bridges. Out of these bridges, one majorly participates in the manganese binding site, which is peculiar and differentiates MnPs from other peroxidases. Oxidation of a variety of phenolic compounds and organic pollutants occurs along with the oxidation of Mn^{2+} to Mn^{3+}. The reaction is initiated by the flow of two electrons from the heme group (Fe^{3+}) to H_2O_2 and is culminated in the reduction of Fe^{3+} of the heme group to Fe^{4+} along with the release of another H_2O_2 molecule. The Mn^{3+}, a highly reactive, small and high redox potential compound, is chelated to oxalic acid and malic acid, and oxidizes many phenolic substrates to generate varied phenol degradation products.

6.3.2.3 Versatile Peroxidase (VP)

VPs belong to a glycosylated class of monomeric enzymes and these are monomeric in nature with a molecular weight averaging around 45 KDa and isoelectric point averaging near 3.6. This class of enzymes has the characteristic feature of possessing multivalent catalytic sites for the oxidation of manganese and various other redox compounds. Their important features are identical to other peroxidases, hence the name versatile peroxidases. Structurally, VPs depict an exposed tryptophan that is engaged in a long-range electron transfer toward the heme group. The catalytic cycle of VPs is much more complicated because of the variety of substrates involved. The molecular mechanism involves the binding of hydrogen peroxide to the heme group of Fe^{3+}, resulting in the formation of the Fe^{4+} oxo-porphyrin radical along with the generation of water molecules. The former then reacts with aromatic compounds, releasing compound II and Mn^{3+}. Compound II goes to the resting enzyme for the oxidation of a second molecule of Mn^{2+}, and water is released. The unstable radical intermediates generated in this manner undergo non-enzymatic reactions leading to cleavage of different bonds, thereby, generating simpler breakdown products. The radical products thus synthesized are highly unsteady and therefore undergo non-enzymatic reactions. The crystal structure of *Pleurotus eryngii* VP has been studied in detail, revealing the presence of an α-helical structure with four disulfide linkages, two Ca^{2+} binding sites, and 1 Mn^{2+} binding site similar to MnPs.

6.3.2.4 Dye Decolorizing Peroxidase (DyP)

DyPs are indexed as a new group of heme-containing ligninolytic peroxidases with EC number 1.11.1.19. Initially, DyPs were believed to originate from fungi, but they were also detected in genomes of various prokaryotes. They are a part of heme-peroxidases that have gained high recognition due to their new substrate specificity and reactivity. DyPs facilitate the oxidation of those substrates that have a large size and are unable to reach the active site of the enzyme. Therefore, the transfer of electrons must take place from the active site to the surface of the enzyme via a long-range electron transfer pathway. Typically, tryptophan and tyrosine residues best perform catalysis at the surface of the enzyme (Uchida et al., 2015). These amino acids can form stable radicals that participate directly in charge transfer. Their catalytic mechanisms are similar to that of other peroxidases and can catalyze the breakdown of those substrates that are too large to fit in their active sites. DyPs have been explored for their potential applications in ligninolysis, decolorization and bioremediation of phenolic pollutants.

DyPs are subcategorized into type A (extracellular function, bacterial origin), B, C (intracellular function, cytoplasmic in both bacteria and fungi) and D (less explored, fungal origin), based upon the primary structure. In general, DyPs have a conserved 3D topology with a non-covalently attached heme group and possess a molecular size of 40 to 60 KDa with the ability to adopt a quaternary configuration. Their C-terminal end has a heme-binding motif possessing highly conserved "histidine residue" and a specific GXXDG motif. Phylogenetically, DyPs have been considered primitive peroxidases in contrast to general peroxidases possessing both hydrolytic and peroxidase activities.

6.3.2.5 Lignin-Degrading Auxiliary (LDA) Enzymes

LDA enzymes catalyze reactions indirectly on the degradation process of lignin or symbiotically with chief lignin-degrading enzymes. These are classified into seven enzymes: aryl alcohol oxidase (LDA1), vanillyl alcohol oxidase (LDA2), glyoxal oxidase (LDA3), pyranose oxidase (LDA4), galactose oxidase (LDA5), glucose oxidase (LDA6) and benzoquinone reductase (LDA7). LDAs produce H_2O_2 from oxidized aryl and vanillyl alcohol, hydrogen peroxide, and aromatic alcohol, which become substrates for peroxidases.

LDA1 (EC.1.1.3.7) is being reported in white rot basidiomycetes such as *Pleurotus* sp., *Bjerkandera adusta* and some ascomycetous fungi. LDA1 is a monomeric glycoflavoprotein with a molecular weight of 69.1 kDa along with flavin adenine dinucleotide (FAD) as a cofactor. The primary structure of LDA1 revealed that it belongs to the glucose–methanol–choline oxidoreductase family. LDA2 (EC.1.1.3.38) is a flavin-containing enzyme and reported for the oxidization of vanillyl alcohol to vanillin (4-methoxymethyl phenol to 4-hydroxybenzaldehyde). LDA2 is considered an industrially important enzyme for the production of various value-added products such as vanillin, 4-hydroxybenzaldehyde and coniferyl alcohol. LDA2 can catalyze several reactions such as deamination, hydroxylation, oxidation, dehydrogenation and demethylation. LDA3 is an extracellular hydrogen peroxide-producing enzyme produced by several lignolytic fungi of *P. chrysosporium*.

It is a monomeric acidic glycoprotein containing a molecular mass of 57 kDa. LDA4 (EC.1.1.3.10; oxygen 2-oxidoreductase), oxidizes the C-2 of several aldopyranoses and D-glucose, and major producers belong to basidiomycetes. LDA5 (EC 1.1.3.9) is also an extracellular enzyme produced by *Fusarium* spp. and belongs to the alcohol oxidoreductase family. It is a monomeric enzyme with one copper ion and catalyzes the oxidation of primary alcohol substrates into aldehydes and hydrogen peroxide. LDA6 (E.C.1.1.3.4) oxidizes β-D-glucose to gluconic acid via molecular oxygen, thus producing H_2O_2. It is a homodimeric glycoprotein linked with two polypeptide chains. LDA6 has a significant role in the food industry, facilitating quality and inhibition of pathogens, such as *Salmonella infantis* and *Staphylococcus aureus*. LDA7 (EC.1.6.5.6) is produced intracellularly by *P. chrysosporium* and may be secreted with a limited source of nitrogen. It is a NADPH-dependent enzyme with a molecular weight of 44 kDa.

6.4 MICROBIAL PRODUCTION OF LIGNIN-DEGRADING ENZYMES

Numerous organisms, ranging from bacteria to insects, are capable of producing lignin-degrading enzymes. Although many ligninolytic enzyme producers are known, white rot fungi are reported as the lead producers of laccase with high redox potential. High-redox potential laccases can break down hard-to-decompose organic compounds. Fungal producers of various lignin-degrading enzymes include *Auricularia auricular, Irpex lacteus* and *Thanatephorus cucumeris* (DyP producers); *Dichomitus squalens* (MnP producer); *Bjerkandera fumosa, Phlebia tremellosa* and *Pleurotus ostreatus* (VP producers); *Ceriporiopsis subvermispora* (laccase and MnP producer); *Phanerochaete eryngii* (laccase, MnP and VP producers); *Phanerochaete chrysosporium, Phanerochaete sordida, Phlebia radiata* laccase, *Bjerkandera adusta* (LiP and MnP producers); and *Tramates hirsute, Tramates ochracea* and *Tramates versicolor* (laccase). Various bacterial strains reported for ligninolytic enzyme production include *Amycolatopsis* sp., *Escherichia coli, Pseudomonas fluorescens, Rhodococcus jostii, Steptomyces viridosporus, Thermobifida fusca* and *Thermomonospora curvata* (DyP producers); *Bacillus subtilis* and *Steptomyces coelicolor* (laccase and DyP producers); *Bacillus atrophaeus, Bacillus licheniformis, Bacillus pumilus, Streptomyces griseus, Streptomyces ipomoea, Streptomyces lavendulae* and *Streptomyces cyaneus* (laccase producers).

For ligninolytic enzyme production, the fungi can be grown both under submerged (SmF) and solid-state (SSF) fermentation (Chmelová et al., 2022). SmF is industrially more suitable due to the ease of downstream processing, but SSF mimics the natural habitat of microbes and, therefore, is most often associated with high enzyme production. SSF involves fungal cultivation on solid support devoid of free water. The solid used as support for fungal growth can either be inert or biological material such as lignocellulosic biomass (LCB). If the former is used, then nutrients are usually added for fungal growth often in the form of solutions. Chemically synthesized media is optimized to intensify laccase production in comparison to cultivation dependent on natural support. SSF using biomass seems more effective for laccase production than inert materials since fungal mycelium flourishes through the biomaterials with a steady release effect, which together in the presence of lignin leads to the efficient production of laccase enzyme. Water is an appropriate agent for moisturizing the solid substrate and can be used with or without nutrients like carbon, nitrogen and various other stimulants. The existence of nutrient compounds boosts the momentum of the initial stages of biomass colonization, which in turn enhances enzyme productivity. As discussed next, several factors influence the overall process of laccase enzyme production.

6.4.1 Choice of Substrate

Lignocellulosic waste biomasses are of great use in laccase enzyme production, as these are cheap and present in abundant quantities. A variety of agricultural residues like wheat, cotton, mustard, corn and rice, in the form of bran, husk or straws are the most prominently utilized solid substrates for laccase production under SSF (Ergun and Urek, 2017). Lignocellulosic waste promotes the yield of desired enzymes, for it provides nearly identical conditions as compared to their natural habitat and may also possess substances that accelerate the yield of the desired enzyme. The commonly used agricultural residues as substrates in SSF possess various lignin derivatives like hydroxycinnamic, ferulic and p-coumaric acids, which are reported to enhance laccase production. One of the crucial characteristics affecting the production of lignolytic enzymes is the carbon-to-nitrogen (C/N) ratio in culture media. Substrates with a C/N ratio less than 16 are best suited for ligninolytic enzyme production. Substrates with a ratio greater than 40 have been found unsuitable for enzyme synthesis by fungi.

6.4.2 Substrate Breakdown

Though lignocellulosic waste tends to have optimal composition, its pretreatment is believed to enhance laccase production. The breakdown of substrate is a normal procedure required prior to its application in the fermentation. The particle size of the substrate can severely influence the efficiency and productivity of the process. In general, reduced particle size is directly proportional to interaction with the enzyme because a smaller size enhances the surface area-to-size ratio, thereby improving enzyme and substrate interactions. On the other hand, fine particles pose a few disadvantages, as these can assemble easily, and reduce oxygen and heat transfer, which may lead to retarded fungal biomass. Particles larger than 500 μm cause a 1.5 times increment in enzyme activity as compared to fragments of a smaller size.

6.4.3 Humidity

It is one of the crucial criteria for the cultivation of white rot fungi under solid-state fermentation. The type of organism and the water binding or retention capacity of the biomass substrate used for SSF regulate the level of humidity. Elevated humidity results in a drastic reduction in gaseous transfer, thereby, resulting in a drop in fungal growth. A minimal optimal volume of the solution of organic and inorganic compounds is best suited for fungal growth and propagation.

6.4.4 Inoculum

The method and age of the inoculum directly impact the growth rate of an organism and also the yield of lignolytic enzymes. Various types of inoculation methods are reported for SSF, such as mycelial disc plugs, homogenized mycelial mat and spawn on grains. The proportion of inoculum quantity and substrate is a vital parameter to achieve high productivity and titers of the lignolytic enzymes. A low ratio can lead to less growth of the organism due to cell death, whereas a large inoculum size causes the optimal utilization of available nutrients and the organism swiftly enters the trophophase.

6.4.5 Temperature and pH

Generally, no considerable alterations between cultivation temperature and pH of various white rot fungi are observed during lignolytic enzyme production. The cultivation temperature of an inoculum media is nearly identical to the production media, ranging around 25°C–30°C, which is most often the room temperature. However, the exact temperature used will be primarily dependent upon the fungal strain used. The pH of the cultivation media is nearly adjusted to between 5.0 and 6.0. In a majority of studies, temperatures around 25°C and pH 5.0–6.0 are reported as suitable physiological parameters for cultivation and lignolytic enzyme production by fungal strains.

6.5 ACTIVITY OF LIGNIN-DEGRADING ENZYMES ON VARIOUS SUBSTRATES

Enzyme activity analysis of lignolytic enzymes, especially laccases, has been a subject of debate due to the availability of several substrates and methods in use. Moreover, many authors have defined the enzyme activity of laccases differently and used different formulae to derive the activities of laccases. As far as the determination of laccase activity on phenolic substrates is concerned, phenolic substrates can be directly oxidized by laccase to phenoxy free radicals, producing end products that cannot be further processed or the phenolic substrate further acts as the mediator compound for the catalysis of other reactions. Laccases are mostly active on phenolic compounds when the pH is between acidic and neutral, and lignin degradation cannot proceed under the basic environment. Concerning the enhancement of laccase activity at the basic pH range, a better understanding of the transfer of electrons and the reduction of oxygen to water via T2/T3 catalysis is needed at a molecular level for improved catalysis of laccases.

ABTS is the substrate of choice for laccase activity and kinetic analysis as it is a strong electron donor that is reduced to radical cation $ABTS^{+}$ (absorption maxima at 734 nm) imparting a blue color to the reaction mixture in correlation to the laccase activity. ABTS donates its electron to the T1 active site of laccase and forms H-linkages and π–π interactions to the histidine residues within the active site, transferring electrons to the active site T3 and finally to the T2 active site, thereby reducing one oxygen to two water molecules. ABTS as a mediator relieves the steric hindrance during substrate–laccase interaction, and being smaller in size, it can easily diffuse to or from enzyme active sites. ABTS can get oxidized and finally reduced by the larger substrates that are not accessible to the enzyme's active site, thereby, finally oxidizing the larger substrate.

In comparison to ABTS, assay substrates like o-dianisidine and guaiacol give lower enzyme activities due to the high sensitivity of ABTS than other substrates due to differences in their oxidation rates and redox potential. Guaiacol is a substituted monophenol that is oxidized to aldehydes. O-Dianisidine is a substituted diamine that is oxidized to disazo-containing groups. ABTS is a non-phenolic heterocyclic compound that is oxidized to cation radical and ABTS di-cation. 2,6-Dimethoxyphenol (DMP) is also being utilized as a substrate but is not well studied. The catalysis of the laccase enzyme takes place in aqueous solutions (Wan et al., 2008). Another reported substrate for laccase is syringaldazine solubilized in 10% ethanol and measured at 530 nm (Perna et al., 2018). Fourier transform infrared spectroscopy (FTIR) is a swift and versatile technique where spectral evolution profiling can be used to assess enzymatic activity when combined with multivariate data analysis. Coupling of FTIR measurements with parallel factor analysis (PARAFAC) is a quantitative assay for assessing enzyme activity, and a multivariate tool capable of comparing and differentiating various enzyme reaction patterns can be developed by combining PARAFAC modeling with principal component analysis.

Other assays for estimation of lignin-degrading enzyme activity are under development. One such assay, named ligNIMS, involves detecting β-O-4 bonds of lignin via nanostructure initiator mass spectrometry (NIMS) analysis. This method was used to assay laccase and MnP enzymes by attaching the NIMS tag to the phenolic and non-phenolic lignin-like model compounds for quantification of β-O-4 link cleavage during enzymatic reaction on two synthesized β-aryl ether substrates. Specific advantages of such an assay in contrast to a spectrophotometric assay involved direct detection/quantification of types of laccase catalysis.

6.6 INDUSTRIAL APPLICATIONS

Lignolytic enzymes have tremendous industrial potential, as these have the capability to participate in oxidation-reduction reactions involving a wide variety of substrates, both phenolic and

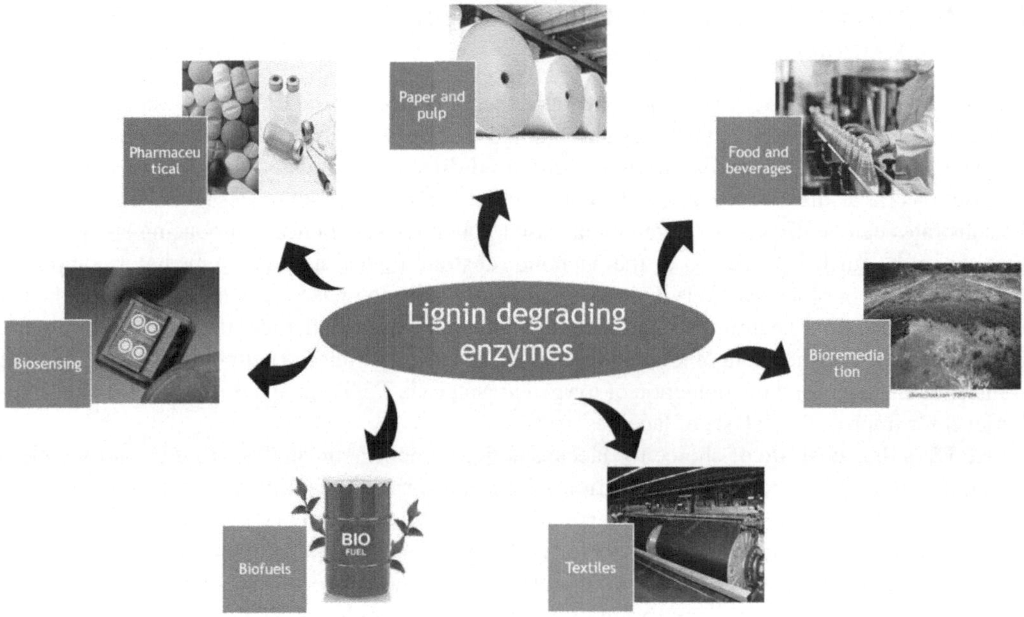

FIGURE 6.2 Important industrial applications of lignin-degrading enzymes.

non-phenolic in nature. The most prominent applications of lignolytic enzymes are shown in Figure 6.2 and discussed next.

6.6.1 Cosmetic Industry

Lignin derived oligomers (LDOs), like lignin peroxidase from *P. chrysosporium*, have been reported for their vital role in the decolorization of melanin and production of numerous skin-lightening products, comparable with that of hydroquinone, without any reported negative impact on the user. Such features of LiPs have been attributed to their capability to oxidize a vast range of structurally distinctive substrates, which makes them a suitable mediator for the oxidation of melanin, which was identical to lignin when compared structurally. This enzyme had a great role in the cosmetic industry due to its melanocytic ability (Falade et al., 2017).

6.6.2 Textile Industry

Synthetic dyes used in the textile industry are one of the major sources of environmental pollution and physicochemical methods employed for dye removal from effluents negatively impact its high operational cost, have low efficiency and excessive sludge is often generated. LDEs have been in use for sustainable industrial effluent treatment for a long time due to their biological origin and high efficiency with lower input costs. For, example, purified LiP from *Kocuria rosea* was successfully employed for decolorization of various dyes belonging to different structural groups, namely, azo, triphenyl-methane, heterocyclic, polymeric and metallic complexes.

6.6.3 Food Industry

The quality of food is not just dependent on nutritional values but also on the presence of various bioactive compounds that benefit human and animal health. Immobilized MnP can positively impact a significant reduction in apple juice color and turbidity. Similar improvement in turbidity reduction in the juices of grapes, oranges and pomegranates has also been reported by the application of such enzymes. Generation of natural aromatic flavors such as vanillin, ferulic, p-coumaric etc. is also possible with the application of MnPs.

6.6.4 Bioremediation of Environmental Pollutants

Rapid industrialization and intense use of pesticides for enhanced agricultural production destructed the quality of air, water and soil, thereby, adversely impacting human and environmental health. Ligninolytic enzymes facilitate oxidation-reduction reactions, thus helping in the biodegradation of potentially active dangerous substances like phenolic compounds, pesticides and pharmaceuticals compounds (Bilal et al., 2019). Immobilized enzyme has the potential to degrade various carcinogenic compounds like bisphenol A, triclosan and other emerging phenolic pollutants.

6.6.5 Beer and Wine Stabilization

Another important application of laccase is in wine stabilization. Characteristics like the aroma, color, taste and texture of wine are directly regulated by the abundance of polyphenols, which in turn is indirectly controlled by particular phenolic composition. Such compounds also affect the maderization process, which results in turbidity, color intensification and aroma/flavor modification. Cutting down the polyphenolic content by laccase enzyme has been reported as the most reliable method to restore wine quality because enzyme utilization would remove target-specific polyphenols without any alteration to wine's organoleptic properties.

6.6.6 Pulp and Paper Industry

The lignin-degrading enzyme, laccase, finds extensive usage in the paper and pulping process where cellulose is extracted from the wooden biomass and the lignin is dissolved and concentrated by the application of the 'Kraft process' for its separation from the pulp. It leads to lignin depolymerization, which is followed by condensation and final extraction of lignin, generating many toxic by-products at various processing steps, which remain suspended/dissolved in water and come out of the industry as effluent. Laccase application allows a sustainable and green alternative to costly and environmentally unfriendly chemical agents including various oxidizing and reducing agents. Furthermore, laccase application during the pulping process enhances brightness due to lignin removal. Post-biobleaching improvement of nearly 12% pulp consistency occurs with the application of laccase of *Trametes villosa* along with mediator compounds. Similarly, *Bacillus* sp.-derived laccase has been found to enhance brightness and whiteness by 13% and 106%, respectively, following biobleaching and deinking of waste newspaper (Curran et al., 2021).

Various industrial applications of lignin-degrading enzymes are summarized in Table 6.1.

6.6.7 Biosensor Development

Lignolytic enzymes play an important role in the identification and monitoring of various chemical compounds, including many phenolics, by application in analytical devices termed 'biosensors'. Such devices transform physicochemical reactions/changes to suitable signals, which upon amplification and processing help in monitoring the targeted compounds. Laccase-based and related enzyme-based biosensors are more efficient, consume less time and provide many other benefits which are highly suited for their applications in the food, environmental and pharmaceutical sectors. Suitable methods of immobilization of these enzymes to the desired matrices are available, including encapsulation, entrapment, covalent interaction, adsorption and cross-linkage. Phenolic

TABLE 6.1
Industrial Applications of Various Lignin-Degrading Enzymes

Enzyme	Source Microorganism	Substrates Acted Upon	Applications	Reference
Lac	*Trametes versicolor*	Protein and hydrogel	Catalyze polymerization of hydroquinone, deliver protein, treat water	Zhang et al., 2020
Lac	*Trametes versicolor*	BPA	86.7% of micropollutant removal by adsorption method using Zeolite Y as immobilization carrier	Taghizadeh et al., 2020
Lac	*Myceliopthora thermophile*	Morphine	100% removal of micro pollutant using polypropylene beads as carrier	Huber et al., 2018
Lac	*Trametes versicolor*	N-formyl-2-bromo-O-methylnorbelladine	Cure Alzheimer's disease	Zippilli et al., 2020
MnP	*Trametes pubescens* strain i8	Acid Blue 158 Poly R-478	95% and 88% dye decolorization respectively	Rekik et al., 2019
Lac, MnP	*Aspergillus terreus* GS28	Direct Blue-1	98.4% dye was decolorized	Singh & Dwivedi, 2020
VP	*Kocuria rosea* MTCC 1532	Methyl orange, cotton blue	100% and 60 % decolorization of dyes	Parshetti et al., 2012
LiP	*Phanerochaete chrysosporium*	Melanin and hydrogen peroxide	Decolorization of melanin	Woo et al., 2004
DyP	*Bjerkandera adusta*	Anthraquinone dye	Decolorization of dye	Kim et al., 1995

DyP = dye decolorizing peroxidase; MnP = manganese peroxidase; Lac = laccase; LiP = lignin peroxidase; VP = versatile peroxidase; BPA= bisphenol A

compounds like triclosan, BPA and other emerging pollutants can be easily detected with the help of laccase-based phenolic biosensors.

6.7 CHALLENGES IN THE PRODUCTION AND APPLICATIONS

Lignolytic enzymes perform the degradation of carbon-rich lignocellulosic biomass, and their oxido-reductive property is a matter of increasing worldwide interest. However, despite their vast industrial potential, the majority of lab-scale studies have not been successfully translated to commercial scale. This is due to a variety of factors involving the economics of enzyme production, the current unavailability of suitable mediators and lignocellulose recalcitrance, catalytic efficiencies and enzyme stabilities under challenging conditions, and recycling for prolonged use.

It has been a challenge to compare the degradation capability of lignolytic enzymes, and very few reports have focused on this important issue. This is mainly due to the difficulty of substrate (lignin) preparation, and the majority of available reports employ technical lignin, which is obtained via chemical processing of the lignocellulosic biomass. The enzymatic degradation of technical lignin differs from that of the natural substrate due to the complex and heterogeneous nature of the latter. Although the rapid degradation of tropical forest litter biomass by lignolytic enzymes was recently studied, similar studies are still needed for degrading industrial lignocellulose-rich biomasses (Fujii et al., 2020).

The economics of the production of lignolytic enzymes represent another bottleneck to their large-scale industrial use. In spite of the progress in genetic engineering-based approaches, the development of a robust lignolytic enzyme is still a huge challenge for researchers. The development of the chimeric enzyme possessing multiple activities and having the ability to completely degrade lignin and its derivatives in a sequential manner could be a great boon for industries. This may require selecting suitable expression vectors or appropriate hosts for expressing such proteins in a heterologous fashion. Rational strategies to improve the catalytic or structural properties of the lignolytic enzymes via protein redesigning or engineering based on the approaches of genetic engineering or evolutionary engineering will certainly improve the catalytic efficiency and substrate specificity as well as the pH and thermostability of such enzymes. Large-scale commercialization and application of lignolytic enzymes require cost reduction, improved catalysis, purity and enhancement in production capacity, as well as exploration of affordable and highly effective mediators. Cost reduction can be achieved by economic use of the growth nutrients, formulation and optimization of enzyme production medium, and downstream processing steps for economic recovery and purification of the enzyme formulation.

Improved catalysis of the enzymes can be achieved by performing the catalysis at higher temperatures, which speeds up the reaction rate along with an evident reduction in microbial contamination. Therefore, thermostable enzymes are considered advantageous over conventional enzymes due to greater hydrophobicity, shortened loop, better atom packing and fewer thermo-labile amino acids. Extremophilic ligninolytic enzymes have been explored from microorganisms thriving under extremophilic conditions. Such enzymes were reported in the past from various microorganisms, which include laccase-producing fungi (*Trametes versicolor*, *Polyporus pinisitus*, *Pleurotus* sp., and *Trichoderma atroviride*) and lignin peroxidase-producing fungi (*Trametes trogii* and *Phlebia tremellosa*). Bioprospecting of such enzymes in the future will certainly help increase the catalytic efficiency and improve the application of lignin-degrading enzymes.

Another major challenge associated with the cost reduction of laccase application is the non-availability of appropriate technology for recycling. Recycling and reuse of lignolytic enzymes has immense potential to improve the cost-effectiveness of their application. Therefore, future studies should be directed to look for viable options for effective recycling of such enzymes for their repeated and prolonged use. Suitable immobilization agents such as iron or magnetic nanoparticles could be explored in the future for their suitability and cost-effectiveness. Further, membrane technology can also be used to enhance the retention or recovery of lignolytic enzymes.

Different lignolytic enzymes have different mechanisms of action that can cumulatively bring about complete biodegradation of the lignocellulosic material and other recalcitrant pollutants of phenolic or non-phenolic origin. However, considering the huge gap in the vast potential of lignolytic enzymes and their current industrial uses, it is obvious that the current proportion of lignolytic enzymes in industrial sectors is still in its infancy. Notwithstanding these obstacles, the research and development efforts in the recent past and currently underway indicate the vast scope and bright future of lignolytic enzyme applications in various industrial sectors.

6.8 CONCLUSIONS AND PERSPECTIVES

The highly complex and recalcitrant nature of lignin poses several challenges in the effective valorization of plant biomass, which has led to many investigations on its biodegradation by several microorganisms and their lignin-degrading enzymes. Laccase, manganese peroxidase, lignin peroxidase and versatile peroxidase play crucial roles in the sequential degradation of lignin owing to their specific structural properties based on the presence of copper, iron, manganese and calcium ions, which also make them highly stable under extreme environments. The combination of these enzymes simultaneously targets multiple bonds, thereby, facilitating effective lignin degradation in comparison to the action of individual enzymes. Excellent oxidative properties of these lignin-degrading enzymes on a wide variety of substrates other than lignin have led to many environment-friendly innovative applications in solving global problems like environmental pollution, global warming and waste treatment. Moreover, these enzymes are excellent catalysts used in the cosmetic, healthcare, textile, food, feed and alcoholic beverage industries. Lignin-degrading enzymes have been extensively used in the degradation of many environmental pollutants like many synthetic dyes as well as many emerging pollutants of environmental and health concerns.

Among the various lignin-degrading enzymes, laccase has emerged as one of the most versatile enzymes suited for many biotechnological applications. However, industrial applications of laccase and other lignin-degrading enzymes face challenges related to mediator cost, low redox potential, stability, etc. The low redox potential of laccase, for example, hinders complete lignin degradation, due to which several mediators are used to enhance the oxidation potential of laccases, which are very costly. In addition, many fungi don't produce a complete set of lignin-degrading enzymes needed to fully degrade lignin. Moreover, the limited knowledge about the chemical properties and exact mode of action of many ligninolytic enzymes further limit many of their industrial applications.

Therefore, better biotechnological tools and innovative bioengineering approaches are needed to make such enzymes industrially more robust and affordable. Specifically, the production and characterization of individual recombinant enzymes via genetic engineering-based and protein engineering-based approaches will be needed to better recognize their mechanisms of action on different substrates. This will further enable more effective and customized degradation of lignin, as well as many other phenolic and non-phenolic pollutants. Tailor-made cocktails of lignin-degrading enzymes will be potentially useful in the targeted degradation of lignin to one or more specific added-value products.

ACKNOWLEDGMENTS

The authors thank the Central University of Haryana (CUH) for administrative and infrastructural support. AK thanks CUH for a PhD fellowship (Non-NET). JKS and H thank DBT for financial support under the twinning program (BT/PR25530/NER/95/1239/2017).

REFERENCES

Bilal, M., Iqbal, H. M., & Barceló, D. (2019). Persistence of pesticides-based contaminants in the environment and their effective degradation using laccase-assisted biocatalytic systems. *Science of the Total Environment, 695*, 133896.

Cagide, C., & Castro-Sowinski, S. (2020). Technological and biochemical features of lignin-degrading enzymes: A brief review. *Environmental Sustainability, 3*(4), 371–389.

Chmelová, D., Legerská, B., Kunstová, J., Ondrejovič, M., & Miertuš, S. (2022). The production of laccases by white-rot fungi under solid-state fermentation conditions. *World Journal of Microbiology and Biotechnology, 38*(2), 1–20.

Curran, L. M. L. K., Sale, K. L., & Simmons, B. A. (2021). Review of advances in the development of laccases for the valorization of lignin to enable the production of lignocellulosic biofuels and bioproducts. *Biotechnology Advances*, 54, 107809.

Ergun, S. O., & Urek, R. O. (2017). Production of ligninolytic enzymes by solid state fermentation using *Pleurotus ostreatus. Annals of Agrarian Science, 15*(2), 273–277.

Falade, A. O., Nwodo, U. U., Iweriebor, B. C., Green, E., Mabinya, L. V., & Okoh, A. I. (2017). Lignin peroxidase functionalities and prospective applications. *Microbiology Open, 6*(1), e00394.

Fujii, K., Nakada, Y., Umezawa, K., Yoshida, M., Shibata, M., Hayakawa, C., Inagaki, Y., Kosaki, T., & Hangs, R. (2020). A comparison of lignin-degrading enzyme activities in forest floor layers across a global climatic gradient. *Soil Ecology Letters, 2*(4), 281–294.

Huber, D., Bleymaier, K., Pellis, A., Vielnascher, R., Daxbacher, A., Greimel, K. J., & Guebitz, G. M. (2018). Laccase catalyzed elimination of morphine from aqueous systems. *New Biotechnology, 42*, 19–25.

Kim, S. J., Ishikawa, K., Hirai, M., & Shoda, M. (1995). Characteristics of a newly isolated fungus, *Geotrichum candidum* Dec 1, which decolorizes various dyes. *Journal of Fermentation and Bioengineering, 79*(6), 601–607.

Parshetti, G. K., Parshetti, S., Kalyani, D. C., Doong, R. A., & Govindwar, S. P. (2012). Industrial dye decolorizing lignin peroxidase from *Kocuria rosea* MTCC 1532. *Annals of Microbiology, 62*(1), 217–223.

Perna, V., Agger, J. W., Holck, J., & Meyer, A. S. (2018). Multiple Reaction Monitoring for quantitative laccase kinetics by LC-MS. *Scientific Reports*, *8*(1), 1–9.

Pollegioni, L., Tonin, F., & Rosini, E. (2015). Lignin-degrading enzymes. *The FEBS journal*, 282(7), 1190–1213.

Ponnusamy, V. K., Nguyen, D. D., Dharmaraja, J., Shobana, S., Banu, J. R., Saratale, R. G., … Kumar, G. (2019). A review on lignin structure, pretreatments, fermentation reactions and biorefinery potential. *Bioresource Technology*, *271*, 462–472.

Rekik, H., Jaouadi, N. Z., Bouacem, K., Zenati, B., Kourdali, S., Badis, A., … Jaouadi, B. (2019). Physical and enzymatic properties of a new manganese peroxidase from the white-rot fungus *Trametes pubescens* strain i8 for lignin biodegradation and textile-dyes biodecolorization. *International Journal of Biological Macromolecules*, *125*, 514–525.

Singh, G., & Dwivedi, S. K. (2020). Decolorization and degradation of Direct Blue-1 (azo dye) by newly isolated fungus *Aspergillus terreus GS28*, from sludge of carpet industry. *Environmental Technology and Innovation*, *18*, 100751.

Taghizadeh, T., Talebian-Kiakalaieh, A., Jahandar, H., Amin, M., Tarighi, S., & Faramarzi, M. A. (2020). Biodegradation of bisphenol A by the immobilized laccase on some synthesized and modified forms of zeolite Y. *Journal of Hazardous Materials*, *386*, 121950.

Uchida, T., Sasaki, M., Tanaka, Y., & Ishimori, K. (2015). A dye-decolorizing peroxidase from *Vibrio cholerae. Biochemistry*, *54*(43), 6610–6621.

Wan, Y., Du, Y., & Miyakoshi, T. (2008). Enzymatic catalysis of 2, 6-dimethoxyphenol by laccases and products characterization in organic solutions. *Science in China Series B: Chemistry*, *51*(7), 669–676.

Woo, S. H., Cho, J. S., Lee, B. S., & Kim, E. K. (2004). Decolorization of melanin by lignin peroxidase from *Phanerochaete chrysosporium. Biotechnology and Bioprocess Engineering*, *9*(4), 256–260.

Zhang, A., Meng, X., Bao, C., & Zhang, Q. (2020). In situ synthesis of protein-loaded hydrogels via biocatalytic ATRP. *Polymer Chemistry*, *11*(8), 1525–1532.

Zippilli, C., Botta, L., Bizzarri, B. M., Baratto, M. C., Pogni, R., & Saladino, R. (2020). Biomimetic synthesis of galantamine via laccase/TEMPO mediated oxidative coupling. *RSC Advances*, *10*(18), 10897–10903.

7 Lytic Polysaccharide Monooxygenases
Production and Applications

Pooja Dixit, Anil Kumar Patel, and Reeta Rani Singhania

7.1 INTRODUCTION

Enzymatic saccharification of cellulose is a heterogeneous system in which cellulases adsorb over the surface of the insoluble cellulose substrate. These enzymes have limited accessibility to cellulose structures inside the lignocellulosic matrix as they can act only on the surface of cellulose microfibrils (Singhania et al., 2021, 2022). Hence, for efficient and complete cellulose hydrolysis, crystalline cellulose microfibrils should be disrupted from their tightly packed structure before the hydrolytic attack of cellulases. The process of disruption and delamination of microfibrils which disrupt the crystalline structure and increase the surface area of cellulose is termed "amorphogenesis" (Arantes & Saddler, 2010).

Many non-hydrolytic proteins are present in cellulolytic organisms which induce amorphogenesis and they represent the C_1 factor of the C_1–C_x model. Reese et al. (1950) first introduced this C_1–C_x model of cellulose degradation that explains the involvement of different kinds of enzymes in cellulose degradation. The authors studied a range of cellulolytic and non-cellulolytic fungi and bacteria, and it was observed that non-cellulolytic organisms can degrade small linear poly-anhydro-glucosyl chains into soluble monomers. However, these non-cellulolytic organisms could not utilize polymeric native cellulose and hence it was proposed that these non-cellulolytic organisms do not possess C_1 enzymes, which are essential to degrade crystalline cellulosic substrates, as presented in Figure 7.1. It was also observed that these C_1 enzymes were present only in the cellulolytic organisms.

Although this theory was accepted as a lignocellulolytic hydrolysis model decades later, this theory indicated the existence of today's non-hydrolytic enzymes involved in the amorphogenesis of lignocelluloses. In 1974, a study led by Eriksson et al. (1974) further validated the theory of the C_1–C_x model and proposed the existence of oxidative enzymes as the C_1 components of this model. The authors described an enzyme secreted from *Sporotrichum pulverulent* that was able to oxidize the cellulose. Later, it was also observed that the secreted oxidative enzyme was also able to enhance the hydrolytic activity of the fungal enzyme mixture twofold. Now, with the current knowledge of lytic polysaccharide monooxygenases (LPMOs), it seems that the authors might be explaining LPMO activity from this fungus. The incorporation of LPMOs has revised the classical hydrolytic system of cellulases. An LPMO-mediated pre-step is required before the action of hydrolytic enzymes (endo-cellulase, exo-cellulase and β-glucosidases) and this synergy ultimately increases the hydrolytic action of canonical cellulases. In Figure 7.1, LPMO-oxidized glucose monomers and native cello-oligomers and glucose moieties produced after the action of hydrolases are presented for better understanding.

DOI: 10.1201/9781003335313-8

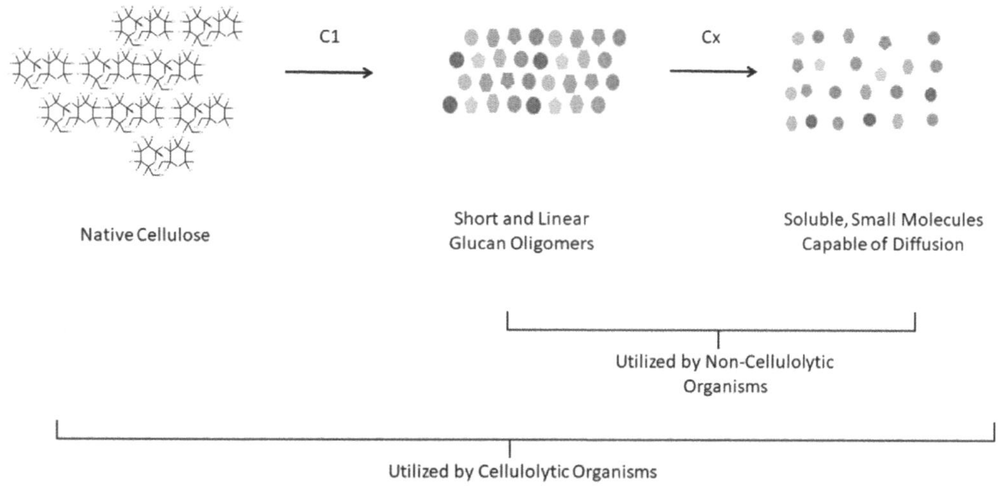

FIGURE 7.1 C_1–C_x model of cellulose conversion.

7.2 LYTIC POLYSACCHARIDE MONOOXYGENASES (LPMOs)

LPMOs, formerly known as glycoside hydrolase family 61 (GH61), are powerful enzymes that can oxidatively cleave glycosidic bonds in polysaccharides, such as cellulose, chitin, starch and xylan (Courtade & Aachmann, 2019; Kracher et al., 2016; Phillips et al., 2011; Quinlan et al., 2011; Vaaje-Kolstad et al., 2010), and are also involved indirectly in the degradation of lignin (Li et al., 2019), suggesting the overlapping role of LPMOs in polysaccharide as well as lignin degradation.

LPMOs are found in different cellulolytic bacteria and fungi and have also been found associated with virulence factors in viruses. LPMOs are also associated with an ancient cellulose-degrading insect (Sabbadin et al., 2018). This finding suggests that these enzymes are spread onto several clades of the Tree of Life and are not just restricted to complex animals but also found in primitive organisms including oomycetes, protists and algae.

LPMOs were recognized as weak endoglucanases due to their weak endoglucanase activity as observed by their ability to reduce viscosity on some cellulose substrates such as CMC and β-glucan and hence were named GH61 enzymes (Karlsson et al., 2001; Saloheimo et al., 1997) However, the first indication for these enzymes as non-hydrolases came through the first crystal structure of GH61 from *Hypocrea jecorina*, TrCel61B (Karkehabadi et al., 2008) From the characterized structure, it was revealed that these enzymes contain a unique metal-binding flat active site on the enzyme surface instead of a cleft or tunnel-like structure conserved in cellulases (Forsberg et al., 2014; Harris et al., 2010; Hemsworth et al., 2015). The conserved carboxylate residue, which is generally found in the active site of cellulases, was found missing from the active site of these enzymes, which indicated a significant structural difference among these enzymes. Later in 2010, a major breakthrough occurred when Vaaje-Kolstad et al. (2010) explained the catalytic action of LPMOs on crystalline chitin, categorizing them as non-hydrolytic accessory oxidative enzymes.

The uniqueness of LPMOs lies in their substrate preference as compared to glycoside hydrolases. Glycoside hydrolases require a single chain stripped out of the polysaccharide matrix and hence these hydrolytic actions are more energy-intensive as compared to the LPMO action. However, LPMO-mediated polysaccharide degradation reactions are carried out on the surface of biomass present in the crystalline lattice. The active site of LPMOs is a flat binding surface that helps in oxidizing glycosidic chains present in a crystalline surface and hence increases substrate porosity. The active site of LPMO contains conserved histidine residues that coordinate with the copper center of the enzymes, and this feature, known as the histidine brace, is the only conserved feature of the whole LPMO superfamily. By oxidizing the glycosidic chain in a crystalline polysaccharide form, these enzymes help in increasing cellulose accessibility to cellulases and enhance the hydrolytic efficiency of cellulases by providing a large number of acting sites for endo-glucanases (Harris et al., 2010; Karnaouri et al., 2017).

Cellulase manufacturer, Novozymes, mined several hundreds of fungal strains for screening synergistic proteins that can help in improving its benchmarked *Trichoderma* cellulase (Johansen, 2016; Teter, 2012). Many fungi were found to synergize and improve the saccharification potential of the benchmark cocktail. After the fractionation of proteins, purified enzymatic components were analyzed for their synergistic action. It was discovered that most of the synergizing proteins were found to be glycoside hydrolases 61, which were later identified as LPMOs (Harris et al., 2010). The authors claimed twofold reductions in required cellulase protein loading for acid-pretreated corn stover hydrolysis by the addition of GH61/LPMOs in cellulase preparations. This report has strongly supported their role in synergism and has spotlighted LPMO proteins as cellulase boosters.

7.3 LPMOs OXIDATIVELY CLEAVE CELLULOSE CHAINS

The structure of LPMOs contains an immunoglobulin or fibronectin-like fold containing a β-sandwich with eight to ten β strands (Harris et al., 2010; Quinlan et al., 2011). Unlike glycoside hydrolases, the active site of LPMO contains a flat binding planar surface with the help of which it interacts with the cellulose structural plane. This means that in the case of glycoside hydrolases, a glucan chain must be abstracted out of the crystalline cellulose so that the action of hydrolases can take place on this chain. This process is energy-intensive, whereas LPMOs do not undergo such energy-draining reactions and can oxidize the substrate at the surface plane only, with the help of their active site present at the surface of the enzyme.

LPMOs abstract the hydrogen from C1 or C4 carbon atoms of glucose and replace it with an oxygen atom to cleave the glycosidic bonds. This hydroxylation reaction demands two protons and two electrons during catalysis. The catalytic action of LPMO starts from a reduction reaction of LPMO-Cu (II) to LPMO-Cu (I), known as a priming reaction. The reduced LPMO-Cu(I) radical further gains two protons and an electron from the enzyme vicinity to carry out the hydroxylation reactions. After the addition of one oxygen in the glycosidic bond, one water molecule is released along with the regeneration of the LPMO-Cu (II) state, which can enter the new catalytic cycle (Beeson et al., 2015; Phillips et al., 2011; Quinlan et al., 2011).

LPMO catalysis reactions involve many LPMO radical intermediates that have been only hypothesized and not yet experimentally proven. Hence, only a little is known about the exact mechanism of LPMO catalysis. Although it has been demonstrated that various electron donors are involved in LPMO catalysis, the exact mechanism of electron supply to the catalytic copper center is not clear. Moreover, the source of protons involved in the LPMO catalysis reaction is also unclear. Therefore, an alternative peroxygenase mechanism of LPMO catalysis has been introduced. In breakthrough research, it was observed that LPMOs might require hydrogen peroxide to drive LPMO catalysis reactions, as given in Figure 7.2. H_2O_2 can act as a source of oxygen, protons and electrons in

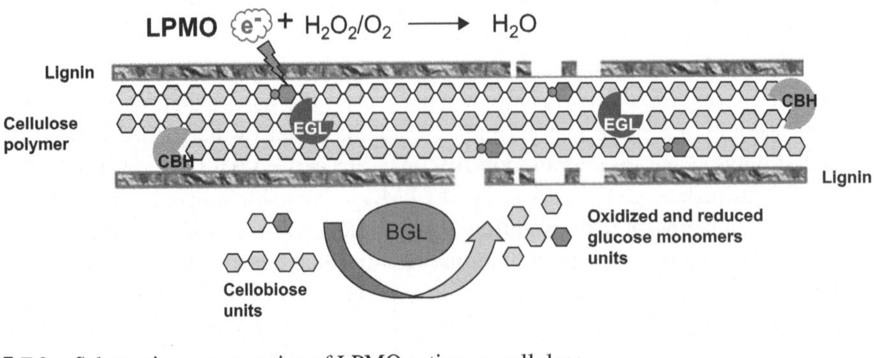

FIGURE 7.2 Schematic representation of LPMO action on cellulose.

LPMO-mediated hydroxylation reactions and hence can be termed as peroxygenases. This mechanism was first proposed by Bissaro et al. in 2016, and they also concluded that H_2O_2 is not an alternative to O_2 but a true co-substrate of LPMOs, as H_2O_2-derived AA9 LPMO and AA10 LPMO reactions have been observed to be faster than oxygen-derived LPMO reactions.

7.4 REGIOSELECTIVITY OF LPMOs

LPMOs have the ability to oxidize the carbon atoms of cellulose chains, either at the C1 or C4 position. This process leads to the production of both oxidized and reduced glucose monomers as end products during hydrolysis (Isaksen et al., 2014). The specific location of oxidation on the cellulose chain determines the classification of the LPMO type. LPMOs are categorized based on their point of action: type I LPMOs oxidize exclusively at the C1 position, resulting in the formation of aldonolactone, which spontaneously converts to aldonic acid. On the other hand, type II LPMOs target the C4 carbon to create non-reducing end C4-ketoladose. These ketoladose molecules typically exist in equilibrium with gemdiols under aqueous conditions. Lastly, type III LPMOs act on both the C1 and C4 carbon atoms within the cellulose chain (Figure 7.3). LPMO's C1/C4 regioselectivity was determined for CBP21 and CelS2 LPMOs that oxidized carbon at the C1 position of chitin and cellulose, respectively (Forsberg et al., 2011; Vaaje-Kolstad et al., 2010). A fungal LPMO TaGH-61 from *Thermoascus aurantiacus* was shown to oxidize both the C1 and C4 end of carbon in a cellulose chain (Quinlan et al., 2011). The regioselectivity of LPMO enzymes is determined experimentally, as such there are some conserved sequence features that can predict the regioselectivity of LPMOs. However, only experimental characterizations of purified LPMOs can accurately predict these regioselectivities in LPMOs.

7.4.1 Electron Donors

A conserved histidine brace contains a copper center at the active site, which allows glycosidic bonds to be cleaved by means of an electron transfer chain. LPMO copper centers require electron donor cofactors to perform this oxidative cleavage. A plethora of such LPMOs which are essential redox molecules have been reported, including small reductants such as ascorbate, gallate, plant

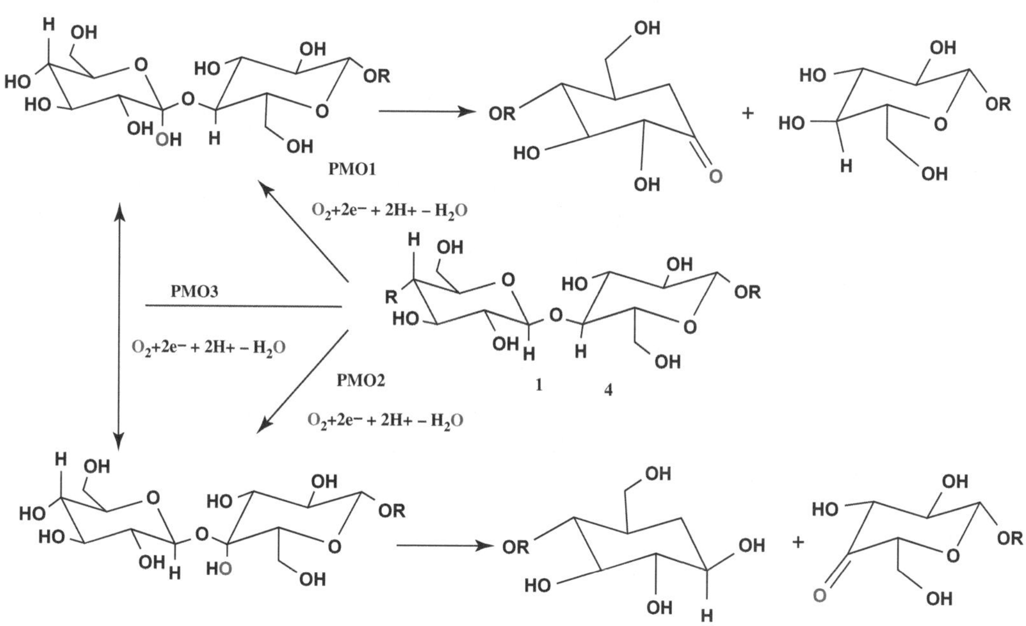

FIGURE 7.3 Scheme of LPMO-catalyzed cellulose oxidation.

phenolics, lignin, lignin-derived aromatic small molecular weight compounds and photosynthetic pigments (Kracher et al., 2016). The reductants directly or indirectly donate electrons to LPMOs, which indicates that these reductants must be present in the natural vicinity of LPMOs.

An important class of LPMO electron donors is GMC (glucose, methanol, choline) oxidoreductases, which were shown to co-secrete with LPMO enzymes in many cellulolytic fungi. These enzymes belong to the AA3 family enzymes in the Carbohydrate-Active Enzymes (CAZy) classification. Cellobiose dehydrogenase was the first member of this enzyme family that was observed as the natural electron donor of LPMOs. Hence, it is clearly reflected that CDH supports the catalytic activity of AA9 LPMO. Many reports have also confirmed the beneficial effect of CDH enzymes on the catalytic activity of GH61 or LPMOs (Couturier et al., 2018; Phillips et al., 2011).

Other members of GMC oxidoreductases such as glucose oxidase, glucose dehydrogenase, pyranose dehydrogenases, aryl alcohol oxidases and aryl alcohol quinone oxidoreductase were also observed to donate the electrons to LPMOs. These enzymes either donate the electrons directly to LPMOs or they are involved in making some lignin-modified compounds that can donate electrons to LPMOs (Garajova et al., 2016; Várnai et al., 2018). Laccase mediator systems that are involved in partial lignin degradation can generate small molecular weight compounds that can donate electrons to LPMOs. LPMO enzymes are also observed to be involved in the lignin-degrading pathway in fungi. Hydrogen peroxide produced from unbound substrate-free LPMO of a white-rot fungus has been observed to fuel the lignin-degrading mechanism of versatile peroxidases. All of these findings have proved that LPMO action is not only limited to cellulose and hemicellulose degradation but can influence the lignin-degradation pathways as well.

7.5 CLASSIFICATION OF LPMOS

Ubiquitous LPMO enzymes are present in nearly all cellulolytic fungi spanning the Basidiomycota and Ascomycota phyla with the sole exception of the anaerobic fungus *Orpinomyces*. The widespread interest and extensive research dedicated to LPMOs have led to their classification as "auxiliary activity" proteins within the CAZy database (www.cazy.org). Currently, LPMOs span eight different classes (AA9, AA10, AA11, AA13, AA14, AA15, AA16 and AA17) in the CAZY database depending upon their substrate of action and organism of origin.

LPMO-AA9, formerly known as GH61, enzymes act on the cellulosic substrate and are derived from fungi. Hence, many times AA9 LPMOs is a term generally associated with fungal LPMOs (Busk & Lange, 2015; Jagadeeswaran et al., 2018). Cleavage of cellulose is carried out by hydroxylation of C1, C4 or both carbon atoms involved in glycosidic bonds of cellulose.

LPMO-AA10 are the oxidative LPMO proteins derived from bacteria and viruses which target chitin and cellulose as the main substrate for oxidative cleavage reactions (Forsberg et al., 2011; Vaaje-Kolstad et al., 2012). These proteins were originally classified as chitin-binding proteins (CBM-33). However, in 2010, these proteins were observed as oxidative chitin-cleaving enzymes, and hence these enzymes were reclassified as AA10 LPMOs.

The LPMO-AA11 family originated after the characterization of LPMO protein from *Aspergillus oryzae*. The characterized LPMO protein possessed chitinolytic activity that oxidized the C1 carbon atom in glycosidic bonds of chitin (Hemsworth et al., 2014).

LPMO-AA13 carries out the cleavage of starch, and some enzymes from the AA11 family have also shown similar behavior. Oxidative cleavage of starch by C1 oxidation was demonstrated for the first time with an LPMO gene NCU8746 from *Neurospora crassa* and this report led to the formation of the AA13 LPMO class (Vu et al., 2014).

LPMO-AA14 acts on xylan and cellulose-associated hemicellulose fractions. This new LPMO family member is prevalent in white-rot and brown-rot basidiomycetes. Two new AA14 family members from *Pycnoporous coccineus* demonstrated enhanced wood saccharification by oxidatively cleaving xylan-coated cellulose fibers (Couturier et al., 2018).

The LPMO-AA15 family was formed after studying two LPMO members from *Thermobia domestica,* a cellulose-degrading insect (Sabbadin et al., 2018). These LPMOs were active on both cellulose and chitin. This finding suggested that LPMOs assist in biomass digestion, development, and metamorphosis in arthropods.

LPMO-AA16 was discovered after studying an unknown protein from *Aspergillus aculeatus.* This protein was found to synergize with cellulase to oxidize cellulosic chains at the C1 position. Bioinformatic analysis and genome mining for other members of this family demonstrated that AA16 LPMOs occur in fungi as well as oomycetes (Filiatrault-Chastel et al., 2019).

The LPMO-AA17 family is the most recently discovered LPMO family in the plant pathogenic oomycetes *Phytophthora infestans* (Sabbadin et al., 2021). The characterized LPMO AA17 protein was found to oxidatively cleave the backbone of pectin, another polysaccharide of lignocellulosic origin. The discovery of this new LPMO family has suggested that LPMOs play an important role in plant infection suggesting that LPMOs are spread onto several clades of the tree of life.

7.6 ANALYTICAL TOOLS AND COLORIMETRIC ASSAYS FOR LPMO ACTIVITY

LPMO-mediated oxidative cleavage of cellulose produces a mixture of oxidized and reduced glucose and cellulosic oligomers. Currently, the most commonly used tools for characterizing LPMO action are chromatographic-based methodology: high-performance anion-exchange chromatography–pulsed amperometric detector (HPAEC-PAD), liquid chromatography–mass spectrometry (LC-MS) and the matrix-assisted laser desorption/ionization–time of flight mass spectrometry (MALDI-TOF MS).

Typically, LPMO enzymatic saccharification reactions involve incubating the LPMO enzyme with a suitable electron donor (such as ascorbate or cellobiose dehydrogenase) and a polymeric cellulosic substrate (such as Avicel, phosphoric acid swollen cellulose [PASC], carboxymethyl cellulose [CMC] and pretreated lignocellulosic biomass). The incubation temperature can range from 37°C to 50°C, and the reactions are usually carried out for several hours, typically up to 48–72 hours. Following the reaction, the resulting products, including oxidized cello-oligosaccharides, glucose, gluconic acid and cellobionic acid, along with native reduced reaction products, are further processed for analysis using HPAEC or mass spectrometry.

To study LPMO enzymes, HPAEC-PAD is the preferred method due to the strong affinity of acidic sugars with anion-exchange columns, which enable clear separation of native and acidic/oxidized cello-oligosaccharides. Additionally, this instrument has a good sensitivity for detecting lower concentrations of gluconic and cellobionic acid (ranging from 0.05 to 0.5 g/L). Researchers have utilized HPAEC-PAD to characterize the oxidative activity of Cellic® CTec2 by measuring both C1 and C4 oxidized cellodextrins produced during cellulose hydrolysis (Cannella et al., 2012).

To achieve the objective of identifying an effective and responsive technique to distinguish between native and oxidized cello-oligosaccharides, various HPLC-based analytical tools were assessed. The findings indicated that HPAEC-PAD emerged as a superior and sensitive method for quantifying oxidized species (Westereng et al., 2016). Additionally, the researchers recommended exploring LC-MS-based approaches, such as hydrophilic interaction chromatography (HILIC-MS) or porous graphitized carbon liquid chromatography (PGC-LC-MS), for simultaneous analysis of C1 and C4 oxidation products.

In the absence of substrates and after activation by a suitable electron donor, LPMOs generate hydrogen peroxide in a futile cycle. This feature of LPMOs has been utilized in many colorimetric assays. In 2012, a colorimetric assay based on the reduction of O_2 to H_2O_2 was employed to indirectly quantify LPMO activity (Figure 7.4). In the presence of hydrogen peroxide, Amplex Red dye is converted to a pink substrate named Resorufin. Kittl and co-workers tested this assay in their work on the expression and purification of LPMOs from *Neurospora crassa*. Using ascorbic acid and CDH as reducing agents, the group showed that the formation of H_2O_2 was directly proportional to the LPMO concentration (Kittl et al., 2012).

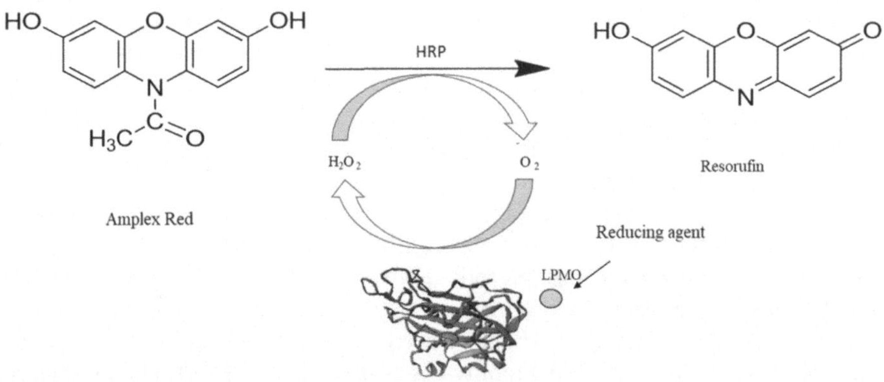

FIGURE 7.4 Schematic representation of Amplex Red assay.

Another fast and robust microplate-based assay for LPMO quantification was designed based on the labeling of aldonic acids with a chromogenic substrate, 2,6-dimethoxyphenol (2,6-DMP) (Breslmayr et al., 2018). Another quick and efficient colorimetric assay for type I LPMO quantification was based on the capability of cation Ni^{2+} ion binding to carboxylate moieties formed after type I LPMO action. Cationic nickel is present as a complex with pyrocatechol violet (PV-Ni^{2+}) dye and reacts with carboxylate moieties formed after type I LPMO action (Wang et al., 2018).

7.7 APPLICATIONS OF LPMOS IN BIOREFINERIES

Despite extensive research focused on enzyme development processes, there is no superior single enzymatic system that can provide cost-effective and satisfactory enzymatic hydrolysis on different lignocellulosic biomass (Bischof et al., 2016). The need to develop more efficient and effective enzymatic preparations has led to the application of sophisticated system biology tools that allow researchers to identify and develop efficient enzyme components. The application of these powerful tools has led to the discovery of novel non-hydrolytic LPMOs, which are reported as "synergistic enzymes" that boost cellulase action (Horn et al., 2012). Over some years, improvements in "cellulolytic cocktails" are getting quite popular in which basic cellulase systems are fortified with accessory enzymes that are directed towards the lignocellulosic composition of biomass. Designing cellulolytic cocktails that are rich in accessory enzymes can significantly enhance the hydrolysis of biomass.

Few recent studies have been conducted which prove LPMOs' significance in biorefineries. A recent study by Hansen et al. (2023) found the potential of externally added H_2O_2 to sustain oxidative cellulose depolymerization by LPMOs during solid-state fermentation (SSF) and to enhance lactic acid production. Bioreactor experiments were conducted using 100 g/L cellulose, and the results demonstrated that continuous addition of small amounts of H_2O_2 (at a rate of 80 μM/h) during SSF enabled LPMO activity and improved lactic acid production. These proof-of-concept findings suggest that modern cellulase cocktails containing LPMOs, such as Cellic CTec2, can be utilized in SSF setups without sacrificing the activity of LPMOs. Further process optimization is necessary to enhance the efficiency of the SSF process. Nonetheless, this study highlights the potential of using externally added H_2O_2 to sustain LPMO activity and improve the conversion of lignocellulosic biomass into valuable products like lactic acid during SSF processes.

Koskela et al. (2022) investigated the effect of hemicellulose content in starting pulp fibers on the oxidation capabilities of cellulose-active LPMO during the production of cellulose nanofibrils (CNFs) from wood pulp fibers. Two types of commercial softwood pulp fibers were evaluated: one with a hemicellulose content of 16% (kraft pulp) and the other with a lower hemicellulose content of 4% (dissolving pulp). The findings revealed that the presence of hemicellulose had a significant impact on the production of LPMO-oxidized CNFs. Kraft pulp, which had a higher hemicellulose content, facilitated better access of LPMO to the fiber cell wall. This enhanced accessibility enabled the production of homogeneous CNFs with a thin width of 3.7 ± 1.7 nm. In contrast, the lower hemicellulose content in dissolving pulp resulted in a partial disintegration of the fibers into thin CNFs, leaving a significant amount of cellulose microfibril aggregates with widths ranging from 50 to 100 nm.

Another study by Koskela et al. (2023) explored the use of LPMOs to modify the cell wall structure of delignified wood. By subjecting delignified wood to LPMO oxidation under mild reaction conditions, nanofibrillation of cellulose microfibril bundles within the wood cell wall is achieved. This enzymatic nanofibrillation enables the densification of delignified wood into transparent anisotropic films using ambient conditions and low pressure. The process of enzymatic nanofibrillation enhances microfibril fusion and improves the adhesion between adjacent wood fiber cells during the densification process. As a result, the mechanical performance of the films is significantly enhanced in both longitudinal and transverse directions. These findings provide a better understanding of the microstructural changes induced by LPMO in wood. Even though we have various biorefinery applications of LPMOs, its significance in the cellulolytic cocktail is immense.

LPMOs have the potential to provide new avenues in the research targeted towards cellulase cost reduction. LPMOs hold great potential to act as accessory enzymes and are capable of significantly improving the efficiency of substrate hydrolysis (Table 7.1). Cellic CTec2 and Cellic CTec3 are

TABLE 7.1
Examples of Synergistic AA9 LPMOs

Organism	Substrate	Base Cellulase	Enhancement	Reference
Thermoascus aurantiacus TaLPMO9	Steam-pretreated Milled Norway spruce (*Picea abies*)	Celluclast + Novozyme 188 and Cellic CTec2	2 fold	(Caputo, F et al., 2023)
Penicillium funiculosum, PfMig1	Acid-pretreated wheat straw	—	20%	(Ogunyewo et al., 2020)
AfAAnB LPMO from *Aspergillus fumigaturs*	Pretreated sugarcane bagasse	Celluclast 1.5L	22%	(de Gouvêa et al., 2019)
GtLPMO from *Gleophyllum trabeum*	Pretreated wheat straw	Xylanase and Celluclast 1.5L	54%	(Sanhueza et al., 2018)
TtAA9 E from *Thelavia terrestris*	Avicel	Celluclast 1.5L	1.9 fold	(Kim et al., 2017)
TaAA9 A from *Thermoascus aurantiacus*	Avicel	Celluclast 1.5L	1.2 fold	(Kim et al., 2017)
PaLPMO 9E from *Podospora amserina*	Chlorite treated miscanthus	Celluclast 1.5L	2 fold	(Chabbert et al., 2017)
AN3046 LPMO from *Aspergillus nidulans*	Sorghum stover	Endoglucanase	2.1 fold	(Jagadeeswaran et al., 2018)
Gt LPMO from *Gleophyllum trabeum*	Pretreated oak and kenaf	Endoglucanase	26%	(Jung et al., 2015)
Chaetomium globosum	Avicel	Celluclast 1.5L	1.7 fold	(Kim et al., 2015)
Ta GH61 A from *Thermoascus aurantiacus*	Pretreated corn stover	Cellobiose dehydrogenase from *Humicola insolens*	2 fold	(Langston et al., 2011)
PcGH61 D *Phanerochaete chrysosporium*	Avicel	Celluclast 1.5L	2 fold	(Westereng et al., 2011)
Ta GH61 A from *Thermoascus aurantiacus*	Pretreated corn stover	Crude cellulase mixture	18%	(Harris et al., 2010)
GH61 E *Thelavia terrestris*	Pretreated corn stover	Crude cellulase mixture	13%	(Harris et al., 2010)

successful commercial synergistic protein-containing cellulase mixtures in which AA9 LPMO has been supplemented (Cannella et al., 2012; Müller et al., 2015; Rodríguez-Zúñiga et al., 2015). Hence, synthesizing customized cellulase preparations fortified with boosting synergistic proteins may help in reducing the cost of cellulases.

7.8 CONCLUSIONS AND PERSPECTIVES

Since their discovery in 2005, LPMOs have come up as tremendously important entities when the cellulase components are considered for biomass hydrolysis. Recently they have grabbed the attention of researchers, scholars and industrialists for biomass biorefinery purposes. The role of LPMOs has been considered so significant that they can increase the rate of hydrolysis of biomass via enzymes twofold or higher, as has been presented in Table 7.1. Though their importance in biomass hydrolysis has been realized, still the analysis of LPMOs has technical challenges, as no specific chromophores determination method is available. Colorimetric assays for LPMOs are an indirect method, hence research in this direction to have a specific colorimetric assay is a necessity. In fact, the late discovery of LPMOs compared to other components of cellulase is attributed to the complexity of their catalytic mechanism as well as analysis. This challenge limits researchers' exploration, but highly sophisticated equipment is being employed for analysis based on determining the product structurally.

The oxidative mechanism of this monocopper enzyme remains enigmatic, although many studies have been conducted in the last decade to uncover it. A variety of electron donors may be used, such as oxygen, hydrogen peroxide, ascorbic acid, glutathione reduced, gallate or lignin. Lignin itself can act as an electron donor in the pretreated biomass, which explains the higher enzymatic hydrolysis efficiency than in pure cellulose. Enzyme manufacturers are adding this booster protein to their enzyme cocktails to enhance saccharification efficiencies, resulting in fewer enzymes to produce equivalent sugar monomers, in turn reducing the cost of cellulases. Due to their addition to cellulase, LPMOs have improved cellulose enzymatic hydrolysis, which has earned them the title booster's protein in the cellulolytic cocktail. They are now a key component of industrial cellulase cocktails for biomass hydrolysis.

REFERENCES

Arantes, V., Saddler, J.N. 2010. Access to cellulose limits the efficiency of enzymatic hydrolysis: The role of amorphogenesis. *Biotechnology for Biofuels*, **3**(1), 4.

Beeson, W.T., Vu, V.V., Span, E.A., Phillips, C.M., Marletta, M.A. 2015. Cellulose degradation by polysaccharide monooxygenases. *Annual Review of Biochemistry*, **84**, 923–946.

Bischof, R.H., Ramoni, J., Seiboth, B. 2016. Cellulases and beyond: The first 70 years of the enzyme producer *Trichoderma reesei*. *Microbial Cell Factories*, **15**(1), 106.

Bissaro, B., Rohr, A.K., Skaugen, M., Forsberg, Z., Horn, S.J., Vaaje-Kolstad, G., Eijsink, V. 2016. Fenton-type chemistry by a copper enzyme: Molecular mechanism of polysaccharide oxidative cleavage. *bioRxiv*, 097022.

Breslmayr, E., Hanžek, M., Hanrahan, A., Leitner, C., Kittl, R., Šantek, B., Oostenbrink, C., Ludwig, R. 2018. A fast and sensitive activity assay for lytic polysaccharide monooxygenase. *Biotechnology for Biofuels*, **11**(1), 79.

Busk, P.K., Lange, L. 2015. Classification of fungal and bacterial lytic polysaccharide monooxygenases. *BMC Genomics*, **16**(1), 368.

Cannella, D., Hsieh, C.W.C., Felby, C., Jørgensen, H. 2012. Production and effect of aldonic acids during enzymatic hydrolysis of lignocellulose at high dry matter content. *Biotechnology for Biofuels*, **5**(1), 26.

Caputo, F., Tõlgo, M., Naidjonoka, P., Krogh, K., Novy, V., Olsson, L. 2023. Investigating the role of AA9 LPMOs in enzymatic hydrolysis of differentially steam-pretreated spruce. *Biotechnology for Biofuels and Bioproducts*, **16**(68), 2–13.

Chabbert, B., Habrant, A., Herbaut, M., Foulon, L., Aguie-Beghin, V., Garajova, S., et al. 2017. Action of lytic polysaccharide monooxygenase on plant tissue is governed by cellular type. *Science Reports*, **7**(1), 17792.

Courtade, G., Aachmann, F.L. 2019. Chitin-active lytic polysaccharide monooxygenases. *Advances in Experimental Medicine and Biology*, **1142**, 115–129.

Couturier, M., Ladeveze, S., Sulzenbacher, G., Ciano, L., Fanuel, M., Moreau, C., Villares, A., Cathala, B., Chaspoul, F., Frandsen, K.E., Labourel, A., Herpoel-Gimbert, I., Grisel, S., Haon, M., Lenfant, N., Rogniaux, H., Ropartz, D., Davies, G.J., Rosso, M.N., Walton, P.H., Henrissat, B., Berrin, J.G. 2018. Lytic xylan oxidases from wood-decay fungi unlock biomass degradation. *Nature Chemical Biology*, **14**(3), 306–310.

Eriksson, K.-E., Pettersson, B., Westermark, U. 1974. Oxidation: An important enzyme reaction in fungal degradation of cellulose. *FEBS Letters*, **49**(2), 282–285.

Filiatrault-Chastel, C., Navarro, D., Haon, M., Grisel, S., Herpoël-Gimbert, I., Chevret, D., Fanuel, M., Henrissat, B., Heiss-Blanquet, S., Margeot, A., Berrin, J.G. 2019. AA16, a new lytic polysaccharide monooxygenase family identified in fungal secretomes. *Biotechnology for Biofuels*, **12**(1), 55.

Forsberg, Z., Mackenzie, A.K., Sorlie, M., Rohr, A.K., Helland, R., Arvai, A.S., Vaaje-Kolstad, G., Eijsink, V.G. 2014. Structural and functional characterization of a conserved pair of bacterial cellulose-oxidizing lytic polysaccharide monooxygenases. *Proceedings of the National Academy of Sciences of the United States of America*, **111**(23), 8446–8451.

Forsberg, Z., Vaaje-Kolstad, G., Westereng, B., Bunæs, A.C., Stenstrøm, Y., MacKenzie, A., Sørlie, M., Horn, S.J., Eijsink, V.G. 2011. Cleavage of cellulose by a CBM33 protein. *Protein Science*, **20**(9), 1479–1483.

Garajova, S., Mathieu, Y., Beccia, M.R., Bennati-Granier, C., Biaso, F., Fanuel, M., Ropartz, D., Guigliarelli, B., Record, E., Rogniaux, H., Henrissat, B., Berrin, J.G. 2016. Single-domain flavoenzymes trigger lytic polysaccharide monooxygenases for oxidative degradation of cellulose. *Scientific Reports*, **6**, 28276.

de Gouvêa, P.F., Gerolamo, L.E., Bernardi, A.V., Pereira, L.M.S., Uyemura, S.A., Dinamarco, T. M. 2019. Lytic polysaccharide monooxygenase from aspergillus fumigatus can improve enzymatic cocktail activity during sugarcane bagasse hydrolysis. *Protein & Peptide Letters*, **26**(5), 377–385.

Harris, P.V., Welner, D., McFarland, K.C., Re, E., Navarro Poulsen, J.C., Brown, K., Salbo, R., Ding, H., Vlasenko, E., Merino, S., Xu, F., Cherry, J., Larsen, S., Lo Leggio, L. 2010. Stimulation of lignocellulosic biomass hydrolysis by proteins of glycoside hydrolase family 61: Structure and function of a large, enigmatic family. *Biochemistry*, **49**(15), 3305–3316.

Hansen, L.D., Eijsink, V.G., Horn, S.J., Várnai, A. 2023. H2O2 feeding enables LPMO-assisted cellulose saccharification during simultaneous fermentative production of lactic acid. *Biotechnology and Bioengineering*, **120**(3), 726–736.

Hemsworth, G.R., Henrissat, B., Davies, G.J., Walton, P.H. 2014. Discovery and characterization of a new family of lytic polysaccharide monooxygenases. *Nature Chemical Biology*, **10**(2), 122–126.

Hemsworth, G.R., Johnston, E.M., Davies, G.J., Walton, P.H. 2015. Lytic polysaccharide monooxygenases in biomass conversion. *Trends in Biotechnology*, **33**(12), 747–761.

Horn, S.J., Vaaje-Kolstad, G., Westereng, B., Eijsink, V. 2012. Novel enzymes for the degradation of cellulose. *Biotechnology for Biofuels*, **5**(1), 45.

Isaksen, T., Westereng, B., Aachmann, F.L., Agger, J.W., Kracher, D., Kittl, R., Ludwig, R., Haltrich, D., Eijsink, V.G., Horn, S.J. 2014. A C4-oxidizing lytic polysaccharide monooxygenase cleaving both cellulose and cello-oligosaccharides. *Journal of Biological Chemistry*, **289**(5), 2632–2642.

Jagadeeswaran, G., Gainey, L., Mort, A.J. 2018. An AA9-LPMO containing a CBM1 domain in *Aspergillus nidulans* is active on cellulose and cleaves cello-oligosaccharides. *AMB Express*, **8**(1), 171.

Johansen, K.S. 2016. Discovery and industrial applications of lytic polysaccharide mono-oxygenases. *Biochemical Society Transactions*, **44**(1), 143–149.

Jung, S., Song, Y., Kim, H.M., Bae, H.J. 2015. Enhanced lignocellulosic biomass hydrolysis by oxidative lytic polysaccharide monooxygenases (LPMOs) GH61 from Gloeophyllum trabeum. *Enzyme and Microbial Technology*, **77**, 38–45.

Karkehabadi, S., Hansson, H., Kim, S., Piens, K., Mitchinson, C., Sandgren, M. 2008. The first structure of a glycoside hydrolase family 61 Member, Cel61B from *Hypocrea jecorina*, at 1.6 angstrom resolution. *Journal of Molecular Biology*, **383**(1), 144–154.

Karlsson, J., Saloheimo, M., Siika-aho, M., Tenkanen, M., Penttilä, M., Tjerneld, F. 2001. Homologous expression and characterization of Cel61A (EG IV) of *Trichoderma reesei*. *European Journal of Biochemistry*, **268**(24), 6498–6507.

Karnaouri, A., Muraleedharan, M.N., Dimarogona, M., Topakas, E., Rova, U., Sandgren, M., Christakopoulos, P. 2017. Recombinant expression of thermostable processive MT EG5 endoglucanase and its synergism with MT LPMO from *Myceliophthora thermophila* during the hydrolysis of lignocellulosic substrates. *Biotechnology for Biofuels*, **10**(1), 126.

Kim, I.J., Nam, K.H., Yun, E.J., Kim, S., Youn, H.J., Lee, H.J., Choi, I-G, Kim, K.H. 2015. Optimization of synergism of a recombinant auxiliary activity 9 from Chaetomium globosum with cellulase in cellulose hydrolysis. *Applied Microbiology Biotechnology*, **99**, 8537–8547.

Kim, I.J., Seo, N., An, H.J., Kim, J.H., Harris, P.V., Ki, K.H. 2017. Type-dependent action modes of TtAA9E and TaAA9A acting on cellulose and differently pretreated lignocellulosic substrates. *Biotechnol Biofuels*, **10**, 46.

Kittl, R., Kracher, D., Burgstaller, D., Haltrich, D., Ludwig, R. 2012. Production of four *Neurospora crassa* lytic polysaccharide monooxygenases in *Pichia pastoris* monitored by a fluorimetric assay. *Biotechnology for Biofuels*, **5**(1), 79.

Koskela, S., Wang, S., Li, L., Zha, L., Berglund, L.A., Zhou, Q. 2023. An oxidative enzyme boosting mechanical and optical performance of densified wood films. *Small*, **19**(17), 2205056.

Koskela, S., Zha, L., Wang, S., Yan, M., Zhou, Q. 2022. Hemicellulose content affects the properties of cellulose nanofibrils produced from softwood pulp fibres by LPMO. *Green Chemistry*, **24**(18), 7137–7147.

Kracher, D., Scheiblbrandner, S., Felice, A.K., Breslmayr, E., Preims, M., Ludwicka, K., Haltrich, D., Eijsink, V.G., Ludwig, R. 2016. Extracellular electron transfer systems fuel cellulose oxidative degradation. *Science*, **352**(6289), 1098–1101.

Langston, J.A., Shaghasi, T., Abbate, E., Xu, F., Vlasenko, E., Sweeney, M.D. 2011. Oxidoreductive cellulose depolymerization by the enzymes cellobiose dehydrogenase and glycoside hydrolase 61. *Applied and Environmental Microbiology*, **77**, 7007–7015.

Li, F., Ma, F., Zhao, H., Zhang, S., Wang, L., Zhang, X., Yu, H. 2019. A lytic polysaccharide monooxygenase from a white-rot fungus drives the degradation of lignin by a versatile peroxidase. *Applied and Environmental Microbiology*, **85**(9), e02803-18.

Müller, G., Várnai, A., Johansen, K.S., Eijsink, V.G.H., Horn, S.J. 2015. Harnessing the potential of LPMO-containing cellulase cocktails poses new demands on processing conditions. *Biotechnology for Biofuels*, **8**(187), 2–9.

Ogunyewo, O.A., Randhawa, A., Gupta, M., Kaladhar, V.C., Verma, P.K., Yazdani, S.S. 2020. Synergistic action of a lytic polysaccharide monooxygenase and a cellobiohydrolase from *Penicillium funiculosum* in cellulose saccharification under high-level substrate loading. *Applied and Environment Microbiology*, **86**(23), e01769-20.

Phillips, C.M., Beeson, W.T., Cate, J.H., Marletta, M.A. 2011. Cellobiose dehydrogenase and a copper-dependent polysaccharide monooxygenase potentiate cellulose degradation by *Neurospora crassa*. *ACS Chemical Biology*, **6**(12), 1399–1406.

Quinlan, R.J., Sweeney, M.D., Lo Leggio, L., Otten, H., Poulsen, J.-C.N., Johansen, K.S., Krogh, K.B., Jørgensen, C.I., Tovborg, M., Anthonsen, A., Tryfona, T., Walter, C.P., Dupree, P., Xu, F., Davies, G.J., Walton, P.H. 2011. Insights into the oxidative degradation of cellulose by a copper metalloenzyme that exploits biomass components. *Proceedings of the National Academy of Sciences of the United States of America*, **108**(37), 15079–15084.

Reese, E.T., Siu, R.G., Levinson, H.S. 1950. The biological degradation of soluble cellulose derivatives and its relationship to the mechanism of cellulose hydrolysis. *Journal of Bacteriology*, **59**(4), 485.

Rodríguez-Zúñiga, U.F., Cannella, D., Giordano, R.D., Giordano, R.D.C., Jørgensen, H., Felby, C. 2015. Lignocellulose pretreatment technologies affect the level of enzymatic cellulose oxidation by LPMO. *Green Chemistry*, **17**(5), 2896–2903.

Sabbadin, F., Hemsworth, G.R., Ciano, L., Henrissat, B., Dupree, P., Tryfona, T., Marques, R.D., Sweeney, S.T., Besser, K., Elias, L., Pesante, G., Li, Y., Dowle, A.A., Bates, R., Gomez, L.D., Simister, R., Davies, G.J., Walton, P.H., Bruce, N.C., McQueen-Mason, S.J. 2018. An ancient family of lytic polysaccharide monooxygenases with roles in arthropod development and biomass digestion. *Nature Communications*, **9**(1), 756.

Sabbadin, F., Urresti, S., Henrissat, B., Avrova, A.O., Welsh, L.R., Lindley, P.J., Csukai, M., Squires, J.N., Walton, P.H., Davies, G.J., Bruce, N.C., Whisson, S.C., McQueen-Mason, S.J. 2021. Secreted pectin monooxygenases drive plant infection by pathogenic oomycetes. *Science*, **373**(6556), 774–779.

Saloheimo, M., Nakari-SetäLä, T., Tenkanen, M., Penttilä, M. 1997. cDNA cloning of a *Trichoderma reesei* cellulase and demonstration of endoglucanase activity by expression in yeast. *European Journal of Biochemistry*, **249**(2), 584–591.

Sanhueza, C., Carvajal, G., Soto-Aguilar, J., Lienqueo, M.E., Salazar, O. 2018. The effect of a lytic polysaccharide monooxygenase and a xylanase from Gloeophyllum trabeum on the enzymatic hydrolysis of lignocellulosic residues using a commercial cellulase. *Enzyme and Microbial Technology*, **113**, 75–82.

Singhania, R.R., Dixit, P., Patel, A.K., Giri, B.S., Kuo, C.H., Chen, C.W., Di Dong, C. 2021. Role and significance of lytic polysaccharide monooxygenases (LPMOs) in lignocellulose deconstruction. *Bioresource Technology*, **335**, 125261.
Singhania, R.R., Patel, A.K., Raj, T., Chen, C.W., Ponnusamy, V.K., Tahir, N., Kim, S.H., Dong, D.C. 2022. Lignin valorisation via enzymes: A sustainable approach. *Fuel*, **311**, 122608.
Teter, S.A. 2012. Decrease, final technical report, development of a commercial ready enzyme application system for ethanol. *Prepared for the U.S. Department of Energy.*
Vaaje-Kolstad, G., Bøhle, L.A., Gåseidnes, S., Dalhus, B., Bjørås, M., Mathiesen, G., Eijsink, V.G. 2012. Characterization of the chitinolytic machinery of Enterococcus faecalis V583 and high-resolution structure of its oxidative CBM33 enzyme. *Journal of molecular biology*, **416**(2), 239–254.
Vaaje-Kolstad, G., Westereng, B., Horn, S.J., Liu, Z., Zhai, H., Sørlie, M., Eijsink, V.G. 2010. An oxidative enzyme boosting the enzymatic conversion of recalcitrant polysaccharides. *Science*, **330**(6001), 219–222.
Várnai, A., Umezawa, K., Yoshida, M., Eijsink, V.G. 2018. The pyrroloquinoline-quinone-dependent pyranose dehydrogenase from *Coprinopsis cinerea* drives lytic polysaccharide monooxygenase action. *Applied and Environment Microbiology*, **84**(11), e00156-18.
Vu, V.V., Beeson, W.T., Span, E.A., Farquhar, E.R., Marletta, M.A. 2014. A family of starch-active polysaccharide monooxygenases. *Proceedings of the National Academy of Sciences of the United States of America*, **111**(38), 13822–13827.
Wang, D., Li, J., Wong, A.C.Y., Aachmann, F.L., Hsieh, Y.S.Y. 2018. A colorimetric assay to rapidly determine the activities of lytic polysaccharide monooxygenases. *Biotechnology for Biofuels*, **11**(1), 215.
Westereng, B., Arntzen, M.O., Aachmann, F.L., Varnai, A., Eijsink, V.G., Agger, J.W. 2016. Simultaneous analysis of C1 and C4 oxidized oligosaccharides, the products of lytic polysaccharide monooxygenases acting on cellulose. *Journal of Chromatography. Part A*, **1445**, 46–54.
Westereng, B., Ishida, T., Vaaje-Kolstad, G., Wu, M., Eijsink, V.G.H., Igarashi, K., Samejima, M., Ståhlberg, J., Horn, S.J., Sandgren, M. 2011. The putative endoglucanase PcGH61D from Phanerochaete chrysosporium is a metal-dependent oxidative enzyme that cleaves cellulose. *PLoS One*, **6**, e27807.

Section B

Advancements and Challenges in Applications

8 Genetic Modification of Enzymes for Biomass Hydrolysis

Aravind Madhavan, K. B. Arun, Bipin G. Nair, Reeta Rani Singhania, Sindhu Raveendran, Mukesh Kumar Awasthi, Ashok Pandey, and Parameswaran Binod

8.1 INTRODUCTION

Biomass hydrolyzing enzymes are the main crux of biomass-to-bioethanol technology. The feeble activity and excessive price of cellulase hinder the economic efficiency of the technology (Madhavan et al., 2023), therefore cellulase production at low cost with enhanced efficacy is an important need. Various attempts have been made to apply gene manipulation techniques in the production of a range of cellulases in a cost-effective manner with high lignocellulose saccharification efficacy (Singh et al., 2017).

Various engineering technologies can be applied to reduce the price of enzymes for biomass hydrolysis (Liu et al., 2013). First, strain improvement with enhanced enzyme production might decrease the price of enzymes. Second, enhancement in the efficacy of enzymes might decrease their requirement. To attain these objectives, production strains need to be efficiently manipulated (Contreras et al., 2020). This chapter describes the developments in manipulating biomass hydrolytic enzymes and enzyme-producing microbial strains and presents future perceptions of cellulase enzyme engineering.

8.2 MAJOR LIGNOCELLULOSE BIOMASS-DEGRADING MICROORGANISMS

The environment is rich with lignocellulosic substrates, and due to their low cost, they are widely used for producing second-generation biofuel (Dey et al., 2020). Biochemical conversion of lignocellulose biomass to biofuel is the preferred approach, which includes a pretreatment step followed by hydrolysis and fermentation steps (Limayem and Ricke, 2012). The biochemical process makes use of various microorganisms and enzymes. Bacteria and fungi are reported to utilize their specific enzymes with cellulolytic, lignolytic, amylolytic, pectinolytic, proteolytic, and lipolytic activity, which ultimately decompose lignocellulose biomass (Ferdeş et al., 2020). Fungal strains are more prominent than bacterial strains in degrading lignocellulose biomass. Lignocellulose biomass is efficiently degraded by white rot and brown rot fungal species. The main white rot fungi currently in use are *Streptomyces viridosporus*, *Irpex lacteus*, *Echinodontium taxodii*, *Trametes versicolor*, *Phanerochaete chrysosporium*, *Pleurotus ostreatus*, *Flammulina velutipes*, *Ceriporiopsis subvermispora*, *Ceriporia lacerata*, *Stereumhirsutum*, *Trametes versicolor*, and *Trichoderma viride* (Tsegaye et al., 2019). *Coniophora puteana*, *Rhodonia placenta* (*Postia placenta*), *Serpula lacrymans*, *Gloeophyllum trabeum*, and *Fomitopsis pinicola* are brown rot fungi reported to degrade lignocellulose biomass (Ray et al., 2010).

DOI: 10.1201/9781003335313-10

The fungal degradation of lignocellulose biomass proceeds through oxidative or hydrolytic mechanisms. The oxidative mechanism is controlled by enzymes such as aryl alcohol oxidase, pyranose 2 oxidase, and glyoxaline oxidase, which initiate the production of free radicals and hydrogen peroxide that work together to decompose lignin (Martínez et al., 2009). During hydrolytic reactions, specific enzymes trigger the breakdown of glycosidic bonds in lignocellulose biomass (Baig, 2020). Enzymes such as cellulases (cellobiohydrolase and β-glucosidase), hemicellulases (β-xylosidase, xyloglucanase, β-mannosidase, and β-1,4-xylanase), and debranching enzymes (α-galactosidase, α-glucuronidase, and glucuronoyl esterase) from fungal sources are widely used for decomposing lignocellulose biomass (Østby et al., 2020).

Bacterial strains from actinobacteria, streptomyces, and Nocardia species are also reported to produce enzymes capable of degrading lignocellulose biomass (Adegboye et al., 2021; Tsegaye et al., 2019).

For efficient production of biomass hydrolyzing enzymes and to have efficient hydrolysis of biomass, genetic modification of naturally occurring microorganisms is inevitable.

8.3 GENETIC ENGINEERING OF *TRICHODERMA REESEI*

Fungi produce cellulase for their own need, and to economize energy, they switch off their cellulase-producing machinery after sensing the availability of glucose. *Trichoderma reesei* is the model fungi for cellulase production and has been the most studied due to its inherent capacity to produce cellulase capable of hydrolyzing crystalline cellulose. It has been applauded by researchers and industrialists due to its capacity to produce all the components of cellulase. *T. reesei* RUT-C30 is a mutant strain having a truncated CRE1 site. CRE1 is a catabolite repressor site where the end product of hydrolysis, i.e., glucose, binds and inhibits further enzyme production. Hence, the strain is known as a hyperproducer of cellulase, and several genetic modification studies have been done on this strain targeting various genes responsible for regulation.

8.3.1 Engineering of Transcriptional Regulators

In *T. reesei*, the gene expression of lignocellulose hydrolytic enzymes is controlled in a tightly regulated manner by cultivation on carbon substrates like lignocellulosic substrates, or certain sugars like lactose or sophorose. The gene expression of cellulases can be controlled by manipulating a small number of transcriptional factors.

In *T. reesei*, unlike in *Neurospora crassa* and *Aspergillus* spp., the transcriptional factor XYR1 (Stricker et al., 2006) is the main transcriptional enhancer of both cellulase and xylanase synthesis. It comes under the fungal transcriptional regulators that comprise a domain that attaches to Zn^{2+} ions by six cysteine amino acids. XYR1 attaches specifically to 5′-GGCW_4-3′ motifs (Furukawa et al., 2008). Mutants of this gene eradicate the induction of cellulase expression by all the inducers, while the overexpression of *xyr1* increases the expression (Gupta et al., 2016). Recently, da Silva Delabona et al. (2017) demonstrated that a constitutive gene expression of *xyr1* in *Trichoderma harzianum* showed the enhanced synthesis of an enzyme complex that resulted in a 25% increased degradation percentage of sugarcane bagasse. Also, Ellilä et al. (2017) recombinantly expressed XYR1 mutant, resulting in decreased glucose repression (Derntl et al., 2013) in *T. reesei* and showed an enhanced hydrolytic enzyme synthesis on sugarcane bagasse.

Three important transcription enhancers—ACE2, ACE3, and the CCAAT-binding factor (HAP2, HAP3, HAP5)—are also present in the control of biomass hydrolytic enzyme synthesis in *T. reesei*. Recombinant expression of *ace3* has been reported to result in increased cellulase synthesis (Häkkinen et al., 2014). Incorporation of several copies of the gene for the hypothetical protein Znf2Cys6 transcription factor Trire2:80291 into *T. reesei* was also reported to enhance cellobiohydrolase 1 (Häkkinen et al., 2014). Additionally, Zhang et al. (2017) recombinantly expressed a hybrid cellulase controller circuit, which comprises a DNA-binding region of the carbohydrate-repressing protein CRE1 combined with the XYR1 protein. Its recombinant expression in *T. reesei* resulted in more than 20-fold enhancement in cellulase and hemicellulase synthesis on glucose substrate. Engineering strategies attempted in *T. reesei* are listed in Table 8.1.

8.3.2 Epigenetic Regulation

Manipulation of epigenetic mechanisms has recently been identified as a possible novel technology for industrial strain engineering in fungi (Aghcheh and Kubicek, 2015). The important epigenetic regulations are DNA methylation and histone modification. Gupta et al. (2016) emphasized the remodelling of chromatin by transcription enhancers and highlighted it as an evolving method for the enhancement of cellulase synthesis. In cellulase-inducing conditions, the chromatin packaging on cellulase-coding genes *cellobiohydrolase 1* and *cellobiohydrolase 2* loosens, but this does not happen in a *xyr1*-deficient strain (Mello-de-Sousa et al., 2016). Subsequently, Mello-de-Sousa et al. (2015) identified various genes which code for chromatin-modifying enzymes, including HAT, whose gene expression is altered in the *xyr1*-deficient strain and its wild-type strain. The histone acetyltransferase GCN5 has been reported to be essential for cellulase expression, and the acetylations of histone H3 at *cbh1* promoter were reduced in the *T. reesei* deficient in GCN5 (Xin et al., 2013). Various engineering strategies applied to *T. reesei* promoter and transcription factors are listed in Table 8.2.

8.3.3 Engineering Promoter Strength

Recombinant protein synthesis depends on the accessibility of effective promoters that permit a precise and higher level of target gene expression (Fitz et al., 2018). Engineering of promoter strength has, thus, been found effective in enhancing gene expression. Researchers engineered *cbh1* promoter of *T. reesei* by substituting the negative regulator of promoter CREI binding sites with the binding sequence of transcriptional activating sequence for ACEII and the HAP2/3/5 protein, which is a positive regulator (Zhou et al., 2016). The incorporation of an α-helix linker in cbh1 from *T. reesei* and endoglucanase from *Acidothermus cellulolyticus* enhanced the efficacy of recombinant protein expression. This heterologously produced enzyme exhibited improved thermal stability as well as approximately 30% enhancement in cellulase activity, respectively. The outcome of the *cis* regulatory element on promoter activity has also been reported (Kiesenhofer et al., 2018). The CBH1 and Xyn1 of *T. reesei* were transcriptionally enhanced by the regulatory protein Xyrl. The binding sites for xyrl (XBS) can be organized in tandem or in inverted repeats. The inverted CIS repeat sequence is reported to play a significant role in inducing *xylanase* and *cbh* promoters that

TABLE 8.1
Improvement of Cellulase Production in *Trichoderma reesei* by Promoter and Transcription Factor Engineering

Gene	Engineering Method	Promoter System	Reference
Ace3	Introduction of multiple copies		Häkkinen et al., 2014
Lae1	Constitutive expression	gpd1	Schuster et al., 2012
Xyr1	Constitutive expression	Ppdc	Xue et al., 2020
Vel1	Constitutive expression	tef1	Karimi Aghcheh et al., 2014
Xyr1	Constitutive expression and downregulation of ace1	Pdc1	Wang et al., 2014
Cbh1	Manipulation of promoter region	Deletion of cre1 binding site and introduction of ace1, hap1, hap2 and hap3	Adnan et al., 2022

TABLE 8.2
Metabolic Engineering for Strain Improvement

Strain	Method	Property Attained	Reference
S. cerevisiae	Xylose isomerase from *C. phytofermentans*	Enhances xylose utilization	Brat et al., 2009
S. cerevisiae	Xylose isomerase from *P. acidipropionici* and chaperonins from *E. coli*	Enhances xylose utilizsation	Temer et al., 2017
S. cerevisiae	5-phospho-ribosyl-1(alpha)-pyrophosphate synthetase overexpression	Helps to survive the stresses during bioethanol production	Cunha et al., 2015
S. cerevisiae	Expressed *XYL*1 and *XYL2*	Enhances xylose utilization	Costa et al., 2017
S. cerevisiae	Overexpress xylose reductase, *ISU*1 (Fe-S cluster synthesis), *IRA*2 and *HOG*1 signalling proteins	Enhances consumption of xylose	Sato et al., 2016
S. cerevisiae	Engineered to produce cellulose-degrading enzymes on the cell surface	Enhances production of ethanol from lignocellulose	Cunha et al., 2021
K. marxianus	Mutagenesis to improve extracellular expression of endoxylanase, endoglucanase and endomannanase	Better utilization of biomass	Zhou et al., 2018
K. marxianus	Hexokinase 1 gene was disrupted, and NAD(P) H-dependent xylose reductase 1 from *Neurospora crassa* was introduced	Enhances production of xylitol	Hua et al., 2019
K. marxianus	Expressing dockerin-fused enzymes containing cellobiose dehydrogenase, lytic polysaccharide monooxygenase and three fungal cellulases	Enhances ethanol production	Anandharaj et al., 2020
K. marxianus	Expressing ribulose 1,5 bis-phosphate carboxylase/oxygenase genes	Enhances ethanol production	Ha-Tran et al., 2021
Penicillium oxalicum	Non-catalytic region of CBH I replaced with the corresponding sequence in *T. reesei*	Improves its activity on crystalline cellulose	Du et al., 2018
T. reesei	Replacement of the native cbh1 gene with homolog from *Chaetomium thermophilum*	Better cellulase activity	Jiang et al., 2020
A. niger	Overexpression of GaaR	Enhances production of pectinases	Alazi et al., 2018

allow enzyme production. Figure 8.1 shows various modifications at the genetic level to enhance enzyme secretion and its ability to act on substrate.

8.3.4 Effects of Genetic Modification

8.3.4.1 Enhancing Catalytic Efficiency of Cellulase

A major drawback of cellulases is their little efficacy in hydrolyzing the natural crystalline cellulose (Gao et al., 2013). Despite various works on the molecular processes of enzyme-substrate interaction and catalysis, little effort has been applied to guide the manipulation of lignocellulose hydrolytic enzymes. In filamentous fungi, cellobiohydrolase I (CBHI) is the crucial enzyme for the hydrolysis of crystalline cellulose. The most efficient technique for the manipulation of this enzyme so far comes from analyzing the cellulase enzyme diversity in nature. For instance, CBHI and various other enzymes from *Penicillium verruculosum* have demonstrated more efficient hydrolytic efficacies than those from *T. reesei* (Morozova et al., 2010). The comparison profile of CBHI from *Penicillium funiculosum*, which exposed a 60% larger activity than *T. reesei* CBHI, the activity of the latter enzyme was enhanced through catalytic domain substitutions or the exact

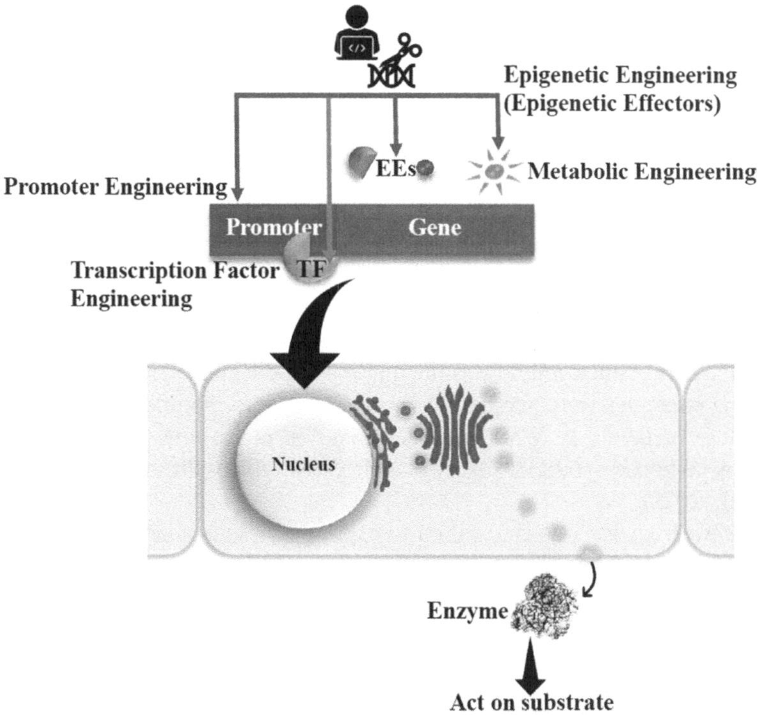

FIGURE 8.1 Engineering strategies for the enhanced production of biomass hydrolytic enzymes.

alteration of two amino acid sites in its catalytic site (Taylor et al., 2018). For the improvement of *Penicillium oxalicum*'s CBHI, substitution of the non-catalytic sequence with the matching sequences in *T. reesei* also enhanced its catalytic activity on cellulose that is in crystalline form (Du et al., 2018). Based on the elucidated multimodular arrangement of the cellulase (CelA) from *Caldicellulosiruptor bescii* (Brunecky et al., 2020), a fusion cellulase consisting of endoglucanase and bacterial cellulose-binding domain and a cellobiohydrolase domain from fungi was created. The fusion enzyme exhibited increased activity in comparison with wild cellulases (Brunecky et al., 2020). Additionally, rational design was applied to enhance the action of endoglucanase by decreasing product-based inhibition. Application of these engineered biomass hydrolytic enzymes to substitute wild enzymes in industrial strains is predictable to synthesize new cellulolytic enzyme platforms with larger efficacies. Certainly, the substitution of the wild *cbh1* gene in *T. reesei* with its homologous gene from *C. thermophilum* resulted in a several-fold enhancement in cellulase action (Jiang et al., 2020).

8.3.4.2 Enhancing Inhibitor Tolerance

Biomass hydrolytic enzymes can be repressed or deactivated by several components synthesized during the pretreatment and fermentation of biomass. The major inhibitors include sugars and their derivatives, and phenolics (Jiang et al., 2020). For instance, the activity of cellulase is inhibited by tannic acid and 60% of the cellulase activity is inhibited (Ximenes et al., 2011). The action of some inhibitors can be complemented with equivalent hydrolases or by the instantaneous fermentation of sugars to other products. Nevertheless, some inhibitors like phenolics are difficult to eliminate, and consequently, the enhancement of enzymatic resistance is important to achieve effective hydrolysis. The β-glucosidase enzyme from *Aspergillus niger* exhibited higher resistance to phenolic compounds than that from *T. reesei*, signifying the likelihood of enhancing phenolic compound resistance by genetic engineering (Ximenes et al., 2011). So far, no such reports have been made on filamentous fungi.

8.4 OTHER GENETIC ENGINEERING STRATEGIES FOR THE SYNTHESIS OF LIGNOCELLULOSE HYDROLYTIC ENZYMES

The genetic engineering of hydrolytic enzymes, especially cellulase and xylanase, is very difficult as the manipulation of these genes leads to unwanted effects on the strains (Hao and Mohnen, 2014). Knowledge of the genome sequence of the organism and its use can overcome the limitations in gene manipulation of hydrolytic enzymes. By implementing metabolic engineering techniques, the lignocellulosic content of the cell wall can be modified for easy access to polysaccharides (Mohapatra et al., 2019). Since cellulose is the major component of lignocellulosic biomass, metabolic engineering strategies were adopted in various studies to modify the biochemical pathways involving cellulose metabolism. Work has been done on modifying the expression of genes of enzymes such as cellulose synthase, glycosyl hydrolases, glycosyltransferases, and sucrose synthase (Joshi et al., 2022).

Saccharomyces cerevisiae was engineered to express xylose isomerase from *Clostridium phytofermentans* and later with xylose isomerase from *Propionibacterium acidipropionici* and chaperonins from *Escherichia coli* to enhance the xylose utilization of yeast (Brat et al., 2009; Temer et al., 2017).

The overexpression of 5-phospho-ribosyl-1(alpha)-pyrophosphate synthetase (PRS3) in CCUG53310 and PE-2 strains of *S. cerevisiae* helps them to survive the stresses taking place in the production of bioethanol from lignocellulosic biomass (Cunha et al., 2015). pMEC1049 vector was engineered to incorporate *XYL*1 (mutated to enhance NADH specificity) and *XYL*2 genes from *Scheffersomyces stipitis* and the endogenous *XKS*1 and *TAL*1 genes under different *S. cerevisiae* promoters *TEF*1, *TDH*3, *PGI*1, and *FBA*1, respectively in *S. cerevisiae* strains. The engineered strains exhibited better xylose consumption and better fermentation of lignocellulosic biomass (Costa et al., 2017). Sato et al. (2016) engineered *S. cerevisiae* strains to express xylose reductase, *ISU*1 (Fe-S cluster synthesis), signalling proteins *HOG*1 for MAP kinase, and *IRA*2 for protein kinase A that showed enhanced consumption of xylose. Cunha et al. (2015) genetically modified an efficient *S. cerevisiae* industrial variant to produce cellulose-degrading enzymes on the cell surface. This strain was subsequently used for the production of ethanol from lignocellulose (Cunha et al., 2015).

The engineered *K. marxianus* strain (expressing ribulose 1,5 bis-phosphate carboxylase/oxygenase genes) was used for the fermentation of hydrolyzed lignocellulosic biomass (rice straw and Napier grass hydrolyzed by *Hungateiclostridium thermocellum*) to produce 10% more ethanol (Ha-Tran et al., 2021). Random mutagenesis of the *K. marxianus* strain improved the extracellular expression of cellulolytic enzymes such as endoxylanase, endoglucanase, and endomannanase (Zhou et al., 2018). In another study, the xylitol production was enhanced using an engineered *K. marxianus* strain in which the endogenous hexokinase 1 gene was disrupted, and NAD(P)H-dependent xylose reductase 1 from *Neurospora crassa* was introduced (Hua et al., 2019). Anandharaj et al. (2020) reported enhanced ethanol production by engineered *K. marxianus* strain expressing dockerin-fused enzymes containing cellobiose dehydrogenase, lytic polysaccharide monooxygenase, and three fungal cellulases along with some cellulosomal genes.

β-Glucosidase (bgl) is a crucial enzyme that breaks the β-1,4 bonding of cellobiose to release glucose. This step is the rate-limiting stage in the conversion of lignocellulosic biomass to sugars due to the production of glucose as the reaction product (Kim, 2018). The presence of glucose inhibits the binding of the substrate to the active site which decreases the reaction rate (Singhania et al., 2013). Several bgls are reported to be repressed by glucose resulting in decreased catalytic efficiency. This might reduce the biotechnological application of these enzymes, therefore, there is a need for bgls with effective enzymatic and catalytic efficiency even at large glucose concentrations. Several genetic engineering techniques have been tried to synthesize enhanced bgls.

A new bgl obtained from a metagenomic clone, manipulated by the replacements of three amino acids resulted in a glucose tolerance with IC50 of 3.0 M. Supplementation of the engineered bgl to cellulose effectively enhanced the glucose synthesis by 30% from the pretreated sugarcane bagasse (Cao et al., 2018). The replacement of two amino acids in β-glucosidase of *T. harzianum* and Cel1A of *T. reesei* were recognized as crucial amino acids in glucose tolerance (Santos et al., 2019).

Cel1A mutant of *T. reesei* also exhibited increased pH stability and thermotolerance and increased substrate specificity in comparison to the wild-type Cel1A (Guo et al., 2016). Two mutations in bgl—D237V and N235S—from *Humicola insolens* resulted in a decreased tolerance to monosaccharides (Meleiro et al., 2017).

Metabolic engineering of hydrolytic enzymes is not much exploited, which may be due to the pleiotropic effect developing after gene manipulation. More research has to be done to manipulate the biochemical pathways involving hydrolytic enzyme production so that lignocellulosic biomass can be efficiently converted to useful products.

8.5 CRISPR-BASED GENOME EDITING FOR CELLULASE PRODUCTION

The CRISPR system is a recently developed method for the metabolic engineering of fungi (Zhao et al., 2018). This system contains two parts: a single chimeric guide RNA and an effector protein like Cas9 endonuclease enzyme. CRISPR is categorized into two major classes: class I (requires more than one effector protein) and class II (requires only one effector protein). Class II comprises a guide RNA and an effector protein as either Cas9 or Cpf1 endonuclease, but in the fungal system, the Cas9 effector protein is widely studied. The guide RNA along with Cas9 binds to DNA at a specific target and inserts a double-strand break. The guide RNA is designed specifically for the target and recognizes a 20bp region situated ahead of the protospacer adjacent motif (PAM) sequence. The double-strand break is repaired by a repair mechanism such as homologous recombination (HR) or NHEJ in cells (Xue and Greene, 2021), leading either to the insertion or deletion of sequences in the target DNA.

The CRISPR system works efficiently when Cas9 (fused with a nuclear localization signal) and guide RNA are expressed properly. The codon optimization of modified Cas9 can be done *in vivo* or *in vitro* and the off-target responses are reduced by introducing inducible promoters to control Cas9 expression. In *T. reesei*, efforts have been made to express guide RNAs *in vitro* and co-transformed with Cas9 (gene or protein) (Zhang et al., 2016). In contrast to other gene editing techniques, CRISPR-Cas is a simple and efficient method and can be used in fungal systems to target different genes without affecting the growth of the fungi. However, in fungal systems, only insertion and deletion effects are achieved by the CRISPR-Cas method; further studies are required for using this method to control the expression of a particular gene in the fungal genome (Zhao et al., 2018).

When it comes to the expression of cellulase systems, gene targeting has always been a tough challenge. *T. reesei*, the most often utilized fungus for commercial cellulase enzyme production, requires an efficient tool for gene editing. *T. reesei* gene targeting efficiency is improved by inhibiting the NHEJ DNA repair mechanism. CRISPR-Cas9 technology was employed to modify the fungus. In fungi, the CRISPR-Cas9 system was performed first in *T. reesei*, where Cas9 codon optimization and *in vitro* transcription of guide RNA was done. The researchers modified the selectable marker by inserting homologous arms of different sizes on both sides of the marker. In *T. reesei*, CRISPR-Cas9-directed gene knockout was achieved by introducing homologous arms (more than 600bp) to attain HR of 100% frequency (Liu et al., 2015). The recombinant guide RNA along with Cas9 thus creates a specific double-strand break in the DNA, allowing the desired gene to be integrated at the target.

8.6 CONCLUSIONS AND PERSPECTIVES

Despite notable progressions in the manipulation of biomass hydrolytic enzymes and the microbes which produce them, there is still an urgent need for future research work to synthesize highly efficient large-scale lignocellulose-degrading enzymes in industry. Additionally, the stability and inhibitor tolerance of lignocellulolytic enzymes need to be enhanced in the background of particular lignocellulose pretreatment techniques. To accomplish these objectives, extensive prospecting of the natural enzyme's diversity and engineering of crucial enzyme machineries will have to be directed. Additionally, existing studies on biomass hydrolytic enzymes need to be unified to develop the biorefinery concept.

The need for enhancing the characteristics of biomass hydrolyzing enzymes such as thermotolerance and enhanced enzyme action for cellulose hydrolysis has led to a spout of interest in genetic manipulation and enzyme engineering. The diverse array of complexities in the biology of various lignocellulose-degrading microorganisms may cause several difficulties in developing new technologies for the synthesis of vigorous cellulases. Technologies to advance efficient cellulase synthesis are needed to comprehend the machinery of cellulase production, progress genetic manipulation techniques along with metabolic and systems engineering, improve parameters for enzyme synthesis, and generate appropriate and beneficial gene cassettes or cellulase synthesis. Computer-assisted data have also been beneficial in this respect because they offer predicted data that cannot be attained through experimental data. Genetic manipulation techniques are proving to be beneficial in advancing cellulase production. Various developments in the analysis of complex metabolic networks, gene assembly techniques, optimized construction of cellulase synthetic pathways, and prospecting and identification of novel highly active cellulases and other carbohydrate-active enzymes may lead to exhilarating advancements in enzyme synthesis and their implications in industrial technologies. Metagenome-based approaches should be emphasized to prospect and cultivate unculturable microorganisms and should be utilized for the synthesis of highly efficient biomass hydrolyzing enzymes. It is one of the most feasible techniques for prospecting novel biosynthetic pathways.

REFERENCES

Adegboye, M.F., Ojuederie, O.B., Talia, P.M., Babalola, O.O., 2021. Bioprospecting of microbial strains for biofuel production: Metabolic engineering, applications, and challenges. *Biotechnol. Biofuels* 14(1), 5. https://doi.org/10.1186/s13068-020-01853-2

Adnan, M., Ma, X., Olsson, S., Wang, J., Liu, G., 2022. Promoter regulation and genetic engineering strategies for enhanced cellulase expression in Trichoderma reesei. *Microbiol. Res.* 259, 127011. https://doi.org/10.1016/j.micres.2022.127011

Aghcheh, R.K., Kubicek, C.P., 2015. Epigenetics as an emerging tool for improvement of fungal strains used in biotechnology. *Appl. Microbiol. Biotechnol.* 99(15), 6167–6181. https://doi.org/10.1007/s00253-015-6763-2

Alazi, E., Knetsch, T., Di Falco, M., Reid, I.D., Arentshorst, M., Visser, J., Tsang, A., Ram, A.F.J., 2018. Inducer-independent production of pectinases in Aspergillus niger by overexpression of the D-galacturonic acid-responsive transcription factor gaaR. *Appl. Microbiol. Biotechnol.* 102(6), 2723–2736. https://doi.org/10.1007/s00253-018-8753-7

Anandharaj, M., Lin, Y.-J., Rani, R.P., Nadendla, E.K., Ho, M.-C., Huang, C.-C., Cheng, J.-F., Chang, J.-J., Li, W.-H., 2020. Constructing a yeast to express the largest cellulosome complex on the cell surface. *Proc. Natl. Acad. Sci.* 117, 2385–2394. https://doi.org/10.1073/pnas.1916529117

Baig, K.S., 2020. Interaction of enzymes with lignocellulosic materials: Causes, mechanism and influencing factors. *Bioresour. Bioprocess.* 7(1), 21. https://doi.org/10.1186/s40643-020-00310-0

Brat, D., Boles, E., Wiedemann, B., 2009. Functional expression of a bacterial xylose isomerase in saccharomyces cerevisiae. *Appl. Environ. Microbiol.* 75, 2304–2311. https://doi.org/10.1128/AEM.02522-08

Brunecky, R., Subramanian, V., Yarbrough, J.M., Donohoe, B.S., Vinzant, T.B., Vanderwall, T.A., Knott, B.C., Chaudhari, Y.B., Bomble, Y.J., Himmel, M.E., Decker, S.R., 2020. Synthetic fungal multifunctional cellulases for enhanced biomass conversion. *Green Chem.* 22(2), 478–489. https://doi.org/10.1039/C9GC03062J

Cao, L., Li, S., Huang, X., Qin, Z., Kong, W., Xie, W., Liu, Y., 2018. Enhancing the thermostability of highly active and glucose-tolerant β-glucosidase Ks5A7 by directed evolution for good performance of three properties. *J. Agric. Food Chem.* 66(50), 13228–13235. https://doi.org/10.1021/acs.jafc.8b05662

Contreras, F., Pramanik, S., Rozhkova, A.M., Zorov, I.N., Korotkova, O., Sinitsyn, A.P., Schwaneberg, U., Davari, M.D., 2020. Engineering robust cellulases for tailored lignocellulosic degradation cocktails. *Int. J. Mol. Sci.* 21(5), 1589. https://doi.org/10.3390/ijms21051589

Costa, C.E., Romaní, A., Cunha, J.T., Johansson, B., Domingues, L., 2017. Integrated approach for selecting efficient Saccharomyces cerevisiae for industrial lignocellulosic fermentations: Importance of yeast chassis linked to process conditions. *Bioresour. Technol.* 227, 24–34. https://doi.org/10.1016/j.biortech.2016.12.016

Cunha, J.T., Aguiar, T.Q., Romaní, A., Oliveira, C., Domingues, L., 2015. Contribution of PRS3, RPB4 and ZWF1 to the resistance of industrial Saccharomyces cerevisiae CCUG53310 and PE-2 strains to lignocellulosic hydrolysate-derived inhibitors. *Bioresour. Technol.* 191, 7–16. https://doi.org/10.1016/j.biortech.2015.05.006

da Silva Delabona, P., Rodrigues, G.N., Zubieta, M.P., Ramoni, J., Codima, C.A., Lima, D.J., Farinas, C.S., da Cruz Pradella, J.G., Seiboth, B., 2017. The relation between xyr1 overexpression in *Trichoderma harzianum* and sugarcane bagasse saccharification performance. *J. Biotechnol.* 246, 24–32. https://doi.org/10.1016/j.jbiotec.2017.02.002

Derntl, C., Gudynaite-Savitch, L., Calixte, S., White, T., Mach, R.L., Mach-Aigner, A.R., 2013. Mutation of the xylanase regulator 1 causes a glucose blind hydrolase expressing phenotype in industrially used Trichoderma strains. *Biotechnol. Biofuels* 6(1), 62. https://doi.org/10.1186/1754-6834-6-62

Dey, P., Pal, P., Kevin, J.D., Das, D.B., 2020. Lignocellulosic bioethanol production: Prospects of emerging membrane technologies to improve the process – A critical review. *Rev. Chem. Eng.* 36(3), 333–367. https://doi.org/10.1515/revce-2018-0014

Du, J., Zhang, X., Li, X., Zhao, J., Liu, G., Gao, B., Qu, Y., 2018. The cellulose binding region in *Trichoderma reesei* cellobiohydrolase I has a higher capacity in improving crystalline cellulose degradation than that of *Penicillium oxalicum*. *Bioresour. Technol.* 266, 19–25. https://doi.org/10.1016/j.biortech.2018.06.050

Ellilä, S., Fonseca, L., Uchima, C., Cota, J., Goldman, G.H., Saloheimo, M., Sacon, V., Siika-aho, M., 2017. Development of a low-cost cellulase production process using *Trichoderma reesei* for Brazilian biorefineries. *Biotechnol. Biofuels* 10, 30. https://doi.org/10.1186/s13068-017-0717-0

Ferdeş, M., Dincă, M.N., Moiceanu, G., Zăbavă, B.S., Paraschiv, G., 2020. Microorganisms and enzymes used in the biological pretreatment of the substrate to enhance biogas production: A review. *Sustainability* 12(17), 7205. https://doi.org/10.3390/su12177205

Fitz, E., Wanka, F., Seiboth, B., 2018. The promoter toolbox for recombinant gene expression in *Trichoderma reesei*. *Front. Bioeng. Biotechnol.* 6. https://doi.org/10.3389/fbioe.2018.00135

Furukawa, T., Shida, Y., Kitagami, N., Ota, Y., Adachi, M., Nakagawa, S., Shimada, R., Kato, M., Kobayashi, T., Okada, H., Ogasawara, W., Morikawa, Y., 2008. Identification of the cis-acting elements involved in regulation of xylanase III gene expression in *Trichoderma reesei* PC-3-7. *Fungal Genet. Biol.* 45(7), 1094–1102. https://doi.org/10.1016/j.fgb.2008.03.006

Gao, D., Chundawat, S.P.S., Sethi, A., Balan, V., Gnanakaran, S., Dale, B.E., 2013. Increased enzyme binding to substrate is not necessary for more efficient cellulose hydrolysis. *Proc. Natl. Acad. Sci. U. S. A.* 110(27), 10922–10927. https://doi.org/10.1073/pnas.1213426110

Guo, B., Amano, Y., Nozaki, K., 2016. Improvements in glucose sensitivity and stability of *Trichoderma reesei* β-glucosidase using site-directed mutagenesis. *PLOS ONE* 11(1), e0147301. https://doi.org/10.1371/journal.pone.0147301

Gupta, V.K., Steindorff, A.S., de Paula, R.G., Silva-Rocha, R., Mach-Aigner, A.R., Mach, R.L., Silva, R.N., 2016. The post-genomic era of *Trichoderma reesei*: What's next? *Trends Biotechnol.* 34(12), 970–982. https://doi.org/10.1016/j.tibtech.2016.06.003

Häkkinen, M., Valkonen, M.J., Westerholm-Parvinen, A., Aro, N., Arvas, M., Vitikainen, M., Penttilä, M., Saloheimo, M., Pakula, T.M., 2014. Screening of candidate regulators for cellulase and hemicellulase production in *Trichoderma reesei* and identification of a factor essential for cellulase production. *Biotechnol. Biofuels* 7(1), 14. https://doi.org/10.1186/1754-6834-7-14

Hao, Z., Mohnen, D., 2014. A review of xylan and lignin biosynthesis: Foundation for studying Arabidopsis irregular xylem mutants with pleiotropic phenotypes. *Crit. Rev. Biochem. Mol. Biol.* 49, 212–241. https://doi.org/10.3109/10409238.2014.889651

Ha-Tran, D.M., Lai, R.-Y., Nguyen, T.T.M., Huang, E., Lo, S.-C., Huang, C.-C., 2021. Construction of engineered RuBisCO Kluyveromyces marxianus for a dual microbial bioethanol production system. *PLoS One* 16, e0247135. https://doi.org/10.1371/journal.pone.0247135

Hua, Y., Wang, J., Zhu, Y., Zhang, B., Kong, X., Li, W., Wang, D., Hong, J., 2019. Release of glucose repression on xylose utilization in Kluyveromyces marxianus to enhance glucose-xylose co-utilization and xylitol production from corncob hydrolysate. *Microb. Cell Fact.* 18, 24. https://doi.org/10.1186/s12934-019-1068-2

Jiang, X., Du, J., He, R., Zhang, Z., Qi, F., Huang, J., Qin, L., 2020. Improved production of majority cellulases in *Trichoderma reesei* by integration of cbh1 gene from *Chaetomium thermophilum. Front. Microbiol.* 11. https://doi.org/10.3389/fmicb.2020.01633

Joshi, A., Verma, K.K., Rajput, V., Minkina, T., Arora, J., 2022. Recent advances in metabolic engineering of microorganisms for advancing lignocellulose-derived biofuels. *Bioengineered* 13, 8135–8163. https://doi.org/10.1080/21655979.2022.2051856

Karimi Aghcheh, R., Németh, Z., Atanasova, L., Fekete, E., Paholcsek, M., Sándor, E., Aquino, B., Druzhinina, I.S., Karaffa, L., Kubicek, C.P., 2014. The VELVET A orthologue VEL1 of *Trichoderma reesei* regulates fungal development and is essential for cellulase gene expression. *PLOS ONE* 9(11), e112799. https://doi.org/10.1371/journal.pone.0112799

Kiesenhofer, D.P., Mach, R.L., Mach-Aigner, A.R., 2018. Influence of cis element arrangement on promoter strength in *Trichoderma reesei. Appl. Environ. Microbiol.* 84(1). https://doi.org/10.1128/AEM.01742-17

Kim, D., 2018. Physico-chemical conversion of lignocellulose: Inhibitor effects and detoxification strategies: A mini review. *Molecules* 23(2), 309. https://doi.org/10.3390/molecules23020309

Limayem, A., Ricke, S.C., 2012. Lignocellulosic biomass for bioethanol production: Current perspectives, potential issues and future prospects. *Prog. Energy Combust. Sci.* 38(4), 449–467. https://doi.org/10.1016/j.pecs.2012.03.002

Liu, G., Qin, Y., Li, Z., Qu, Y., 2013. Development of highly efficient, low-cost lignocellulolytic enzyme systems in the post-genomic era. *Biotechnol. Adv.* 31(6), 962–975. https://doi.org/10.1016/j.biotechadv.2013.03.001

Liu, R., Chen, L., Jiang, Y., Zhou, Z., Zou, G., 2015. Efficient genome editing in filamentous fungus Trichoderma reesei using the CRISPR/Cas9 system. *Cell Discovery* 1, 15007. https://doi.org/10.1038/celldisc.2015.7

Madhavan, A., Arun, K.B., Sindhu, R., Nair, B.G., Pandey, A., Awasthi, M.K., Szakacs, G., Binod, P., 2023. Design and genome engineering of microbial cell factories for efficient conversion of lignocellulose to fuel. *Bioresour. Technol.* 370, 128555. https://doi.org/10.1016/j.biortech.2022.128555

Martínez, Á.T., Ruiz-Dueñas, F.J., Martínez, M.J., del Río, J.C., Gutiérrez, A., 2009. Enzymatic delignification of plant cell wall: from nature to mill. *Curr. Opin. Biotechnol.* 20(3), 348–357. https://doi.org/10.1016/j.copbio.2009.05.002

Meleiro, L.P., Salgado, J.C.S., Maldonado, R.F., Carli, S., Moraes, L.A.B., Ward, R.J., Jorge, J.A., Furriel, R.P.M., 2017. Engineering the GH1 β-glucosidase from *Humicola insolens*: Insights on the stimulation of activity by glucose and xylose. *PLOS ONE* 12(11), e0188254. https://doi.org/10.1371/journal.pone.0188254

Mello-de-Sousa, T.M., Rassinger, A., Derntl, C., Poças-Fonseca, M.J., Mach, R.L., Mach-Aigner, A.R., 2016. The relation between promoter chromatin status, Xyr1 and cellulase expression in *Trichoderma reesei. Curr. Genomics* 17(2), 145–152. https://doi.org/10.2174/1389202917666151116211812

Mello-de-Sousa, T.M., Rassinger, A., Pucher, M.E., dos Santos Castro, L., Persinoti, G.F., Silva-Rocha, R., Poças-Fonseca, M.J., Mach, R.L., Nascimento Silva, R., Mach-Aigner, A.R., 2015. The impact of chromatin remodelling on cellulase expression in *Trichoderma reesei. BMC Genomics* 16(1), 588. https://doi.org/10.1186/s12864-015-1807-7

Mohapatra, S., Mishra, S.S., Bhalla, P., Thatoi, H., 2019. Engineering grass biomass for sustainable and enhanced bioethanol production. *Planta* 250, 395–412. https://doi.org/10.1007/s00425-019-03218-y

Morozova, V.V., Gusakov, A.V., Andrianov, R.M., Pravilnikov, A.G., Osipov, D.O., Sinitsyn, A.P., 2010. Cellulases of *Penicillium verruculosum. Biotechnol. J.* 5(8), 871–880. https://doi.org/10.1002/biot.201000050

Østby, H., Hansen, L.D., Horn, S.J., Eijsink, V.G.H., Várnai, A., 2020. Enzymatic processing of lignocellulosic biomass: Principles, recent advances and perspectives. *J. Ind. Microbiol. Biotechnol.* 47(9–10), 623–657. https://doi.org/10.1007/s10295-020-02301-8

Ray, M.J., Leak, D.J., Spanu, P.D., Murphy, R.J., 2010. Brown rot fungal early stage decay mechanism as a biological pretreatment for softwood biomass in biofuel production. *Biomass Bioenergy* 34(8), 1257–1262. https://doi.org/10.1016/j.biombioe.2010.03.015

Santos, C.A., Morais, M.A.B., Terrett, O.M., Lyczakowski, J.J., Zanphorlin, L.M., Ferreira-Filho, J.A., Tonoli, C.C.C., Murakami, M.T., Dupree, P., Souza, A.P., 2019. An engineered GH1 β-glucosidase displays enhanced glucose tolerance and increased sugar release from lignocellulosic materials. *Sci. Rep.* 9(1), 4903. https://doi.org/10.1038/s41598-019-41300-3

Sato, T.K., Tremaine, M., Parreiras, L.S., Hebert, A.S., Myers, K.S., Higbee, A.J., Sardi, M., McIlwain, S.J., Ong, I.M., Breuer, R.J., Avanasi Narasimhan, R., McGee, M.A., Dickinson, Q., La Reau, A., Xie, D., Tian, M., Reed, J.L., Zhang, Y., Coon, J.J., Hittinger, C.T., Gasch, A.P., Landick, R., 2016. Directed evolution reveals unexpected epistatic interactions that alter metabolic regulation and enable anaerobic xylose use by *Saccharomyces cerevisiae. PLoS Genet* 12, e1006372. https://doi.org/10.1371/journal.pgen.1006372

Schuster, A., Bruno, K.S., Collett, J.R., Baker, S.E., Seiboth, B., Kubicek, C.P., Schmoll, M., 2012. A versatile toolkit for high throughput functional genomics with *Trichoderma reesei. Biotechnol. Biofuels* 5(1), 1. https://doi.org/10.1186/1754-6834-5-1

Singh, A., Patel, A.K., Adsul, M., Mathur, A., Singhania, R.R., 2017. Genetic modification: A tool for enhancing cellulase secretion. *Biofuel Res. J.* 4(2), 600–610. https://doi.org/10.18331/BRJ2017.4.2.5

Singhania, R.R., Patel, A.K., Sukumaran, R.K., Larroche, C., Pandey, A., 2013. Role and significance of beta-glucosidases in the hydrolysis of cellulose for bioethanol production. *Bioresour. Technol.* 127, 500–507. https://doi.org/10.1016/j.biortech.2012.09.012

Stricker, A.R., Grosstessner-Hain, K., Würleitner, E., Mach, R.L., 2006. Xyrl (xylanase regulator 1) regulates both the hydrolytic enzyme system and D-Xylose metabolism in *Hypocrea jecorina. Eukaryot. Cell* 5(12), 2128–2137. https://doi.org/10.1128/EC.00211-06

Taylor, L.E., Knott, B.C., Baker, J.O., Alahuhta, P.M., Hobdey, S.E., Linger, J.G., Lunin, V.V., Amore, A., Subramanian, V., Podkaminer, K., Xu, Q., VanderWall, T.A., Schuster, L.A., Chaudhari, Y.B., Adney, W.S., Crowley, M.F., Himmel, M.E., Decker, S.R., Beckham, G.T., 2018. Engineering enhanced cellobiohydrolase activity. *Nat. Commun.* 9(1), 1186. https://doi.org/10.1038/s41467-018-03501-8

Temer, B., dos Santos, L.V., Negri, V.A., Galhardo, J.P., Magalhães, P.H.M., José, J., Marschalk, C., Corrêa, T.L.R., Carazzolle, M.F., Pereira, G.A.G., 2017. Conversion of an inactive xylose isomerase into a functional enzyme by co-expression of GroEL-GroES chaperonins in Saccharomyces cerevisiae. *BMC Biotechnol* 17, 71. https://doi.org/10.1186/s12896-017-0389-7

Tsegaye, B., Balomajumder, C., Roy, P., 2019. Microbial delignification and hydrolysis of lignocellulosic biomass to enhance biofuel production: An overview and future prospect. *Bull. Natl. Res. Cent.* 43(1), 51. https://doi.org/10.1186/s42269-019-0094-x

Wang, W., Meng, F., Liu, P., Yang, S., Wei, D., 2014. Construction of a promoter collection for genes co-expression in filamentous fungus *Trichoderma reesei. J. Ind. Microbiol. Biotechnol.* 41(11), 1709–1718. https://doi.org/10.1007/s10295-014-1508-2

Ximenes, E., Kim, Y., Mosier, N., Dien, B., Ladisch, M., 2011. Deactivation of cellulases by phenols. *Enzyme Microb. Technol.* 48(1), 54–60. https://doi.org/10.1016/j.enzmictec.2010.09.006

Xin, Q., Gong, Y., Lv, X., Chen, G., Liu, W., 2013. *Trichoderma reesei* histone acetyltransferase Gcn5 regulates fungal growth, conidiation, and cellulase gene expression. *Curr. Microbiol.* 67(5), 580–589. https://doi.org/10.1007/s00284-013-0396-4

Xue, C., Greene, E.C., 2021. DNA Repair Pathway Choices in CRISPR-Cas9-Mediated Genome Editing. *Trends in Genetics: TIG* 37(7), 639–656. https://doi.org/10.1016/j.tig.2021.02.008

Xue, Y., Han, J., Li, Y., Liu, J., Gan, L., Long, M., 2020. Promoting cellulase and hemicellulase production from *Trichoderma orientalis* EU7-22 by overexpression of transcription factors Xyrl and Ace3. *Bioresour. Technol.* 296, 122355. https://doi.org/10.1016/j.biortech.2019.122355

Zhang, C., Meng, X., Wei, X., Lu, L., 2016. Highly efficient CRISPR mutagenesis by microhomology-mediated end joining in Aspergillus fumigatus. *Fungal Genetics and Biology: FG & B* 86, 47–57. https://doi.org/10.1016/j.fgb.2015.12.007

Zhang, X., Li, Y., Zhao, X., Bai, F., 2017. Constitutive cellulase production from glucose using the recombinant *Trichoderma reesei* strain overexpressing an artificial transcription activator. *Bioresour. Technol.* 223, 317–322. https://doi.org/10.1016/j.biortech.2016.10.083

Zhao, X.Q., Zhang, X.Y., Zhang, F., Zhang, R., Jiang, B.J., Bai, F.W., 2018. Metabolic engineering of fungal strains for efficient production of cellulolytic enzymes. In Fang, X., Qu, Y. (eds.) *Fungal Cellulolytic Enzymes*. Springer, Singapore. https://doi.org/10.1007/978-981-13-0749-2_2

Zhou, J., Zhu, P., Hu, X., Lu, H., Yu, Y., 2018. Improved secretory expression of lignocellulolytic enzymes in *Kluyveromyces marxianus* by promoter and signal sequence engineering. *Biotechnol Biofuels* 11, 235. https://doi.org/10.1186/s13068-018-1232-7

Zhou, Y., Wang, X., Wei, W., Xu, J., Wang, W., Xie, Z., Zhang, Z., Jiang, H., Wang, Q., Wei, C., 2016. A novel efficient β-glucanase from a paddy soil microbial metagenome with versatile activities. *Biotechnol. Biofuels* 9, 36. https://doi.org/10.1186/s13068-016-0449-6

9 Protein Engineering for Biomass Hydrolyzing Enzymes

Manisha Sharma, Amit K. Rai, and Sudhir P. Singh

9.1 INTRODUCTION

Enzymes are the biological catalysts that accelerate the rates of various biochemical reactions without actually getting utilized themselves. In order to meet the requirements of industrial biotechnology, it is somehow essential to alter the critical regions of the enzymes to enhance their specific functions. This could be possible through enzyme engineering. Due to the increasing demand for enzymes in the food, chemical, and pharmaceutical industries, the employment of enzyme engineering tools to develop novel enzyme variants with enhanced or improved catalytic properties has gained much attention (Robinson, 2015; Singh et al., 2013; da Silva Amatto et al., 2022).

Enzyme engineering is the designing of variant(s) of the protein sequence that leads to betterment in the enzyme's activity, stability, and kinetic parameters favorable to industrial applications of the biocatalyst. The protein sequence variants of the targeted enzyme are expressed in the heterologous host by the recombinant method, and screening is performed on the basis of the desirable catalytic trait, such as thermal stability, cold activity, temperature activity, acidic or alkaline stability and activity, turnover number, substrate binding, catalytic efficiency, and reaction specificity. The objective behind enzyme engineering is to synthesize an enzyme such that it is more beneficial from the industrial point of view.

The principle of enzyme engineering is to enhance or modify the kinetic properties of the enzyme by altering its amino acid sequence. However, it is not evident that a change in any amino acid will affect the properties or functioning of protein; only the critical amino acids that change the functionality or structural stability of the enzyme should be altered. Thus, it is essential to identify and mark the critical amino acids responsible for the particular parameters' function. Various databases have been developed providing information regarding the relationship between amino acid sequences and the 3D structure of enzymes (Hinz et al., 2010). Various strategies are currently available for enzyme engineering, including rational design, *de novo* design, directed evolution, and semi-rational design (Figure 9.1) (Steinera and Schwaba, 2012).

i. The two complementary strategies—rational design and directed evolution—are combined in a semi-rational design approach. In rational design, precise changes in amino acid sequence are introduced by site-specific mutagenesis on the basis of the structure, mechanism, function, and biochemical characteristics of the protein. The desired properties of commercial enzymes are optimized through this method.
ii. Directed evolution is considered in cases when limited information is available regarding biochemical, structural, and sequence conservation data. Directed evolution mimics natural evolution and performs the high-throughput assay. This can be achieved by DNA shuffling and error-prone PCR.
iii. Among all the approaches for enzyme engineering, *de novo* design is the computational methods-based optional strategy. All these strategies, with their own characteristics, improve the enzyme properties. The *de novo* approach's main aim is to identify amino acid

 DOI: 10.1201/9781003335313-11

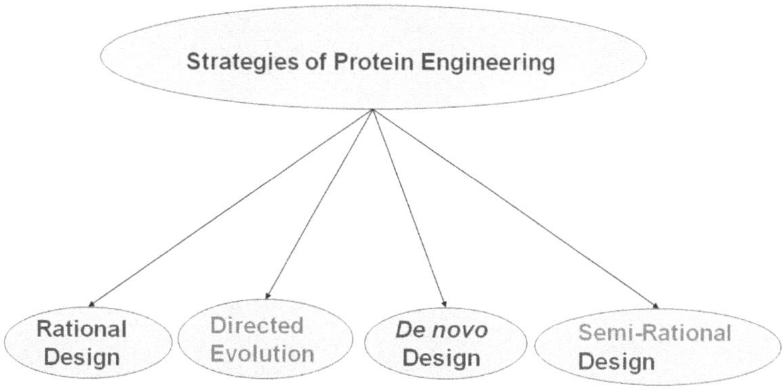

FIGURE 9.1 Different strategies of protein engineering.

sequences that fold into proteins with desired functions. Comparative modeling and fold recognition are certain ways to predict protein structure. The designing of proteins through the *de novo* approach is usually done in such a way that they can often appropriately fold in a faster way. The proteins designed by the *de novo* approach are generally stable and possibly have functional conformations.

9.2 BIOMASS HYDROLYZING ENZYMES

Biomass hydrolyzing enzymes can be classified based on their specificity of catalytic reaction, e.g., EC 3.2.1 glycosidases, EC 4.2.2 lyases, EC 3.1.1 esterases, EC 1.11.1 peroxidases, EC 1.1.3 carbohydrate oxidases, EC 1.10.3 phenol oxidase, and other EC classes. The primary enzyme substrate shared by each class and subclass has facilitated the enzyme with a feature for selecting and targeting biomass materials (Sweeney and Xu, 2012).

On the basis of sequence and structural homology, lignocellulose-degrading enzymes have been classified under the following enzyme families: glycoside hydrolases (GH), polysaccharide lyases (PL), carbohydrate esterases (CE), and lignin oxidases (LO). Many of the lignocellulose-degrading enzymes, in addition to the catalytic core, possess non-catalytic but functionally important domains, such as fibronectin 3-like modules, carbohydrate-binding modules, immunoglobulin-like domains, and dockerins. These domains are found to help the enzymes target carbohydrate molecules present in either clustered or individual polysaccharide forms (Sweeney and Xu, 2012).

The hydrolysis of cellulose and/or hemicellulose involves hydrolytic reactions, whereas lignin degradation involves an oxidoreductive reaction. Almost all cellulases and hemicellulases proceed through either a "retaining" mechanism or an "inverting" mechanism.

Figure 9.2 shows various biomass hydrolyzing enzymes, which are detailed as follows.

9.2.1 Cellulases

Cellulases are the group of enzymes that led to the formation of glucose (Glc) and short cellodextrins by hydrolyzing $\beta(1\rightarrow4)$ glucosidic bonds in cellulose. The cellulolytic enzymes are mainly comprised of cellobiohydrolase or exo-1,4- β -D-glucanase (CBH; EC 3.2.1.91), carboxymethyl cellulase or endo-1,4-β-D-glucanase (EG; EC 3.2.1.4), and β-glucosidase or 1,4- β -D-glucosidase (BG; EC 3.2.1.21) (Poidevin et al., 2013). A vast number of cellulases are available with different modes of reaction. Cellobiohydrolase (CBH) is an important industrial enzyme required for the degradation of crystalline cellulose. It mainly belongs to the GH6, GH7, and GH48 families. CBH of family GH6 and GH7 are found in many cellulolytic fungi. CBH of GH7 is specific to the reducing end of a cellulose chain, whereas CBH of GH6 can be specific to the non-reducing end. Unlike

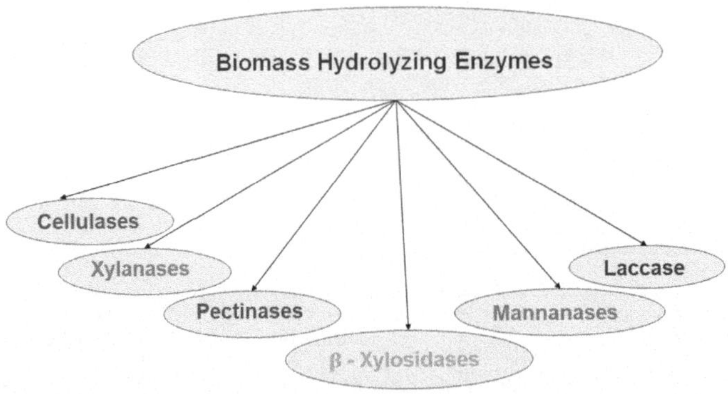

FIGURE 9.2 Flow chart depicting biomass hydrolyzing enzymes.

CBH, endo-1,4-β-Glucanases (EC 3.2.1.4) are responsible for the degradation of amorphous cellulose by randomly hydrolyzing internal glycosidic bonds in cellulose. Both CBH and EG are in significant synergism that provides an efficient system for biomass valorization. Endo-1,4-β-Glucanase have been studied from several GH families, e.g., GH5–10, GH12, GH44–45, GH48, GH51, GH74, GH124, and GH131 (Joshi et al., 2021). β-Glucosidases or cellobiose hydrolases (EC 3.2.1.21) are responsible for the degradation of cellobiose and other cellodextrins. They act on the non-reducing glucose unit from cellobiose or cellodextrin. β-Glucosidases are mainly classified into five classes of enzyme family: GH1, GH3, GH5, GH9, and GH30 (Kaushal et al., 2021).

9.2.2 β-Xylosidases and Xylanases

Xylan is the main constituent of hemicellulose and is categorized into glucuronoxylan, arabinoxylan, glucuronoarabinoxylans, and galactoglucuronoarabinoxylans. It could be present in the form of O-acetyl-4-O-methylglucuronoxylan and arabino-4-O-methylglucuronoxylan (Burlaco et al., 2016; Malgas et al., 2019). Xylanases (EC 3.2.1.8) are the enzymes responsible for the degradation of xylan by cleaving between xylopyranosyl residues on β-1,4-glycosidic linkages. They belong to the GH families 8, 10, 11, 30, and 43 families. Xylanases of GH10 and GH11 families differ in substrate specificity. GH10 acts on a wider range of biomass xylan (Joshi et al., 2020). Although GH11 may be kinetically better, GH10 xylanase produces shorter xylooligosaccharides. Xylanases involve either an "inverting" or a "retaining" mechanism for their catalysis. β-xylosidases (EC 3.2.1.37) belong to GH families 3, 30, 39, 43, 52, and 54 families. They are found to hydrolyze xylobiose or other xylo-oligosaccharides, produced from xylan through xylanase (Burlaco et al., 2016; Malgas et al., 2019).

9.2.3 α-Glucuronidase, α-Galactosidase, and α-L-Arabinofuranosidase

α-Glucuronidases (EC 3.2.1.139) belong to the GH67 and GH115 families and led to the removal of α(1→2)-linked glucuronoyl or its methyl ester in xylan or other hemicellulose components. Some of these enzymes are specific to glucuronated xylooligosaccharides, whereas some have specificity to polymeric glucuronoxylan (Méndez-Líter et al., 2021). The removal of galactose linked to galactomannan, pectin, or other hemicelluloses via α-glycosidic bonds could be carried out using α-galactosidase (EC 3.2.1.22). They belong to GH families 4, 27, 36, 57, and 110, wherein α-galactosidases of GH4 and GH110 families need NAD+ cofactor and a redox mechanism for hydrolytic function (Bhatia et al., 2020). The α-L-arabinofuranosidase (EC 3.2.1.55) of GH families 3, 43, 51, 54, and 62 carry out the removal of arabinose by cleaving the alpha-L-arabinose from the non-reducing ends of α-L-arabinofuranosidase, α-L-arabinans with (1, 3)-linked and/or (1, 5)-linked arabinoxylans and arabinogalactans. This enzyme helps in improving the hydrolysis of lignocellulosic biomass (Poria et al., 2020).

9.2.4 Pectinase, Mannanase, and Glucanase

Pectin is a heteropolysaccharide, predominantly consisting of α(1→4)-linked galacturonic acids as the backbone. The unbranched chain of galacturonic acid residues is called homogalacturonan (HG). The galacturonic acid residues in HG are modified by methylation, esterification, and O-acetylation reactions. The presence of side chains of different sugar moieties on HG results in heteropolysaccharides, such as rhamnogalacturonan I (RGI), rhamnogalacturonan II, rabinan, arabinogalactan I, arabinogalactan II, and xylogalacturonan (Sharma et al., 2022). Pectin is mainly comprised of HG (65%) and RGI (20%–30%), and the remaining 10% are the other heteropolysaccharide components (rhamnogalacturonan II, rabinan, arabinogalactan I, arabinogalactan II, and xylogalacturonan) (Harholt et al., 2010). HG backbone of pectin from different plants varies in the proportion of methylation. The demethylated form of pectin is called pectate.

The common degrading enzymes include polygalacturonases (EC 3.2.1.15, 67, 82) belonging to GH28; pectin/pectate lyases (EC 4.2.2.2, 6, 9, 10) belonging to PL families 1, 2, 3, 9, and 10; and pectin methyl esterase (EC 3.1.1.11) belonging to CE8. The pectin and pectate are hydrolyzed by pectin hydrolases and pectate lyases, respectively (Zheng et al., 2021; Satapathy et al., 2020). Mannans are naturally found polysaccharides that generally exist in four forms, i.e., linear mannan, glucomannan, galactomannan, and galactoglucomannan.

The mannanases (EC 3.2.1.78) from GH5, GH26, and GH113 degrade galacto- and glucomannans (β(1→4)-D-mannosyl or manno/glucopyranosyl polymers with variable α(1→6) galactose side chain and acetylation at O2 and/or O3 position). Another enzyme, β-mannosidases (EC 3.2.1.25) of GH1, GH2, and GH5 could further degrade the mannooligosaccharides produced by mannanase (Sweeney and Xu, 2012; Hlalukana et al., 2021; Burchill et al., 2022). Glucans are a group of polysaccharides linked together by β-1,3 and β-1,4 glycosidic bonds. β-glucanases are endo- or exo-acting enzymes hydrolyzing glycosidic bonds. They belong to GH families 3, 5, 12, 16, 17, 55, 64, and 81. They catalyze the degradation of β(1→3), (1→4), or (1→6) glucan (Chaari and Chaabouni, 2019).

9.2.5 Lignocellulose Oxidoreductases

Lignin is the second most abundant aromatic polymer after cellulose in nature. It is also considered the second most organic polymer after cellulose (Christopher et al., 2014). Lignin is comprised of complex aromatic polymers constructed from syringyl, guaiacyl, and hydroxyphenyl, which are phenylpropanoid monomers linked together via chemical linkages of alkyl–alkyl, alkyl–aryl, and aryl–aryl groups. These three phenylpropanoid monomers (syringyl, guaiacyl, and hydroxyphenyl) are derived from p-coumaryl, coniferyl, and sinapyl alcoholic precursors, respectively (Lu et al., 2017). The lignin-degradation process is basically enzyme-mediated oxidation of lignin polymers. Oxidoreductases mainly catalyze the degradation of lignin.

Lignin peroxidase (EC 1.11.1.14), Mn peroxidase (EC 1.11.1.13), and versatile peroxidase (EC 1.11.1.16) are potential biocatalysts that oxidatively degrade lignin (Ferreira et al., 2021). Laccases (EC 1.10.3.2) are copper-containing phenol oxidase enzymes from the LO1 family. Laccases are not only involved in catalyzing cross-linking reactions among lignin monomers but also in the degradation of lignin polymers (Narnoliya et al., 2019). It can oxidize the phenolic and non-phenolic components of lignin. The degradation of natural lignocellulose can also be carried out using aryl-alcohol oxidase (EC 1.1.3.7), glyoxal oxidase (EC 1.2.3.5), and various carbohydrate oxidases (EC 1.1.3.4, 9, 10). These enzymes belong to the LDA1–6 families that can lead to H_2O_2 production from O_2, with concomitant oxidation of glyoxal and aromatic alcohol. Many lignocellulolytic bacteria and fungi are responsible for the production of cellobiose dehydrogenase (EC 1.1.99.18). The cellobiose or other cellodextrins may get dehydrogenated or oxidized to corresponding aldonolactones using lavoheme enzyme (belonging to the LO3 family) (Sweeney and Xu, 2012; Ferreira et al., 2021; Østby et al., 2020).

9.3 PROTEIN ENGINEERING OF BIOMASS HYDROLYZING ENZYMES

9.3.1 Cellulase Engineering

Enzymes require consistent improvement to make biomass processing economical by increasing the biocatalysts' yield, stability, efficiency, and, subsequently, biotransformation of biomass into value-added products. The betterments in the biochemical properties of enzymes possibly lead to minimizing the number of enzymes required for the degradation of the lignocellulose. The enzymatic improvements could be achieved through protein engineering, involving rational, semi-rational design, and random mutagenesis to develop cellulases with improved characteristics. In a study, mutations were introduced in the outer channel of the active site of β-glucosidase of *Trichoderma reesei* (TrBgl2). Site-directed mutagenesis was performed (W120F, W417F, P338G, T315F, Y298F, H297Y, F250A, D227S, F179H, T178V, P172L, C168V, L167W) replacing the non-conserved amino acids with the conserved residues identified by aligning amino acid sequences of β-glucosidases from diverse bacterial and fungal sources. The catalytic efficiency of the mutants P172L and P172L/F250A toward 4-nitrophenyl-β-D-glucopyranoside (*p*-NPG) substrate was noted to be significantly (more than fivefold) improved as compared to the wild type. Thermal tolerance was also substantially improved in the mutants L167W and P172L/F250A, exhibiting the optimum temperature of 50°C, whereas the wild-type enzyme was inactive at 50°C (Lee et al., 2012). In another approach, the structure-based rational design was carried out in β-1,4-endoglucanase from *Chaetomium thermophilum* (CTendo45). The crystal structure of β-1,4-endoglucanase from *Thielavia terrestris* TtCel45A (a homologous protein of CTendo45) was studied in complex with cellotriose and cellotetraose (PDB: 5GLY). Four single mutants (R29H, Y30F, W31S, and Y173F) and one double mutant (Y30F/Y173F) were generated based on the selection of different conserved and non-catalytic residues. This led to the increase in the specific activity of two single mutants, Y30F and Y173F, toward the substrates carboxymethylcellulose sodium (CMC-Na) and β-D-glucan. Further, improvement was recorded in the thermal stability of double mutant Y30F/Y173F (Chen et al., 2018). In the case of β-glucosidase 1 from *Aspergillus aculeatus* (AaBGL1), the amino acids constituting subsite +1, responsible for β-glucodisaccharides discrimination, were targeted for site-saturation mutagenesis. Based on the crystal structure of AaBGL1, the amino acids interacting with thiocellobiose (TCB) were identified and selected for mutation: W68, R98, D99, R200, Q201, E204, F305, D437, and Y511. The mutant AaBGL1 (Q201E) showed more than a higher efficiency (about 2.7-fold) toward cellobiose than the wild type (Baba et al., 2016). The endoglucanase from *Escherichia coli*, Cel8M, underwent error-prone PCR (99% identity) with the reported endoglucanase from *E. coli* K12. After screening around 10,000 variants of Cel8M, two variants (Cel8ME15 and Cel8ME18) were obtained with 1.42-fold and 1.61-fold enhanced activities, respectively. Further sequence analysis indicated that in Cel8ME15 two residues, Ala9 and Glu353, were substituted for the residues Val9 and Lys353 of Cel8M; whereas in Cel8ME18, Ser117 residue was replaced with Gly117 of Cel8M (Lin et al., 2016). In a study, directed evolution was employed to enhance the thermal stability of cellulases Cel7A from *Hypocrea jecorina*. The library of variants of *H. jecorina* Cel7A mutants expressed heterologously and screened for higher thermal stability. An enzyme variant, FCA398, with 18 mutated sites, was obtained with an improvement in melting temperature by 10.4°C and half-life by 44-fold, as compared to the wild-type enzyme (Goedegebuur et al., 2017).

Enzymes are required to be tolerant to the presence of high salt, organic solvents, ionic liquids, etc. This is even more important when enzymes are employed for the valorization of agri-industrial biomass because of the complexity and pretreatment parameters of biomass. Therefore, strategies were designed to develop cellulase variants that could function even in non-conventional media for their application in the degradation of biomass. The cellulase cocktail from *T. reesei* was succinylated by using succinic anhydride, which resulted in an increase in cellulose conversion in 15% (*v/v*) 1-butyl-3-methylimidazolium chloride by twofold. This might be due to the change in the density of positive to negative surface charge because of the conversion of positively charged primary amine groups to negatively charged and neutral groups, respectively (Nordwald et al., 2014).

Furthermore, pH is another important parameter for the activity and stability of enzymes. Certain protein engineering investigations have been instrumental in maintaining the activity and stability of different cellulases over a wide range of pH (from acidic to alkaline), which is a valuable enzymatic property for the degradation of agri-industrial biomass. The directed evolution of endo-β-1,4-glucanase III from *Trichoderma reesei* QM9414 was conducted by using error-prone PCR for the improvement. This led to the development of a mutant (2R4) with a wide range of pH stability (from pH 4.4–8.8) and higher thermal stability (at 55°C). The mutant enzyme 2R4 exhibited about 1.4-fold higher turnover number (k_{cat}) toward carboxymethyl-cellulose as compared to the wild type (Nakazawa et al., 2009).

9.3.2 Xylanase Engineering

Xylanases play an essential role in industries that produce value-added products from agri-biomass. Protein engineering is a promising approach for the improvement of physico-kinetic properties of xylanases, including activity, pH and thermal stability, and turnover number, as per industrial demands. A study was conducted to enhance the thermostability of a mesophilic xylanase PjxA from *Penicillium janthinellum* MA21601. This enzyme was found to have high specific activity under acidic conditions. On the basis of sequence alignment and structural analysis, five amino acid residues at the N-terminus of PjxA were substituted with that of thermophilic xylanase TfxA from *Thermomonospora fusca* (T10Y, N11H, N12D, Y15F, N30L) to construct a mutant, PTxA. This mutation was introduced to form hydrophobic interactions in the enzyme. Further, an additional disulfide bridge was constructed by replacing two residues of PjxA (threonine-2 and threonine-29) into cysteines by using site-directed mutations to construct PjxA-DB. Then, both mutational strategies were combined to construct a mutant, PTxADB, that led to the improvement in its Tm value from 21.3°C to 76.6°C, and its half-life was increased by 107-fold as compared to the wild type (Xiong et al., 2019). In another study, the catalytic activity of xylanase XynLC9 of the GH11 family from *Bacillus subtilis* was improved through protein engineering.

On the basis of sequence and structural comparisons among XynLC9, xylanases Xyn2 from *Trichoderma reesei* and Xyn11A from *Thermobifida fusca*, and the N-terminal residues YWQN in XynLC9 were identified as engineering hotspots and subjected to site-saturation and iterative mutagenesis. The mutants generated were W6F/Q7H and N8Y; wherein W6F/Q7H and N8Y were thermostable and showed catalytic activity higher than XynLC9 by 2.6- and 1.8-fold, respectively. Even the production yield of oligosaccharides from xylan extracted from corn cob was 160% higher in the case of both mutants (Wang et al., 2021). The three-dimensional structures of xylanase *mxynB* from the *Aspergillus niger* nl-1 were constructed, and then the presence of disulfide in the xylanase structure was studied to engineer mutants with improved optimum temperature and thermostability. The mutant MxynB-116-135, generated with the substitution of amino acid residues G116C+Y135C, showed an increase in 10°C optimum temperature as compared with the wild-type enzyme. Moreover, another mutant (MxynB-G116C-Y135C-D76R-N28H-N29D-Y45M-N47L) generated using site-directed mutagenesis led to the enhancement of thermostability by 570% as compared with the wild-type enzyme. The mutants MxynB-G116C-Y135C-D76R-N28H-N29D-Y4 5M-N47L and MxynB-G116C-Y135C-S58H-D76R-N28H-N29D-Y45M-N47L showed increases in pH optima from 4.5 to 6.0 and pH stability from 120% to 194% as compared to the wild type under pH 5.0 to 7.0 (Li et al., 2019).

9.3.3 Laccase Engineering

Enzyme engineering of laccase enzymes has been carried out to improve their physico-kinetic properties. Researchers employed rational and random mutagenesis to develop engineered laccases (Stanzione et al., 2020). Directed evolution was performed by Liu et al. (2013) to develop an improved variant of laccase from *Trametes versicolor*, making it resistant to ionic liquids. This helped in enhancing the application of laccase for lignin degradation. The combination of computational design with directed evolution created a laccase library in order to enhance the activity of enzymes toward high-redox potential mediators (Mateljak et al., 2019). This led to the improvement in thermal and pH (acidic) stability of the enzyme due to an increase in T1 redox potential. Several other methods, including evolution and hybrid approaches, were employed to generate different laccase variants to meet the harsh conditions required during processing (Pita et al., 2014; Lopes et al., 2019).

9.3.4 Pectinase Engineering

Pectinases are employed in food bioprocessing industries for the clarification aspects during the preparation of juices. The industry demands pectin-hydrolyzing enzymes with significant stability at different temperatures and pH with magnificent catalytic activity. In a previous study, a structure-based rational design approach was employed on endopolygalacturonase (*pg8fn)* (wild type) from *Achaetomium* sp. strain Xz8 in which single and double mutations at two residues D244 and D299 led to the generation of different variants—D244A, D299R, and D244A/D299R. In comparison to the wild type, all mutants showed an improvement in thermal properties, with the maximum level of improvement in a double mutant, D244A/D299R, in terms of enzymatic activity, melting temperature, and catalytic activity (Tu et al., 2015).

Xu et al. (2022) developed a thermostable variant from a wide-type pectate lyase (Pel419) from *Dickeya Dadantii* DCE-01 for degumming ramie through a rational design based on structural analysis. In this study, five non-catalytic mutation sites—V52, K99, A282, N284, and N294—were selected and mutated into V52A, K99R, A282V, N284D, and N294D. A mutant, V52A, showed a significant improvement in enzyme activity and thermal stability, important for industrial-scale degumming of ramie. The pectate lyase PEL168 from *Bacillus subtilis* was modified through various approaches by Wang et al. (2018). First, a mutant V132F was developed through rational design that showed an increase in activity by 1.7-fold with wider pH stability. Afterward, a random mutagenesis library was constructed for the selection of a mutant K47E that depicted an increase in activity and half-life at 50°C by 1.8-fold and 2.0-fold, respectively. The combined effect of these two mutants K47E/V132F led to an improvement in activity by 2.2-fold as compared to the wild type. Pectate lyase from *Bacillus* sp. strain N16-5 (BspPelA) has high specific activity under extremely alkaline conditions but is a thermolabile enzyme. Directed evolution was performed to improve the thermal stability of this enzyme. It was carried out through two rounds of error-prone PCR followed by a screening of more than 12,000 mutants, out of which 10 mutants with improved thermostability were obtained. It was concluded through sequence analysis and site-directed mutagenesis that E124I, T178A, and S271G could be critical for the improvement of thermostability. The combined mutant, EAET (E124I/T178A/N186D/I211V/A254T/S271G/N289H), was found to be the most thermostable variant, exhibiting 140-fold improvement in half-life at 50°C as compared to the wild type. However, a decrease was noted in the activity by 84.3% as compared with that of wild-type BspPelA. In contrast, another combined mutant, EA (N186D/I211V/A254T/S271G/N289H), showed an increase of 24-fold in the half-life at 50°C, with an increase in activity by 23.3% as compared to wild-type BspPelA (Zhou et al., 2015).

9.3.5 β-Xylosidase Engineering

Xylosidases are inactivated at high temperatures, and the reaction product xylose inhibits the activity. This negatively affects the application of xylosidase in biomass processing. Variants of β-xylosidases have been developed through protein engineering approaches. In a study, on the basis of a homologous model and prediction of the protein's active pocket (by using I-TASSA and PyMOL), site-directed mutations were done in two putative xylose tolerance relevant sites (283 and 284) present at the bottom of the protein's active pocket, which led to improvement in the xylose sensitivity and thermostability of *Dictyoglomus thermophilum* β-xylosidase Xln-DT. The xylose tolerance in mutants Xln-DT-284ASP and Xln-DT-284ALA were increased by 9%–35%, with the *Ki* values of 4602 mM and 3708 mM, respectively, as compared to the wild type, whereas the thermal stability of mutant Xln-DT-284ASP was significantly improved at 75°C and 85°C, with no change in activity (Li et al., 2021). Another approach of gene shuffling in β-xylosidase XylBH43 led to the identification of three mutations—R45K, M69P, and L186Y—that affected the thermostability parameter and enhanced Kt0.5 (the temperature at which enzyme activity is reduced by half) by −1.8°C ± 0.1°C, 1.7°C ± 0.3°C, and 3.2°C ± 0.4°C, respectively (Singh et al., 2014). In a recent study, xylosidase Xyl21 was subjected to directed evolution in order to improve thermostability and product tolerance. The triple mutant Xyl21-A16 (K16R, L94I, and K262N) depicted remarkable improvement in xylose tolerance, ethanol tolerance, and thermostability without any change in the optimum pH, temperature, and specific activity (Zhang et al., 2022).

9.4 CONCLUSION AND PERSPECTIVES

The agriculture sector generates huge agri-biomass that have limited uses. Biotransformation of the abundant agri-biomass into value-added bioproducts is for sustainable agricultural prospects. The breakdown of agri-industrial biomass is a complicated process, and it requires the involvement of a cocktail of biocatalysts to hydrolyze lignocellulosic macromolecules. The native biomass hydrolyzing enzymes often exhibit lower efficiency and inferior tolerance to industrial bioprocessing reaction conditions. This has encouraged researchers to develop protein engineering approaches to improve the biocatalysts' expression, secretion, kinetic properties, reaction condition tolerance, and stability parameters.

Protein engineering has contributed a lot to the studying of structural details of enzymes and improving the characteristics of enzymes in terms of their catalytic activity, pH activity and stability, thermal activity and stability, optimum activity, and turnover number. This has provided a platform to researchers in exploring improvements in the catalytic performance of various enzymes, such as cellulases, xylanases, pectinases, and xylosidases, to meet the industrial demands of agro-industrial biomass processing. These engineered enzymes have enormous scope in food, beverages, paper, textile, and other industries. Nevertheless, more intense and aggressive research attempts are required to meet the cost-effectiveness of the biocatalysts useful in the industrial bioprocessing of agri-biomass.

ACKNOWLEDGMENTS

The authors acknowledge the Department of Biotechnology, Government of India for its support.

REFERENCES

Baba Y, Sumitani J, Tanaka K, Tani S and Kawaguchi T. (2016). Site-saturation mutagenesis for β-glucosidase 1 from *Aspergillus aculeatus* to accelerate the saccharification of alkaline-pretreated bagasse. *Appl Microbiol Biotechnol*, 100(24):10495–10507. https://doi.org/10.1007/s00253-016-7726-y.

Bhatia S, Singh A, Batra N and Singh J. (2020). Microbial production and biotechnological applications of α-galactosidase. *Int J Biol Macromol*, 150:1294–1313. https://doi.org/10.1016/j.ijbiomac.2019.10.140.

Chen X, Li W, Ji P, Zhao Y, Hua C and Han C. (2018). Engineering the conserved and noncatalytic residues of a thermostable β-1,4-endoglucanase to improve specific activity and thermostability. *Sci Rep*, 8(1):2954. https://doi.org/10.1038/s41598-018-21246-8.

Christopher L P, Yao B and Ji Y. (2014). Lignin biodegradation with laccase-mediator systems. *Front Energy Res*, 2. https://doi.org/10.3389/fenrg.2014.00012.

Burchill L, Males A, Kaur A, Davies G J and Williams S J. (2022). Structure, function and mechanism of N-Glycan processing enzymes: Endo-α-1,2-Mannanase and endo-α-1,2- mannosidase. *Isr J Chem*, e202200067. https://doi.org/10.1002/ijch.202200067.

Burlaco A, Cornea C P and Israel-Roming F. (2016). Microbial xylanase: A review project: Lignocellulose biodegradation. University of Agronomic Sciences and Veterinary Medicine of Bucharest, 59 Marasti Blvd, District 1, Bucharest, Romania.

Chaari F and Chaabouni S E. (2019). Fungal -1,3-1,4-glucanases: Production, properties and biotechnological applications. *J Sci Food Agric*, 99(6):2657–2664.

da Silva Amatto I V, da Rosa-Garzon N G, de Oliveira Simões F A, Santiago F, da Silva Leite N P, Martins J R and Cabral H. (2022). Enzyme engineering and its industrial applications. *Biotechnol Appl Biochem*, 69(2):389–409. https://doi.org/10.1002/bab.2117.

Ferreira R G, Azzoni A R and Freitas S. (2021). On the production cost of lignocellulose-degrading enzymes. *Biofuels Bioprod Bioref*, 15(1):85–99. https://doi.org/10.1002/bbb.2142.

Goedegebuur F, Dankmeyer L, Gualfetti P, Karkehabadi S, Hansson H, Jana S, Huynh V, Kelemen B R, Kruithof P, Larenas E A, Teunissen P J M, Ståhlberg X J, Payne C M, Mitchinson C and Sandgren M. (2017). Improving the thermal stability of cellobiohydrolase Cel7A from *Hypocrea jecorina* by directed evolution. *J Biol Chem*, 292(42):17418–17430. https://doi.org/10.1074/jbc.M117.803270.

Hlalukana N, Magengelele M, Malgas S and Pletschk B I. (2021). Enzymatic conversion of mannan-rich plant waste biomass into prebiotic mannooligosaccharides. *Foods*, 10(9):2010. https://doi.org/10.3390/foods10092010.

Harholt J, Suttangkakul A and Vibe Scheller H. (2010). Biosynthesis of pectin. *Plant Physiol*, 153(2):384–395.

Hinz U and The UniProt Consortium. (2010). From protein sequences to 3D-structures and beyond: The example of the UniProt KnowledgeBase. *Cell Mol Life Sci*, 67(7):1049–1064. https://doi.org/10.1007/s00018-009-0229-6.

Joshi N, Sharma M and Singh S P. (2020). Characterization of a novel xylanase from an extreme temperature hot spring metagenome for xylooligosaccharide production. *Appl Microbiol Biotechnol*, 104(11):4889–4901. https://doi.org/10.1007/s00253-020-10562-7.

Joshi N, Kaushal G and Singh S P. (2021). Biochemical characterization of a novel thermo-halo-tolerant GH5 endoglucanase from a thermal spring metagenome. *Biotechnol Bioeng*, 118(4):1531–1544. https://doi.org/10.1002/bit.27668.

Kaushal G, Rai A K and Singh S P. (2021). A novel β-glucosidase from a hot-spring metagenome shows elevated thermal stability and tolerance to glucose and ethanol. *Enzyme Microb Technol*, 145:109764. https://doi.org/10.1016/j.enzmictec.2021.109764.

Lee H L, Chang C K, Jeng W Y, Wang A H J and Liang P H. (2012). Mutations in the substrate entrance region of β-glucosidase from *Trichoderma reesei* improve enzyme activity and thermostability. *Protein Eng Des Sel*, 25(11):733–740. https://doi.org/10.1093/protein/gzs073.

Li Q, Tong X, Jiang Y, Li D and Zhao L. (2021). Improvements in xylose stability and thermal stability of GH39 β-xylosidase from *Dictyoglomus thermophilum* by site-directed mutagenesis and insights into its xylose tolerance mechanism. *Enzyme Microb Technol*, 151:109921. https://doi.org/10.1016/j.enzmictec.2021.109921.

Li Q, Wu T, Duan Y, Pei J and Zhao L. (2019). Improving the thermostability and pH stability of *Aspergillus niger* xylanase by site-directed mutagenesis. *Appl Biochem Microbiol*, 55(2):136–144, ISSN 0003-6838.

Lin L, Fu C and Huang W. (2016). Improving the activity of the endoglucanase, Cel8M from *Escherichia coli* by error-prone PCR. *Enzyme Microb Technol*, 86:52–58. https://doi.org/10.1016/j.enzmictec.2016.01.011.

Liu H, Zhu L, Bocola M, Chen N, Spiess A C and Schwaneberg U. (2013). Directed laccase evolution for improved ionic liquid resistance. *Green Chem*, 15(5):1348–1355. https://doi.org/10.1039/c3gc36899h.

Lopes P, Koschorreck K, Nedergaard Pedersen J, Ferapontov A, Lörcher S, Skov Pedersen J, Urlacher V B and Ferapontova E E. (2019). Bacillus licheniformis CotA laccase mutant: Electrocatalytic reduction of O2 from 0.6 V (SHE) at pH 8 and in seawater. *Chem Electro Chem*, 6(7):2043–2049. https://doi.org/10.1002/celc.201900363.

Lu Y, Lu Y C, Hu H Q, Xie F J, Wei X Y and Fan X. (2017). Structural characterization of lignin and its degradation products with spectroscopic methods. *J Spectrosc*, 2017: 1–15.

Mateljak I, Rice A, Yang K, Tron T and Alcalde M. (2019). The generation of thermostable fungal laccase chimeras by SCHEMA-RASPP structure-guided recombination in vivo. *ACS Synth Biol*, 8(4):833–843. https://doi.org/10.1021/acssynbio.8b00509.

Malgas S, Mafa M S, Mkabayi L and Pletschke B I. (2019). A mini review of xylanolytic enzymes with regards to their synergistic interactions during heteroxylan degradation. *World J Microbiol Biotechnol*, 35(12):187. https://doi.org/10.1007/s11274-019-2765-z.

Mendez-Líter J A, de Eugenio L I, Nieto-Domínguez M, Prieto A and Martínez M J. (2021). Hemicellulases from Penicillium and Talaromyces for lignocellulosic biomass valorization: A review. *Bioresour Technol*, 324:124623. https://doi.org/10.1016/j.biortech.2020.124623.

Nakazawa H, Okada K, Onodera T, Ogasawara W, Okada H and Morikawa Y. (2009). Directed evolution of endoglucanase III (Cel12A) from Trichoderma reesei. *Appl Microbiol Biotechnol*, 83(4):649–646.

Narnoliya L K, Agarwal N, Patel S N and Singh S P. (2019). Kinetic characterization of laccase from *Bacillus atrophaeus*, and its potential in juice clarification in free and immobilized forms. *J Microbiol*, 57(10):900–909.

Nordwald E M, Brunecky R, Himmel M E, Beckham G T and Kaar J L. (2014). Charge engineering of cellulases improves ionic liquid tolerance and reduces lignin inhibition. *Biotechnol Bioeng*, 111(8):1471–1698. https://doi.org/10.1002/bit.25216.

Pita M, Mate D M, Gonzalez-Perez D, Shleev S, Fernandez V M, Alcalde M and De Lacey A L. (2014). Bioelectrochemical oxidation of water. *J Am Chem Soc*, 136(16):5892–5895. https://doi.org/10.1021/ja502044j.

Poidevin L, Feliu J, Doan A, Berrin J G, Bey M, Coutinho P M, Henrissat B, Record E and Heiss-Blanquet S (2013). Insights into exo- and endoglucanase activities of family 6 glycoside hydrolases from *Podospora anserina*. *Appl Environ Microbiol*, 79(14):4220–4229. https://doi.org/10.1128/AEM.00327-13.

Poria V, Saini J K, Singh S, Nain L and Kuhad R C. (2020). Arabinofuranosidases: Characteristics, microbial production, and potential in waste valorization and industrial applications. *Bioresour Technol*, 304:123019. https://doi.org/10.1016/j.biortech.2020.123019.

Østby H, Hansen L D, Horn S J, Eijsink V G H and Várnai A. (2020). Enzymatic processing of lignocellulosic biomass: Principles, recent advances and perspectives. *J Ind Microbiol Biotechnol*, 47(9–10):623–657. https://doi.org/10.1007/s10295-020-02301-8.

Robinson P K. (2015). Enzymes: Principles and biotechnological applications. *Essays Biochem*, 59:1–41. https://doi.org/10.1042/bse0590001.

Satapathy S, Rout J R, Kerry R G, Thatoi H and Sahoo S L. (2020). Biochemical prospects of various microbial pectinase and pectin: An approachable concept in pharmaceutical bioprocessing. *Front Nutr*, 7:117.

Sharma N, Sahoo D, Rai A K and Singh S P. (2022). A highly alkaline pectate lyase from the Himalayan hot spring metagenome and its bioscouring applications. *Process Biochem*, 115:100–109.

Singh R K, Tiwari M K, Singh R and Lee J K. (2013). From protein engineering to immobilization: Promising strategies for the upgrade of industrial enzymes. *Int J Mol Sci*, 14(1):1232–1277. https://doi.org/10.3390/ijms14011232.

Singh S K, Heng C, Braker J D, Chan V J, Lee C C, Jordan D B, Yuan L and Wagschal K. (2014). Directed evolution of GH43 β-xylosidase XylBH43 thermal stability and L186 saturation mutagenesis. *J Ind Microbiol Biotechnol*, 41(3):489–498. https://doi.org/10.1007/s10295-013-1377-0.

Stanzione I, Pezzella C, Giardina P, Sannia G and Piscitelli A. (2020). Beyond natural laccases: Extension of their potential applications by protein engineering. *Appl Microbiol Biotechnol*, 104(3):915–924.

Steinera K and Schwaba H. (2012). Recent advances in rational approaches for enzyme engineering. *Comput Struct Biotechnol J*, 2:e201209010. https://doi.org/10.5936/csbj.201209010.

Sweeney M D and Xu F. (2012). Biomass converting enzymes as industrial biocatalysts for fuels and chemicals: Recent developments. *Catalysts*, 2(2):244–263. https://doi.org/10.3390/catal2020244.

Tu T, Luo H, Meng K, Cheng Y, Rui M, Shi P, Huang H, Bai Y, Wang Y, Zhang L and Yao B. (2015). Improvement in thermostability of an *Achaetomium* sp. strain Xz8 endopolygalacturonase via the optimization of charge-charge interactions. *Appl Environ Microbiol*, 81(19):6938–6944. https://doi.org/10.1128/AEM.01363-15.

Xu H, Feng X, Yang Q, Zheng K, Yi L, Duan S and Cheng L. (2022). Improvement on thermostability of pectate lyase and its potential application to ramie degumming. *Polymers (Basel)*, 14(14):2878. https://doi.org/10.3390/polym14142878.

Zhou C, Ye J, Xue Y and Ma Y. (2015). Directed evolution and structural analysis of alkaline pectate lyase from the alkaliphilic *Bacterium bacillus* sp. strain N16-5 to improve its thermostability for efficient ramie degumming. *Appl Environ Microbiol*, 81(17):5714–5723. https://doi.org/10.1128/AEM.01017-15.

Zheng L, Xu Y, Li Q and Zhu B. (2021). Pectinolytic lyases: A comprehensive review of sources, category, property, structure, and catalytic mechanism of pectate lyases and pectin lyases. *Bioresour Bioprocess*, 8(1):79. https://doi.org/10.1186/s40643-021-00432-z.

Zhang Z, Zhang Z, Yu Z, Chen S, Zhang M,Zhang T, Luo X, Zhao J and Li Z. (2022). Simultaneous improvement of final product-tolerance and thermostability of GH39 xylosidase for prebiotic production by directed evolution. *Foods*, 11(19):3039. https://doi.org/10.3390/foods11193039.

Xiong K, Hou J, Jiang Y, Li X, Teng C, Li Q, Fan G, Yang R and Zhang C. (2019). Mutagenesis of N-terminal residues confer thermostability on a *Penicillium janthinellum* MA21601 xylanase. *BMC Biotechnol*, 19(1):51. https://doi.org/10.1186/s12896-019-0541-7.

Wang L, Cao K, Pedroso M M, Wu B, Gao Z, He B and Schenk G. (2021). Sequence- and structure-guided improvement of the catalytic performance of a GH11 family xylanase from Bacillus subtilis. *J Biol Chem*, 297(5):101262. https://doi.org/10.1016/j.jbc.2021.101262.

Wang X, Lu Z, Xu T, Selvaraj J N, Yi L and Zhang G. (2018). Improving the specific activity and thermostability of alkaline pectate lyase from *Bacillus subtilis* 168 for bioscouring. Biochem Eng J, 129:74–83. https://doi.org/10.1016/j.bej.2017.11.001.

10 Inhibitory Effects in Corn Stover Pellet Enzymatic Liquefaction versus Monomeric Sugar Formation after Pretreatment

Luana Assis Serra, Antonio Jose Goncalves da Cruz, Rosineide Gomes da Silva Cruz, Diana M. R. Gutierrez, Cristiane Sanchez Farinas, João Ricardo Moreira de Almeida, Eduardo Ximenes, and Michael R. Ladisch

10.1 INTRODUCTION

Lignocellulosic biomass is a renewable energy source that has great potential to produce biofuels and other bioproducts. This feedstock can come from different sources such as household food waste, forests, and crop residues. Corn stover, sugarcane bagasse and straw, and wheat are abundant and cheap agricultural residues that can be used as energy sources for industrial processes. Using this kind of biomass is advantageous because it reduces the need to expand cultivation areas, lowering the emission of gases that contribute to global warming. Furthermore, the sugars (especially C6 sugars [glucose, galactose, and mannose] and C5 sugars [xylose and arabinose]) present in the lignocellulosic biomass can be converted into high value-added products through the integration of processes for the production of fuels, energy, and chemical compounds (Agrawal et al., 2021; Kim, 2018).

Corn stover is a raw material composed mainly of lignocellulose, a complex material that forms the plant cell wall consisting mostly of cellulose (~37%), hemicellulose (~23 %), and lignin (~19 %). These proportions can vary depending on the corn stover origin and state (pelleted or unpelletized) (Agrawal et al., 2021; dos Santos et al., 2021; Kim et al., 2016). Table 10.1 illustrates the chemical composition of the corn stover from different sources, before and after pretreatment, and pelletized or unpelletized. Additionally, the biomass may contain smaller and variable amounts of pectin, phenolic, inorganic compounds, ashes, and extractives. The interaction between the lignocellulosic biomass components provides the cell wall with an extremely organized, stable, and complex structure, strongly held together by covalent bonds and hydrophobic interactions. This organization guarantees a natural resistance to microbial activity and enzymatic deconstruction, named biomass recalcitrance (Agrawal et al., 2021).

Biomass handling is one of the many challenges in corn stover bioprocessing. Limiting factors such as lower mass and energy densities and dispersed feedstock lead to higher transportation costs. One way to overcome these difficulties is densifying the biomass using equipment such as pellet mills, briquette presses, screw presses, or agglomerators, thereby converting agricultural residues into more uniform and dense commodity-type products that are easier to handle (Tumuluru, 2014).

DOI: 10.1201/9781003335313-12

TABLE 10.1
Compositional Analysis of Untreated (No Pretreatment Was Done) Pellet and Unpelletized Corn Stover

Biomass Composition (%)	Untreated Pelletized Corn Stover	Untreated Pelletized Corn Stover	Pretreated Pelletized Corn Stover	Untreated Corn Stover Unpelletized*	Pretreated Corn Stover Unpelletized†
Glucan	23.8 ± 0.5	35.0 ± 2	50.0 ± 1	37 ± 0.06	35.0 ± 0.06
Hemicellulose	—	20.2 ± 1	10.0 ± 2	—	—
Xylan	12.1 ± 0.1	—	—	22.7 ± 0.14	5.6 ± 0.62
Arabinan	1.2 ± 0.0	—	—	—	—
Acetyl	1.3 ± 0.0	—	—	—	—
Total lignin	11.1 ± 0.7	18.5 ± 0.5	35.0 ± 0.4	—	—
Acid-soluble lignin	—	2.2 ± 0.1	1.7 ± 0.3	1.0 ± 0.03	0.8 ± 0.1
Acid insoluble lignin	—	16.5 ± 0.5	33.5 ± 0.5	17.6 ± 0.02	17.1 ± 1.3
Extractives	21.8 ± 0.0	15.5 ± 1.5	—	11.1 ± 0.06	37.6 ± 0.1
Ash	17.4 ± 0.0	3 ± 1	5.1 ± 0.1	7.5 ± 0.8	2.4 ± 0.8
Water	6.9 ± 0.0	—	—	—	—
Mass	95.6 ± 1.2	91.1 ± 1	99.0 ± 3	96.4 ± 0.19	98.5 ± 0.72
Reference	(dos Santos et al., 2021)	(dos Santos et al., 2020)		(Kim et al., 2016)	

* Dry basis (raw corn stover).
† Dry basis (LHW pretreated corn stover).

Biomass storage losses and transportation costs are lower since this material's processing limits residual microorganisms that overwise feed on the corn stover. According to Tumuluru (2014), these densification processes have been applied for many years in several countries to produce densified biomass for fuel applications (Tumuluru, 2014).

As the pelleted biomass enters the biorefinery, we envision the first operation as the formation of a slurry from the pellets (i.e., liquefaction). Liquefaction (transformation of solids into liquid slurries) allows better manipulation of the feedstock, facilitating transportation inside the unit operations in biorefineries. The enzyme-based liquefaction is important for the preprocessing of high biomass-solids concentrations into slurries. In this process, the enzymes are responsible for the partial degradation of the biomass structure turning the solid material into a pumpable slurry (dos Santos et al., 2021). In this step, some inhibitors could be already released, hampering the enzyme activities and compromising the liquefaction. However, there are few studies regarding enzyme-based liquefaction inhibitors describing the main inhibitors generated in this process. Liquefaction experiments carried out at high-solids loadings can release glucose, xylose, oligomers, and phenolic compounds that accumulate and could act as enzyme inhibitors. A well-mixed bulk phase can lead to changes in the interactions among enzymes, substrates, products, inhibitors, and air, which increases inhibition effects on cellulases and hemicellulases (dos Santos et al., 2021, 2020).

Pretreatment disrupts the biomass cell wall, breaking down the external layer of lignin and exposing the innermost carbohydrate chains for hydrolytic enzyme action. In sequence, the hydrolysis phase is responsible for the deconstruction of the complex carbohydrates into monomeric sugars including glucose, mannose, xylose, arabinose, and galactose. Finally, the fermentation step by microorganisms transforms sugar into bioproducts and is followed by purification and finally distribution to the consumer (Østby et al., 2020). Pretreatment methods and technologies include biological (enzymes and microorganisms), chemical (acids and bases), physical (milling and pyrolysis), physical-chemical (steam explosion, and ammonia-based), and hydrothermal (liquid hot water)

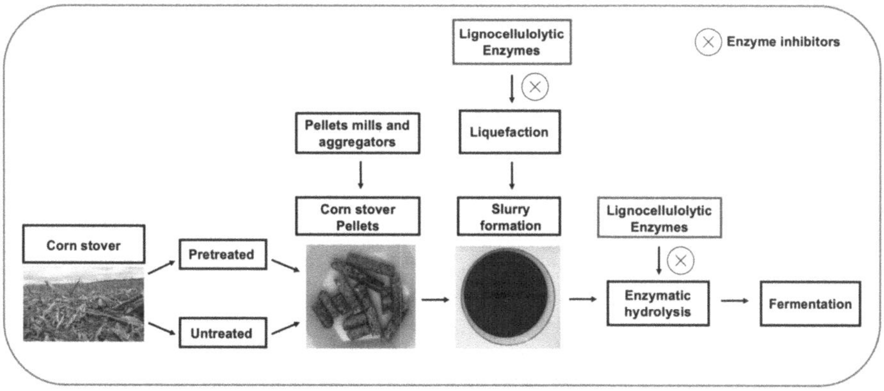

FIGURE 10.1 Schematic representation of the steps of biomass handling from harvest to fermentation. The presence of enzyme inhibitors (represented by the circle) is noticed in the enzyme-based liquefaction and enzymatic hydrolysis of the biomass.

(Chen et al., 2020). Figure 10.1 illustrates the steps starting with the harvest, pretreatment, pelletization, enzyme-based liquefaction, enzymatic hydrolysis, and the fermentation phase, including the presence of enzyme inhibitors during the process.

Among the most used pretreatment methods for corn stover are liquid hot water (LHW), steam explosion (SE), alkaline (AL), and diluted acid (DA). Several inhibitors can be released during this process, compromising the enzymatic hydrolysis and fermentation phases (Agrawal et al., 2021). Depending on the method of pretreatment and the biomass (due to differences in chemical composition and solid concentration), different toxic and inhibitory compounds will be released and generated. In addition to the desired sugars, the most common compounds released and formed from the corn stover are phenolics (ferulic, vanillin, tannic, ρ-coumaric, phenol, and gallic acids), soluble sugars (xylo- and other oligosaccharides, cellobiose, xylose, and glucose), furan aldehydes (5-hydroxymethylfurfural and furfural), and organic acids (acetic, formic, and levulinic acids). Besides the lignin, the phenolics can come from the extractive constituents of biomass, for instance, terpenes and tannins (Kim, 2018; Kim et al., 2011; Tramontina et al., 2020; Ximenes et al., 2011). Table 10.2 demonstrates the different concentrations of these inhibitors, according to the pretreatment method used (Cao et al., 2015, 2013; Chen et al., 2020; Kim et al., 2016).

Using high-solids loadings (>25%) for both processes (enzyme-based liquefaction and pretreatment) can make the process more economically viable and competitive (dos Santos et al., 2021, 2020). However, as previously mentioned, high biomass loads also lead to high concentrations of inhibitory compounds that could accumulate, compromising the enzyme activity during the liquefaction and saccharification steps (Agrawal et al., 2021). A fed-batch strategy for biomass deconstruction can be an affordable strategy for conducting hydrolysis at high-solids loadings. Dividing the addition of the biomass into different time intervals throughout the process gives the enzymes more time to liquefy the substrate, and water can be released from the biomass, although small amounts are consumed by the reaction itself (Cardona et al., 2015; Corrêa et al., 2016). Furthermore, adding surfactants such as Tween 20 and Tween 80 to the reaction could improve enzymatic hydrolysis by reducing the adsorption of cellulases to lignin. Also, biosurfactants, like rhamnolipids, and other chemical compounds, for instance, polyethylene glycols, could improve the enzyme action in high-solids loadings environments by preventing adsorption (Agrawal et al., 2021).

In this chapter, the main inhibitory compounds in enzyme-based liquefaction and pretreated corn stover will be addressed. The literature is vast regarding the compounds derived after the pretreatment of corn stover, but when it comes to liquefaction few studies are available. This chapter will provide some insights into these inhibitory compounds and their effects on the enzyme activity in enzyme-based liquefaction and hydrolysis of corn stover biomass.

TABLE 10.2
Soluble Inhibitor in Different Pretreated Hydrolysates

		Soluble Inhibitors in Hydrolysates				
Pretreatment	**Solid Loading (%)**	**Phenols**	**Oligosaccharides**	**Furan Derivatives**	**Organic Acids**	**Reference**
LWH	20	246 AU*	Xylo-oligomers: 21.7 g/L	Furfural: 7.25 mM HMF: 1.12 mM	Acetic acid: 2.8 g/L	(Kim et al., 2016)
LHW	10	181 AU*	Xylo-oligomers: 9.71 g/L Gluco-oligomers: 1.89 g/L	Furfural: 0.66 g/L HMF: 0.081 g/L	Acetic acid: 2.0 g/L	(Cao et al., 2015)
DA	10	81 AU*	Xylo-oligomers: 0.19 g/L	Furfural: 0.76 g/L HMF: 0.099 g/L	Acetic acid: 3.18 g/L	(Cao et al., 2013)
LWH	10	100 AU*	Xylo-oligomers: 10.7 g/L	Furfural: 0.66 g/L HMF: 0.08 g/L	Acetic acid: 2.24 g/L	
AL†	10	5.92 g/L	Xylo-oligomers: 0.32 g/L Gluco-oligomers: 0.01 g/L	Furfural: 0.66 g/L HMF: 0.08 g/L	Acetic acid: 0.12 g/L	(Chen et al., 2020)
DA†	10	3.33 g/L	Xylo-oligomers: 0.10 g/L Gluco-oligomers: 0.10 g/L	Furfural: 0.0 g/L HMF: 0.01 g/L	Acetic acid: 0.13 g/L	

* Absorbance unit (280 nm).
† Extracted water-soluble phenolic compounds.

10.2 LIGNOCELLULOLYTIC ENZYMES

Several microorganisms produce enzymes that act on the degradation of the plant's cell wall. Among these, filamentous fungi can use decomposing organic material, such as lignocellulosic biomass, to obtain chemical energy. Because of the wide range of enzymes produced by these fungi, some industrially mesophilic relevant species, such as *Trichoderma* spp., *Penicillium* spp., *Aspergillus* spp., and *Neurospora* spp., are used to produce enzymes that are used in commercial enzymatic cocktails for the deconstruction of biomass. Furthermore, thermophilic fungi, such as *Humicola grisea* var. *thermoidea*, *Thielavia terrestris*, *Myceliophthora thermophila*, and *Thermomyces lanuginosus*, have also been exploited as enzyme producers (Bala and Singh, 2019).

Enzyme formulations are in constant development to achieve better results regarding enzyme-based liquefaction and hydrolysis of the biomass. The main components of these consortiums are the enzymes responsible for the degradation of cellulose and hemicellulose, such as exoglucanases, endoglucanases, and xylanases. For cellulose saccharification, the main classes of proteins that act in a coordinated and/or synergistic manner are endoglucanases, exoglucanases, and β-glucosidases. This is the classic concept of cellulose hydrolysis, where the fibers are cleaved by endoglucanases in the internal and amorphous regions. Thus, smaller fragments with reducing and non-reducing ends are released and hydrolyzed at the ends by cellobiohydrolases, which form oligosaccharides, such as cellobiose, which is the substrate for β-glucosidase. These enzymes are classified as glycoside hydrolases (GHs) (Agrawal et al., 2021; Østby et al., 2020).

Recently discovered oxidative enzymes called lytic polysaccharide monooxygenases (LPMOs) aid the degradation of cellulose. These enzymes are classified as proteins with auxiliary activity (AA) and are defined as redox proteins that act with other enzymes, such as GHs. In this way, the hydrolytic activity of GH enzymes, such as endoglucanases and cellobiohydrolases, can be optimized in the presence of AA proteins, as they release new access points on the substrate surface (Agrawal et al., 2021).

Hemicellulose branched chains are made up of different types of sugars, thus requiring specific enzymes for the degradation of the main chains and side groups in question. For example, xylan is one of the main components of hemicellulose chains, and an efficient deconstruction of it requires the synergistic action of endo-β-1,4-xylanase, β-xylosidase, and auxiliary enzymes such as carbohydrate esterases, which hydrolyze the side groups. The endo-β-1,4-xylanase degrades the xylan chain, generating oligosaccharides that are cleaved by β-1,4-xylosidases, releasing the xylose. The enzymes considered auxiliary for the hydrolysis of the side groups are diverse and among them are the acetyl-xylan esterases, feruloyl esterases, and ζ-arabinofuranosidases. The interaction between GH enzymes and auxiliary proteins is essential for the efficient hydrolysis of this substrate (Agrawal et al., 2021; Østby et al., 2020).

10.3 MAIN ENZYME INHIBITORS DERIVED FROM PRETREATED CORN STOVER

10.3.1 Phenols

Phenolic compounds are released from the degradation of the lignin during the corn stover pretreatments, and its main constituents are p-coumaroyl, coniferyl, and sinapyl alcohols. Also, the phenols can come from corn stover extractives. Among the main phenolics in this biomass are the tannic, gallic, cinnamic, ferulic, ζ-coumaric, sinapic, 4-hydroxybenzoic acids, vanillin, and syringaldehyde (Ximenes et al., 2011). Studies have been reporting that phenolics can be much more inhibitory for enzymes than other compounds, such as furan derivatives, oligosaccharides, and organic acids. The most common effects attributed to phenolics are the precipitation and irreversible inhibition of the affected enzymes (Agrawal et al., 2021; Kim, 2018; Ximenes et al., 2011). Vanillin, syringaldehyde, 4-hydroxybenzoic acid, cinnamic acid, ζ-coumaric, ferulic, and tannic acid are reported as being cellulases (endoglucanases, cellobiohydrolases, and ζ-glucosidases) and hemicellulases (xylanases and ζ-xylosidase) inhibitors during hydrolysis, as illustrated in Figure 10.2 (Kim et al., 2011; Ximenes et al., 2011).

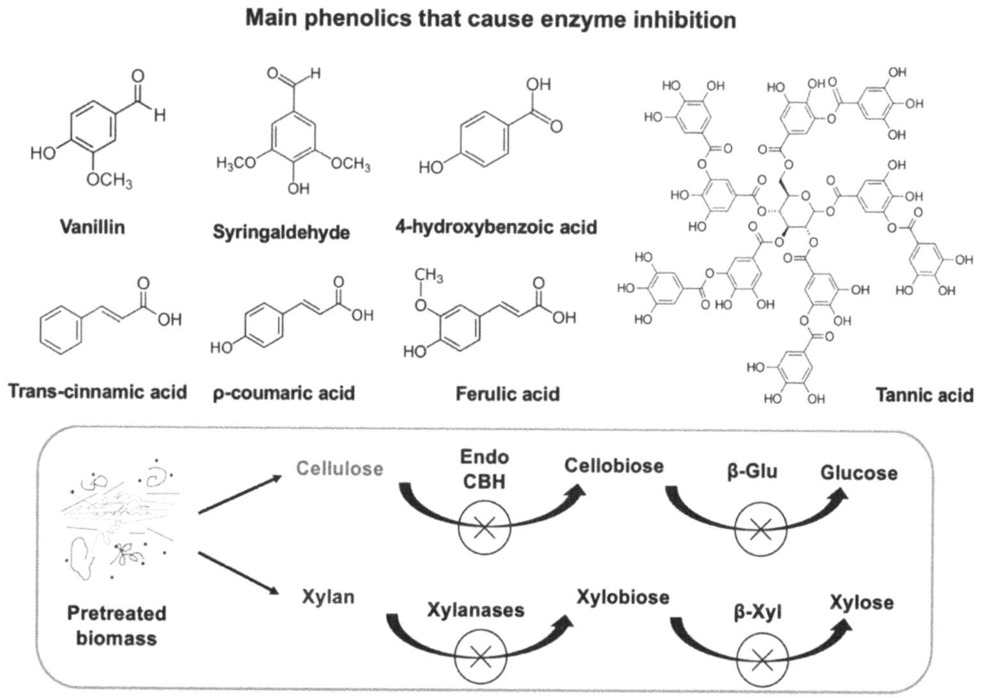

FIGURE 10.2 Main phenolic compounds that cause inhibition in cellulases and hemicellulases. Endo: endoglucanases, CBH: cellobiohydrolases, β-Glu: β-glucosidase, β-Xyl: β-xylosidase.

Vanillin can also lead to enzyme deactivation and precipitation. The work of Ximenes et al. (2011) demonstrated that vanillin and other phenolic compounds (tannic, gallic, trans-cinnamic, and hydroxybenzoic acids) harmed cellulases, such as exo- and endoglucanases and ζ-glucosidase, deactivating theses enzymes by 20%–80%. Tannic acid was the most damaging among phenolic compounds, resulting in deactivation as well as reversible losses of all types of enzyme activities, although some enzymes are more resistant to inhibitors depending on the microbial source of the enzymes. β-glucosidase from *A. niger* resists inhibition up to fivefold and deactivation at tenfold higher phenolic concentrations when compared to the same levels for this enzyme from *T. reesei* (Ximenes et al., 2011).

The inhibitory effect of phenolics on enzymatic hydrolysis can be considerably related to which type of pretreatment is chosen since the concentration of the phenolic compound will vary between different pretreatment methods. Four types of pretreatments (LHW, AL, DA, and ammonia fiber expansion [AFEX]) released higher amounts of phenolics with AL > DA > AFEX > LHW. However, AFEX had the largest inhibitory effect on enzyme hydrolysis followed by LHW, DA, and AL pretreatments (Tramontina et al., 2020). At a solids loading of 20% (w/w) of pretreated corn stover, AFEX generated a phenolic concentration of 6 g/L and decreased the glucose formed (concentration) from 105.1 to 79.9 g/L. This correlates to a 24% reduction in enzymatic hydrolysis. Concerning xylose conversion, the phenolics originating from the four pretreatments had similar inhibitory results, since the xylose values did not change when the phenolics concentrations were varied (Chen et al., 2020). The type of pretreatment has a significant impact on the properties of generated phenolics and, consequently, different levels of inhibition during enzyme hydrolysis.

10.3.2 Oligosaccharides

The sugars generated due to the saccharification, for instance, soluble hydrolysis intermediates, and end products of cellulose and hemicellulose breaking down can have an inhibitory action in lignocellulolytic enzymes. As an example, the monomeric sugars (glucose and xylose) and oligosaccharides, such as xylo-oligomers (Xylo-O) and gluco-oligomers (Gluco-O) are generated and can accumulate throughout the hydrolysis, compromising the enzyme activity, especially the cellulases. It is well-studied that these sugars bind to the active site of the enzymes, lowering and even deactivating their activity (Agrawal et al., 2021; Kim, 2018; Tramontina et al., 2020).

According to some studies in the literature (Agrawal et al., 2021; Kim et al., 2011; Xin et al., 2019), the monomeric sugars (especially glucose and xylose) and cellobiose have a strong effect on cellulases and ζ-glucosidases activities, whereas oligomers can be more detrimental due to inhibition during hydrolysis. The work of Kim et al. (2011) investigated the effect of Xylo-O and xylose in the inhibition of cellulases. They demonstrated that using 21 g/L of xylose resulted in a 10% lower glucose yield. The research found that Xylo-O is more inhibitory than xylose alone. When 8 g/L was added to the reaction, the glucose yield was reduced to 20% (Kim et al., 2011).

Since the hemicellulose has a heterogeneous chain structure, different xylo-oligomers can be generated from it, for example, the arabinoxylooligomers (AXOS). In the work from Xin et al. (2019), they used an ammonia aqueous-pretreated corn stover generating around 15% of the xylan converted into AXOS, and among the AXOS approximately 12% was substituted with arabinose. They investigated the role of the ζ-arabinofuranosidases (ARA) that remove the arabinose branched in the AXOS oligomers. They found that the less substituted AXOS inhibited cellulase activity by 45.7% higher than the ones substituted with arabinose, especially the activity of a cellobiohydrolase (Xin et al., 2019).

10.3.3 Furan Derivatives

Furaldehydes are organic molecules that can be originated from biomass. During the pretreatment step, the furaldehydes furfural ($C_5H_4O_2$) and 5-hydroxymethylfurfural (5-HMF, $C_6H_6O_3$) are formed by dehydration of pentoses and hexoses, respectively, at low pH, although hexoses require higher temperatures and more severe conditions for the formation of hydroxymethyl furfurals compared to pentoses (from hemicellulose) that form furfural degradation products. These two compounds are normally found in biomass hydrolysates when pretreatment or hydrolysis is carried out under acidic conditions. The aldehydes are inhibitory during the microbial fermentation of sugars to ethanol by *Saccharomyces cerevisiae* (Kim, 2018). The furaldehydes act synergistically and affect yeast metabolism and membrane structures, damaging DNA and causing other toxic effects. They especially inhibit enzymes involved in the metabolism, including those in the glycolytic activity and the tricarboxylic acid cycle. Therefore, aldehydes can compromise yeast growth, and at high concentrations cause death to these microorganisms and decrease ethanol production (Kim, 2018).

Although furfural and HMF do not have a strong inhibitory effect on hydrolytic enzymes, in some cases the activity of these proteins can be reduced at higher concentrations of these compounds. For instance, in the work of Mhlongo et al. (2015), the activity of cellobiohydrolase and endoglucanase was affected by HMF using synthetic substrates. For the endoglucanase, after the exposure of these compounds using a 7.5 mM HMF concentration, the residual enzyme activity dropped almost 50% and when 30 mM HMF was used there was no activity at all (Mhlongo et al., 2015). For the cellobiohydrolase, the enzyme had a residual activity of 50% and 40% after immediate exposure with a concentration of 30 mM of HMF and furfural, respectively. However, these furaldehydes derivatives concentrations are lower in natural substrates, for instance, in corn stover, than those tested in the literature.

In the study of Kim et al. (2011), HMF and furfural were tested at a concentration of 4 g/L to evaluate the inhibition of cellulases (commercial cocktail Spezyme CP and Novozyme 188), however, no inhibition was detected. Furfural and HMF were found to have little effect on inhibiting enzyme hydrolysis of cellulose, while the hydrolysis products themselves (glucose and cellobiose) exert a significant inhibitory effect on the hydrolytic enzymes (Kim et al., 2011).

10.3.4 Organic Acids

Organic acids such as acetic, formic, lactic, and levulinic acids can be generated/liberated after pretreatment, and their formation depends strongly on how conditions of this process are performed. Because hemicellulose has the main sugar chain branched with different side groups, some acids can be formed from it, for example, acetic acid is formed from the deacetylation of acetyl groups. Other weak acids are derived from the degradation reactions of sugars and furaldehydes. For instance, levulinic and formic acid can be formed from the breakdown of HMF, which in turn is derived from glucose. Also, formic acid is formed from furfural under high temperatures and acidic conditions, such as those in acid pretreatments or LHW. These weak organic acids can be very toxic for microorganisms in the fermentation phase, affecting the cell membrane and hampering ion transportation and, therefore, interfering with cell metabolism and growth (Kim, 2018; Tramontina et al., 2020).

It is known that organic acids exhibit a small effect on cellulases. For instance, in the work of Kim et al. (2011), 13 g/L of acetic acid did not inhibit cellulases, whereas higher concentrations of acetic and formic acids inhibit enzymes, reducing the rate of hydrolysis by lignocellulolytic enzymes (Kim et al., 2011). The study of Mhlongo et al. (2015) showed that with a concentration of 18.75 mM of formic acid, β-glucosidase was strongly affected, reducing its activity by 80%, whereas, for cellobiohydrolase and endoglucanase, the residual activity was reduced to approximately 40% immediately after exposure to acetic acid at 75 mM. When cellobiohydrolase was tested with formic acid (18.75 mM), the remaining activity was 60% and for endoglucanase around 40%. The addition of 75 mM of acetic and formic acids decreases the pH from 5.2 to 4.37 and 3.69, respectively, and then lowers enzyme activity since pH shifts away from the pH optimum (Mhlongo et al., 2015).

10.4 MAIN ENZYME INHIBITORS FROM ENZYME-BASED LIQUEFACTION

The enzyme inhibitors generated after pretreatment are very well described and documented in the literature. Different studies report the effects of inhibitory compounds in the hydrolysis phase, as described earlier in this chapter. However, when it comes to enzyme-based liquefaction, there are just a few studies that have explored the inhibitors present in this step. The work of dos Santos et al. (2021) presented a new strategy for enzyme-based liquefaction using a high-solids loading (30% w/v) without the need for biomass pretreatment. The corn stover pellet was used as substrate and the study explored three possible pathways for liquefaction: first, applying maleic acid (that acts as an enzyme mimetic), second using enzymes, and third combining both maleic acid and the enzyme in sequence. They tested different concentrations of maleic acid and concluded that at 40 and 100 mM of this acid, liquefaction occurred. Commercial Celluclast 1.5L (concentration of 2.2 mg per g of the substrate) decreased the yield stress from 6806 ± 737 Pa (control, without enzyme) to 298 ± 27 (enzyme in an agitated reactor) after six hours, for a biomass loading of 300 g/L.

Mimetic (i.e., maleic acid at 40 mM) followed by the enzyme treatment improved the results with a 40× decrease (to 171 Pa) in the yield stress, generating pumpable slurries (dos Santos et al., 2021). When combining maleic acid with the enzymes, it was observed that endoglucanases and xylanases were more stable when compared to enzyme treatment alone (Figure 10.3). Also, the complete hydrolysis of hemicellulose into monosaccharides with minimal xylose degradation was verified (Cao et al., 2015; Kim et al., 2016).

The use of maleic acid prior to hydrolysis has been demonstrated to have a beneficial effect on depolymerizing oligosaccharides that may act as inhibitors for cellulases. The work of Cao et al. (2015) used maleic acid to lower the oligosaccharide concentration that could hamper the hydrolysis and observed that all of them were hydrolyzed in xylose, glucose, and arabinose (Cao et al.,

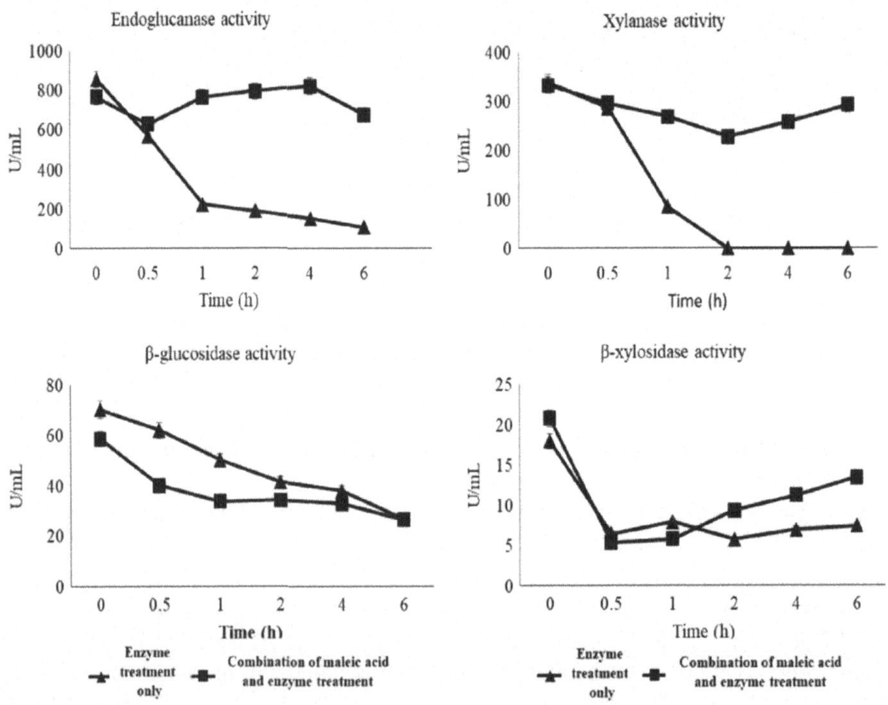

FIGURE 10.3 Endoglucanase, xylanase, β-glucosidase, and β-xylosidase activities measured over six hours during liquefaction of 30% (w/v) corn stover pellets. Triangles represent enzyme treatment only (1 FPU or 2.2 mg protein/g solids), and squares combined maleic acid (40 mM) and enzyme (1 FPU or 2.2 mg protein/g solids) treatments. (From Dos Santos et al., 2021. With permission.)

2015). Following this result, Kim and collaborators (2016) evaluated the effect of hemicellulases and maleic acid to hydrolyze Xylo-O in sugar monomers before the hydrolysis phase in corn stover liquors. The result indicated that maleic acid was more effective in hydrolyzing the substrate, and higher amounts of glucose and xylose were released, while the formation of degradation products was decreased (Kim et al., 2016).

These studies suggest that the maleic acid used in the enzyme-based liquefaction may have the same effect of hydrolyzing the oligomers into monomers, lowering the inhibitory process during the liquefaction of corn stover pellets. More studies are needed to assess the effects of other potential inhibitors, such as phenolic compounds, furan derivatives, and organic acids.

10.4.1 Influence of Operational Conditions on Enzyme-Based Liquefaction, the Lignin, and the Air-interface Effects

Good mixing conditions are important to increase the rate of biomass liquefaction and also to minimize the heat and mass transfer limitations that impair enzymatic cellulose hydrolysis. The overall hydrolysis reaction rate has three steps: (1) external mass transfer of enzymes, (2) diffusion/adsorption of the enzymes on the cellulose surface, and (3) enzyme catalytic action. The process can be governed by either the external mass transfer or the internal diffusion. The design of reactors and impellers has been investigated to improve the progress of the reaction. The selection of the agitation and mixing system in addition to a suitable stirring speed is crucial to ensure that the first effect can be neglected, and the process can operate at the optimum conditions (Corrêa et al., 2016).

In the study by Corrêa et al. (2016), higher cellulose-to-glucose conversion (75.7% in 72 h) and better energetic efficiency, measured by the ratio between cellulose-to-glucose conversion and total energy consumption (78.9 %/MJ), resulted when using steam-exploded sugarcane bagasse (10% w/v and 10 FPU per gram of biomass) in a stirred tank reactor with a dual impeller system consisting of two elephant ear impellers. The results showed that the negative effects caused by soluble inhibitors/deactivators of cellulolytic enzymes could be reduced since the shear stress generated by the elephant ear impellers is lower compared to the system composed of two Rushton impellers (Corrêa et al., 2016).

The other point that deserves attention is the enzyme deactivation by shear stress and contact with the air–liquid interface (Almeida et al., 2021; Bhagia et al., 2018; dos Santos et al., 2020). This effect plays an important role in enzymatic hydrolysis. The results obtained by dos Santos et al. (2020) showed that the cellulases are deactivated by contact with oxygen (dos Santos et al., 2020). In a bioreactor processing corn stover pretreated by LHW in concentrations of 200 g/L and using a low enzyme dosage (3.6 mg protein per gram of biomass), glucose yields of 34% (with no agitation), 45% (with 290 rpm), and 55% (with 290 rpm and nitrogen atmosphere) were obtained. According to Bhagia et al. (2018), the mechanical agitation of the slurry, besides improving the contact between solid and liquid phases, brings about the contact of the third phase, air with protein, thus deactivating cellulolytic enzymes (Bhagia et al., 2018). This may help to explain the use of nitrogen to blanket the air–liquid interface and exhibit a positive impact on glucose yields. The results presented by dos Santos et al. (2020) showed that keeping good mixing conditions throughout the time course of the process enhanced the biomass/enzyme contact and dispersed sugars released during hydrolysis that cause product inhibition (dos Santos et al., 2020).

A further understanding of the effects of the air–liquid and lignin interfaces on endoglucanase activity is presented by Almeida et al. (2021). These authors concluded that the air–liquid interfacial area plays an important role in endoglucanase activity and that the use of non-catalytic proteins such as bovine serum albumin and soy proteins can stabilize endoglucanase activity minimizing the effects of the contact with oxygen in the liquid–air interface. The other protective effect of these non-catalytic proteins is to alleviate the non-specific adsorption of endoglucanases on lignin. From a thermodynamic analysis of the enzyme/protein adsorption, the authors showed that the process is spontaneous, dependent on temperature, endothermy, and entropy. Therefore, lignin hydrophobicity is important to the adsorption of endoglucanase (Almeida et al., 2021).

10.5 CONCLUSIONS AND PERSPECTIVES

The enzyme inhibitors released during enzyme-based liquefaction or generated in subsequent pre-treatment may reduce yields of the released sugars that would further be fermented to desired value-added bioproducts. Phenolics, oligosaccharides, furan derivatives, and organic acids are the main inhibitors present in corn stover pretreatment liquids if dilute acid pretreatment is used. The effect of these compounds on enzyme activity is known and studied extensively, with several approaches being developed to diminish their effect. In addition, poor mixing, shear stress, and the air–liquid interfacial area could also interfere with the success of the liquefaction, leading to deactivation or a decrease in enzyme activity.

The enzyme-based liquefaction is an alternative to preprocessing the biomass facilitating the handling inside the biorefinery. Few studies provide insights into the enzyme inhibitors that could hamper liquefaction, therefore, further elucidation of the severity of these effects is needed, as this is a new field of study. A deeper understanding of the mechanisms behind these effects will help to optimize enzyme formulations for lignocellulose biomass liquefaction and the formation of monomeric sugars after pretreatment of corn stover and liquefied slurries. Also, studies regarding the mixing of high-solids content, shear stress, and other factors are needed to optimize the whole process, opening opportunities for innovative research.

REFERENCES

Agrawal, R., Verma, A., Singhania, R.R., Varjani, S., Di Dong, C., Kumar Patel, A., 2021. Current understanding of the inhibition factors and their mechanism of action for the lignocellulosic biomass hydrolysis. *Bioresour. Technol.* 332, 125042. https://doi.org/10.1016/j.biortech.2021.125042

Almeida, R.M.R.G., Pimentel, W.R.O., Santos-Rocha, M.S.R., Buffo, M.M., Farinas, C.S., Ximenes, E.A., Ladisch, M.R., 2021. Protective effects of non-catalytic proteins on endoglucanase activity at air and lignin interfaces. *Biotechnol. Prog.* 37(4), 1–12. https://doi.org/10.1002/btpr.3134

Bala, A., Singh, B., 2019. Cellulolytic and xylanolytic enzymes of thermophiles for the production of renewable biofuels. *Renew. Energy* 136, 1231–1244. https://doi.org/10.1016/j.renene.2018.09.100

Bhagia, S., Dhir, R., Kumar, R., Wyman, C.E., 2018. Deactivation of cellulase at the air-liquid interface is the main cause of incomplete cellulose conversion at low enzyme loadings. *Sci. Rep.* 8(1), 1–12. https://doi.org/10.1038/s41598-018-19848-3

Cao, G., Ximenes, E., Nichols, N.N., Frazer, S.E., Kim, D., Cotta, M.A., Ladisch, M., 2015. Bioabatement with hemicellulase supplementation to reduce enzymatic hydrolysis inhibitors. *Bioresour. Technol.* 190, 412–415. https://doi.org/10.1016/j.biortech.2015.04.064

Cao, G., Ximenes, E., Nichols, N.N., Zhang, L., Ladisch, M., 2013. Biological abatement of cellulase inhibitors. *Bioresour. Technol.* 146, 604–610. https://doi.org/10.1016/j.biortech.2013.07.112

Cardona, M.J., Tozzi, E.J., Karuna, N., Jeoh, T., Powell, R.L., McCarthy, M.J., 2015. A process for energy-efficient high-solids fed-batch enzymatic liquefaction of cellulosic biomass. *Bioresour. Technol.* 198, 488–496. https://doi.org/10.1016/j.biortech.2015.09.042

Chen, X., Zhai, R., Li, Y., Yuan, X., Liu, Z.H., Jin, M., 2020. Understanding the structural characteristics of water-soluble phenolic compounds from four pretreatments of corn stover and their inhibitory effects on enzymatic hydrolysis and fermentation. *Biotechnol. Biofuels* 13, 1–13. https://doi.org/10.1186/s13068-020-01686-z

Correâ, L.J., Badino, A.C., Cruz, A.J.G., 2016. Mixing design for enzymatic hydrolysis of sugarcane bagasse: Methodology for selection of impeller configuration. *Bioprocess Biosyst. Eng.* 39(2), 285–294. https://doi.org/10.1007/s00449-015-1512-6

dos Santos, A.C.F., Overton, J.C., Szeto, R., Patel, M.H., Gutierrez, D.M.R., Eby, C., Martínez Moreno, A.M., Erk, K.A., Aston, J.E., Thompson, D.N., Dooley, J.H., Sharma, P., Mosier, N.S., Ximenes, E., Ladisch, M.R., 2021. New strategy for liquefying corn stover pellets. *Bioresour. Technol.* 341, 125773. https://doi.org/10.1016/j.biortech.2021.125773

dos Santos, A.C.F., Ximenes, E., Thompson, D.N., Ray, A.E., Szeto, R., Erk, K., Dien, B.S., Ladisch, M.R., 2020. Effect of using a nitrogen atmosphere on enzyme hydrolysis at high corn stover loadings in an agitated reactor. *Biotechnol. Prog.* 36, 1–10. https://doi.org/10.1002/btpr.3059

Kim, D., 2018. Physico-chemical conversion of lignocellulose: Inhibitor effects and detoxification strategies: A mini review. *Molecules* 23(2). https://doi.org/10.3390/molecules23020309

Kim, D., Ximenes, E.A., Nichols, N.N., Cao, G., Frazer, S.E., Ladisch, M.R., 2016. Maleic acid treatment of biologically detoxified corn stover liquor. *Bioresour. Technol.* 216, 437–445. https://doi.org/10.1016/j.biortech.2016.05.086

Kim, Y., Ximenes, E., Mosier, N.S., Ladisch, M.R., 2011. Soluble inhibitors/deactivators of cellulase enzymes from lignocellulosic biomass. *Enzyme Microb. Technol.* 48(4–5), 408–415. https://doi.org/10.1016/j.enzmictec.2011.01.007

Mhlongo, S.I., den Haan, R., Viljoen-Bloom, M., van Zyl, W.H., 2015. Lignocellulosic hydrolysate inhibitors selectively inhibit/deactivate cellulase performance. *Enzyme Microb. Technol.* 81, 16–22. https://doi.org/10.1016/j.enzmictec.2015.07.005

Østby, H., Hansen, L.D., Horn, S.J., Eijsink, V.G.H., Várnai, A., 2020. Enzymatic processing of lignocellulosic biomass: Principles, recent advances and perspectives. *J. Ind. Microbiol. Biotechnol.* https://doi.org/10.1007/s10295-020-02301-8

Tramontina, R., Brenelli, L.B., Sodré, V., Franco Cairo, J.P., Travália, B.M., Egawa, V.Y., Goldbeck, R., Squina, F.M., 2020. Enzymatic removal of inhibitory compounds from lignocellulosic hydrolysates for biomass to bioproducts applications. *World J. Microbiol. Biotechnol.* 36(11), 1–11. https://doi.org/10.1007/s11274-020-02942-y

Tumuluru, J.S., 2014. Effect of process variables on the density and durability of the pellets made from high moisture corn stover. *Biosyst. Eng.* https://doi.org/10.1016/j.biosystemseng.2013.11.012

Ximenes, E., Kim, Y., Mosier, N., Dien, B., Ladisch, M., 2011. Deactivation of cellulases by phenols. *Enzyme Microb. Technol.* 48(1), 54–60. https://doi.org/10.1016/j.enzmictec.2010.09.006

Xin, D., Chen, X., Wen, P., Zhang, J., 2019. Insight into the role of α-arabinofuranosidase in biomass hydrolysis: Cellulose digestibility and inhibition by xylooligomers. *Biotechnol. Biofuels* 12, 1–11. https://doi.org/10.1186/s13068-019-1412-0

11 Research Progress on Enzyme–Substrate Mixing by Periodic Forces during High-Solids Enzymatic Hydrolysis of Lignocellulose

Xin Shi, Lan Wang, and Hongzhang Chen

11.1 INTRODUCTION

The high-solids biological reaction of natural bioresources as a green route is a development trend in biomanufacturing. Typically, conventional bioreaction systems based on enzyme catalysis and microbial fermentation have a low solids content, generally no more than 5%. Low-concentration bioreaction systems imply low substrate concentrations and low product concentrations, as well as lower apparent efficiency and higher operational energy consumption. In addition, low product concentrations increase the cost of downstream product separation and generate large amounts of biological wastewater. If the substrate concentration of the reaction system is increased, the problems of high water and energy consumption, and high wastewater generation can be effectively mitigated. It was reported that about 10% of overall operation costs might be reduced when the solids loading is increased from 20% to 30% (Lu et al., 2010). In the case of ethanol production from starch, for example, the wastewater is reduced from nine tons to seven tons in a high-solids system, reducing emissions by more than 20% and reducing steam consumption by one-third.

For lignocellulose, a process is considered "high solids" if the ratio of solids to liquid is such that very little to no free water is present in the slurry or roughly a solids loading ≥15% (w/w) (Modenbach & Nokes, 2013). The system is called a high-solids and multiphase bioprocess system, which is composed of solid, liquid and gas phases with the solid phase as the continuous phase. High-solids enzymatic hydrolysis of lignocellulose, which is a necessary step for the preparation of bio-based chemicals and materials, has many advantages such as high product concentration, high equipment utilization, low product separation cost, and low water and energy consumption. In the traditional lignocellulosic ethanol industry, the low fermentation concentration of ethanol would lead to high operation costs, making it difficult to achieve the economic operation target of ethanol distillation (Ramos et al., 2015). Increasing the initial ethanol concentration from 1% to 5% in a single distillation unit with a final concentration of 94.5% (w/w) has been reported to reduce energy demand by about two-thirds (Zacchi & Axelsson, 1989). In the production process of lignocellulosic ethanol, a fermented ethanol concentration for distillation must be higher than 4% (w/w), requiring the solids loading to be more than 20% (Zacchi & Axelsson, 1989; Larsen et al., 2008) to ensure that a corresponding sugar concentration is produced.

Although high-solids and multiphase bioprocess systems have many advantages compared with low-solids systems, new problems arise when the solids loading increases. Lignocellulose at a high-solids loading causes a "solid effect" in the multiphase system, which hinders the mass

 DOI: 10.1201/9781003335313-13

transfer and enzymatic hydrolysis efficiency. Water does not only act as a reactant in the high-solids enzymatic process but also plays an important role as a lubricant. The water content of the reaction system determines the mixing process. When the solids loading increases, the lignocellulosic mixture exhibits shear thinning of a non-Newtonian fluid, and its viscosity increases with the increase in solids loading (Russ et al., 2015). When the viscosity of the reaction system increases, its mass transfer efficiency decreases (Du et al., 2017). This is because the increase in viscosity makes the contact between the cellulase and the substrate more difficult. Also, the products accumulate in the vicinity of the cellulase, which affects the catalytic efficiency of the enzyme. Kristensen et al. (2009) demonstrated that mass transfer efficiency is a key factor limiting the high-solids enzymatic process by comparing the enzymatic efficiency at different solids loadings. Meanwhile, Du et al. (2017) observed that higher solids content led to a decrease in enzymatic hydrolysis rate at the beginning of the enzymatic process, even at very low glucose concentrations. This is further evidence that mass transfer efficiency is indeed a key limiting factor in the high-solids enzymatic hydrolysis process. Therefore, to improve the efficiency of high-solids enzymatic hydrolysis, targeted enhancement of the mass transfer process of high-solids enzymatic hydrolysis is required.

Recognizing the mass transfer law of a high-solids reaction system and clarifying the rate-limiting steps of kinetics, we can design targeted enhancement measures to improve the reaction efficiency. Taking the lignocellulosic high-solids enzymatic hydrolysis process as an example, the limiting rate of enzymatic hydrolysis of cellulase is 14 glycosidic bonds/s, which means that its catalytic reaction is completed at the moment of contact with the substrate. Therefore, cellulose breakage is not the rate-limiting step of the enzymatic hydrolysis process (Luterbacher et al., 2013; Arantes & Saddler, 2011). The true rate-limiting step of the enzymatic hydrolysis process is the entry of the enzyme through the dense spatial structure of the lignocellulose to the catalytic point on the fiber surface (Arantes & Saddler, 2011). By intensifying the mixing process between enzyme and substrate, the time required for the two to come into contact can be shortened, thus increasing the efficiency of enzymatic hydrolysis. Therefore, cognitive enzyme mass transfer laws in high-solids reaction systems, improving reaction mixing efficiency, shortening enzyme–substrate mixing time and developing effective process measures have become the keys to improving the efficiency of high-solids enzymatic hydrolysis of lignocellulose. This chapter addresses the progress of the aforementioned research.

11.2 ENZYME–SUBSTRATE MIXING CHARACTERISTICS DURING HIGH-SOLIDS ENZYMATIC HYDROLYSIS OF LIGNOCELLULOSE

Enzyme–substrate mixing is a process that makes enzymes dispersed in the substrate to achieve uniform distribution. Mixing is a common process in chemical engineering, and the pharmaceutical and food industries. The process is divided into liquid–liquid, liquid–solid, gas–liquid, solid–solid, gas–solid and gas–gas mixing. In the pharmaceutical industry, small amounts of active drug ingredients are mixed with large quantities of excipients for a pharmaceutical dosage (Portillo et al., 2008). Content uniformity in drugs is vital to ensure safety and efficacy (Staniforth & Rees, 2011; Bánfai et al., 2007). Inefficient blending can lead to increased variability in active components in the final dosage form, which is not potentially beneficial to patient health (Freeman et al., 2012). For lignocellulose, the enzymes and active drug ingredients are small in quantity, and the substrates are large in amount like excipients. Enzyme–substrate mixing is similar to trace liquid and bulk liquid mixing at a low-solids loading, while at a high-solids loading, it is similar to trace liquid and bulk solid mixing due to the lack of free water in the reaction system. The high-solids enzymatic process is a very complex biological reaction process but also the result of the interaction of the solid–liquid–gas phase. To better analyze the high-solids enzymatic process, we divide it into a mixing stage and an enzymatic hydrolysis phase. However, in the actual process, when the enzyme is added to the substrate, the two processes, mixing and enzymatic hydrolysis, occur almost simultaneously and the two stages affect each other.

11.2.1 Characteristics of Mixing Trace Enzyme and Bulk Substrate

Mixing usually starts with large-scale movements to homogenize the material on a large scale, while the small-scale structure allows the contact surface between the materials to increase rapidly. The final mixing is completed by molecular diffusion or the diffusion and permeation of different components of the material to produce the desired chemical reaction.

A mixing index that can characterize enzyme and substrate mixing is important. This can provide a theoretical basis for the development of effective process measures to achieve uniform mixing. Generally speaking, the value of a mixing index is between 0 and 1, with 0 representing no mixing process occurring and 1 representing complete homogeneous mixing (Cho et al., 2017). Various types of mixing indices can be indicative of the status of mixing, which are primarily classified into two categories (Cho et al., 2017). The first category is based on statistical analysis using standard deviation from spot samplings (Williams, 1968; Lacey, 1954; Feng et al., 2004), while the other category uses all particles in a system as opposed to a few particles from samplings (Siraj et al., 2011; Chandratilleke et al., 2012; Asmar et al., 2002). Many mixing indices based on statistical analysis considering the system of binary mixtures were studied by a number of researchers, including Lacey (1954), Kramer (1968) and Ashton and Valentin (1966). The Lacey mixing index was introduced in 1943 by P.M.C. Lacey, and is well-known as a classical mixing index.

Lacey (1954; Chou et al., 2017) divided the area of interest into N cells and assumed that $\varnothing_i$ is the concentration of the reference component in the ith cell and $\varnothing_m$ is the overall concentration of the reference component. The variance of σ^2 for the concentration of a reference component in each cell can be expressed by

$$\sigma^2 = \sum_{i=1}^{N} \frac{\left(\varnothing_i - \varnothing_m\right)^2}{N-1} \tag{11.1}$$

where N is the total number of cells occupied by the total particles, $\varnothing_i$ is the local concentration of tracer particles in each cell and $\varnothing_m$ is the average concentration of tracer particles in the whole system. Theoretically, for a binary-mixture solid system, the maximum value of mixture variance, σ_0, can be expressed by

$$\sigma_0^2 = \varnothing_m\left(1 - \varnothing_m\right) \tag{11.2}$$

The possible minimum variance in a cell, σ_r, is defined by

$$\sigma_r^2 = \frac{\varnothing_m\left(1 - \varnothing_m\right)}{n} \tag{11.3}$$

where n denotes the possible number of particles in a cell. Then the Lacey mixing index M can be defined by

$$M = \frac{\sigma_0^2 - \sigma^2}{\sigma_0^2 - \sigma_r^2} \tag{11.4}$$

Given the concentrations of the reference component in each cell, the mixing index can be defined and calculated by the statistical method. The Lacey index has an availability problem in response to the size dependence of the extracted sample, which leads to erroneous results when the geometrical size of the components of a binary mixture is not consistent (Cho et al., 2017). The critical disadvantage of mixing indices based on standard deviation or variance is that they lack statistical accuracy in systems without enough samples. In some cases, the Lacey mixing index may be greater than 1, causing several problems (Chou et al., 2017). In addition, they vary with sampling size and sampling methods (Holdich, 2020). Therefore, mixing indices based on statistics suffer from the aforementioned limitations. During the high-solids enzymatic hydrolysis process, it may not be necessary that the substrate particles are uniformly distributed throughout the reactor, but only that the enzyme molecules are homogeneously distributed inside the particles. Therefore, the Lacey mixing index may not be suitable for describing the high-solids enzymatic hydrolysis process.

The experimental method of examining the particle distribution using the coefficient of variation gives an idea of the mixing degree of the system (McFarland et al., 1999; Balami et al., 2013; Van Leeuwen et al., 1996; Egermann et al., 1985). For the system with lignocellulose as the substrate and cellulase as the enzyme, the mixing process is the mixing of trace cellulase with bulk substrate. To distinguish the mixing and enzymatic stages, the mixing of cellulase and substrate can be simulated by a cold-mode test. Bovine serum albumin (BSA) was chosen as a model substance for the test due to its single component suitable for the determination of protein content. Also, BSA as a non-catalytic protein has a similar molecular weight to the major cellulases (e.g., Cel7A, Cel6A and Cel7B) expressed by *Trichoderma reesei*, the industrial cellulase production source (Foreman et al., 2003; Rosgaard et al., 2007b; Lavenson et al., 2011), which was suitable for the cold-mode test. Shi et al. (2022) used BSA instead of cellulase to mix with lignocellulosic substrate at different solids loadings, and multiple samples of fixed mass were randomly taken at different mixing times to determine the amount of BSA in the substrate, and the amount of BSA in different samples was statistically analyzed to calculate the coefficient of variation (*Cv*). The coefficient of variation can well reflect the mixing degree of enzyme and substrate. Shi et al. (2022) divided the mixing process into two stages according to the decreasing trend of the *Cv*. In the first stage, the value of the *Cv* decreases quickly, which means that the enzyme and the substrate mix well at the beginning, while in the second stage, the value of the *Cv* tends to slow with time, which means that the mixing effect is poor at this stage. The trend of the value of the *Cv* is similar but the change rate of the value of the *Cv* is quite different for different solids loadings. The higher the solids loading, the smaller the change rate of the *Cv* value, indicating that the enzyme and substrate mixing process is more difficult. The mixing of enzyme and substrate is mainly divided into two stages: macroscopic mixing and microscopic mixing. Macroscopic mixing refers to reactor-scale mixing, while microscopic mixing includes pore-scale mixing and molecular-scale mixing (Dutta & Chakraborty, 2019). The molecular-scale mixing is the same at different solids loadings at a constant temperature. Based on the fact that lignocellulose as a substrate is a porous material, in the absence of free water, it is hard for the enzyme to diffuse out due to van der Waals forces of the pore wall at high-solids loading (Singh et al., 2019). For low-solids loading systems, water is a continuous phase, which facilitates mass transfer. For high-solids loading systems, the substrate as a solid phase is a continuous phase and water is the discontinuous phase, which is not conducive to mass transfer. It can be seen that microscopic mixing, also known as pore-scale mixing, is the rate-limiting step. To achieve the same mixing degree, it took more than 50 times longer for a system at high-solids loading compared to a low-solids loading system.

11.2.2 Characteristics of Enzymatic Hydrolysis at High-Solids Loadings

11.2.2.1 Rheological Properties at High-Solids Loadings

To enhance mass transfer in high-solids systems, it is important to improve the rheological properties of the system, which is beneficial to enzymatic hydrolysis efficiency, process parameters and equipment selection (Lou et al., 2014). Rheology is the study of the deformation and flow of objects under the action of external forces and is mainly concerned with fluids, as well as soft solids or solids that can flow under certain conditions rather than elastic deformation. It applies to substances with complex structures, including slurries, sludges, suspensions, polymers, foods and other biological materials. The rheological properties during the high-solids enzymatic hydrolysis of lignocellulose are affected by the particle size and distribution, fiber flexibility, physical and chemical characteristics of solid, and pretreatment conditions. Viscosity and yield stress are the two important parameters in the rheological properties of high-solid enzymatic hydrolysis. Many rheological models, such as the Bingham, Herschel–Buckley, power law, Wildemuth–Williams, Casson and Vipulanandan models (Wildemuth & Williams, 1984; Roche et al., 2009; Viamajala et al., 2009; Vipulanandan & Mohammed, 2014; Sharoba et al., 2005; Sahin & Ozdemir, 2004), have been developed to describe the non-Newtonian behavior of high-solids mixtures.

The Wildemuth–Williams semi-empirical model (Wildemuth & Williams, 1984; Roche et al., 2009) is used to describe the relationship between yield stress (τ_y) and the volume fraction (ϕ)

$$\tau_y(\phi) = [A\left(\frac{\phi}{\phi_{m0}} - 1\right) / (1 - \phi / \phi_{m\infty})]^{\frac{1}{m}} \tag{11.5}$$

where ϕ_{m0} is the maximum volume fraction without shear stress, $\phi_{m\infty}$ is the maximum packing fraction in the limiting of infinite shear stress, and A and m are related to microstructure changes of the system under mechanical shearing.

The Casson model (Viamajala et al., 2009) is an empirical model given by

$$\tau^{0.5} = \tau_{Cy}{}^{0.5} + K_C \gamma^{0.5} \tag{11.6}$$

where τ_{Cy} is the apparent Casson yield stress (Pa), γ is the shear rate and K_C is a Casson constant (Pa^2s^2).

The Bingham model (Viamajala et al., 2009) is given as

$$\tau = \tau_{By} + K_B \gamma \tag{11.7}$$

where τ_{By} is the Bingham yield stress, γ is the shear rate and K_B is the Bingham viscosity.

The Vipulanandan model proposed by Vipulanandan and Mohammed (2014) is

$$\tau = \tau_0 + \frac{\gamma\cdot}{A + D\gamma\cdot} \tag{11.8}$$

where τ_0 is the yield stress (Pa), $\gamma\cdot$ is the shear rate (s^{-1}), and A ($[Pas]^{-1}$) and D (Pa^{-1}) are model parameters or constants, respectively.

The power law model (Sahin & Ozdemir, 2004) is given as

$$\tau = k\gamma\cdot^{n} \tag{11.9}$$

where τ is the shear stress (Pa), τ_0 is the yield stress (Pa), $\gamma\cdot$ is the shear rate (s^{-1}) and k is the consistency coefficients (Pa s^n).

The Herschel–Buckley (Sharoba et al., 2005) model modifies the power law model, which is

$$\tau = \tau_0 + k^{'\gamma\cdot^{n}} \tag{11.10}$$

where τ is the shear stress, $\gamma\cdot$ is the shear rate (s^{-1}) and k' is the consistency coefficients (Pa s^n).

11.2.2.2 Water Properties at High-Solids Loadings

The viscosity is closely related to the water properties within the high-solids system. When water migrates to lignocellulose, it is influenced by the chemical components and physical structure of different lignocelluloses and interacts with them differently, and the state and location of water differs (Felby et al., 2008). The water molecules associated with the characteristics of lignocellulose are again very dissimilar to the properties of free water in the system. Compared to free water, they become localized, structured and relatively limited in terms of available degrees of freedom, kinetic motion and the ability to exchange with other water molecules (Selig et al., 2014). According to the definition of high solids, when the solid–liquid ratio exceeds 15%, the water bound by lignocellulose becomes bound and capillary water, resulting in the absence of free water in the system, which is known as the "water-binding effect" of high-solids enzymatic hydrolysis (Roche et al., 2009; Deng et al., 2015; Nicholls, 2015). In the high-solids system, there are five main forms of water: primary bound water, secondary bound water, capillary water, gravimetric water and free water. The binding effect of water is closely related to the biological properties, chemical components and physical structure of lignocellulosic materials. The decrease of free water changes the rheological properties such as the viscosity of the system, which in turn decreases the diffusion efficiency of the enzyme. The reduced water content limits the diffusion of sugar from the active site of the enzyme, inhibiting glycation and preventing the enzyme from reaching the new catalytic site (Himmel et al., 2007), which greatly reduces the enzymatic hydrolysis efficiency. At the same time, the increase in the viscosity increases the requirement for mixing yield stress and the energy consumption of mixing (Nicholls, 2015; Deng et al., 2015; Viamajala et al., 2009).

Roberts et al. (2011) investigated the effect of the combination of water and cellulose on mass transfer and enzymatic hydrolysis efficiency. As the solids loading increased, the physical constraint of water was increased. The addition of glucose and mannose increased the water-binding effect. The inhibition of the enzyme hydrolysis rate and the increase in water binding caused by monosaccharides showed the same trend. The relationship between the water-binding effect induced by macromolecular polymers and its inhibition of enzyme hydrolysis was studied by Selig et al. (2014). The relaxation time profile (T2) of the added polymer system measured with low-field nuclear magnetic resonance approached zero for a short time and the conversion of cellulose decreased due to the added polymer. This trend is consistent with the reduction in the water-binding effect and conversion, demonstrating that the water-binding effect reduces the efficiency of high-solids enzymatic hydrolysis (Selig et al., 2014).

According to the law of bound water changes, Zhang et al. (2020) divided the high-solids enzymatic hydrolysis process of lignocellulose into three stages: the first phase is the solid stage, the second phase is the transition stage and the third phase is the liquid stage. In the first stage, the solid phase in the high-solids system is continuous, and water exists in the form of bound water inside the porous medium, which cannot form a continuous phase, leading to poor mass transfer. Strengthening the mass transfer with bound water as a carrier is the key to improving the efficiency of high-solids enzymatic hydrolysis. In the second stage, the system releases part of bound water and a small amount of free water exists in the system, making the whole substrate present a slurry state. In this stage, it is necessary to continue to strengthen the mass transfer of bound water and enhance the mass transfer of free water in the system at the same time. In the third stage, the lignocellulosic substrate is completely liquefied and almost all bound water is converted to free water. The porous structure of the substrate is destroyed and water becomes a continuous phase, and the mass transfer of free water transfer needs to be enhanced at this stage. From the division of the three stages of high-solids enzymatic hydrolysis, it can be seen that the initial period of enhancing the mass transfer of bound water is important during the high-solids enzymatic hydrolysis. Wang et al. (2022) used the inflection point where the substrate particle size, viscosity and yield stress of the high-solids enzymatic hydrolysis system decreased with time as the starting point for the system to enter the liquid stage, i.e., the pumpable state of the substrate. Similarly, Shi et al. (2022) determined the inflection point of the hardness of the system with time as the liquefaction point, which is also the transition point from high-solids enzymatic hydrolysis to low-solids enzymatic hydrolysis. According to the division of mixing stages of a high-solids system, Shi et al. (2022) examined that the lignocellulosic high-solids enzymatic hydrolysis process was similarly divided into two stages at the reactor scale and pore scale, and its enzymatic efficiency showed a positive correlation with mixing efficiency, i.e., enhanced accessibility of enzyme-substrate contact was the core element while keeping the enzyme activity constant. This division specifies the concept of microscopic mixing for lignocellulose with an intrinsic property of substrate porosity and suggests that the limiting factor of mass transfer lies in the pore-scale mass transfer. The definition of enzyme accessibility to the substrate focuses on the enzyme–substrate interfacial interaction, but even if the two are accessible, cellulase and hydrolysate do not migrate efficiently due to the physical spatial structure of lignocellulose as a substrate, based on the capillary effect it produces, which can lead to a significant decrease in cellulase utilization. At the same time, the hydrolysis products do not migrate in time, which in turn inhibits the activity of cellulase and the enzymatic reaction. Lignocellulose is a natural class of porous media (Zhao & Chen, 2013). The relevant properties of porous media are presented next to further illustrate how they affect mixing efficiency.

11.2.2.3 Porous Media Properties at High-Solids Loadings

In simple terms, porous media is a solid with pores. The pore is the distinctive feature of porous media, which is the existence of porous media form. The degree of pore development is measured by porosity. Generally, a porosity between 0 and 0.3 is a dense medium; a porosity between 0.3 and 0.9 is a loose medium. The tiny voids within a porous medium may be interconnected, or partially connected and partially unconnected. As a representative class of lignocellulose, Zhao and Chen (2013) summarized the highly ordered and compact structure of corn stover formed by porous structure at three levels: the tissue level, cell level and cell wall level. Pores at the tissue level consist mainly of different cellular tubular lumens and intercellular spaces (1–130 μm). Pores at the cellular level point to pits in the cell wall, especially in the vessel cell wall (30 nm–50 μm). The diffusion of fluid throughout plant tissues and cells is controlled by the arrangement of vascular bundles and the pits connecting the cells to each other. The pores of the cell wall layer consist of spaces between polymers, including lignin, hemicellulose, cellulose and ultrafine fibers. Atomic force microscopy images showed that the pores in the primary and secondary walls ranged from 1 to 30 nm (Ding et al., 2012). Meanwhile, Zhao and Chen (2013) analyzed the relationship between porous properties of lignocellulose and mass transfer and the relationship between porous structure and the efficiency of enzymatic hydrolysis.

11.2.2.3.1 Relationships between Porous Properties and Percolation

Hyperosmosis is the science of probabilistic theory and methods to study and characterize the structural shape and phenomena associated with the critical conditions and critical states of porous media from complete impermeability to permeability. Enzymatic hydrolysis first requires that the water be able to osmose through the porous structure of lignocellulose substrate and reach saturation. The percolation probability can be determined from the critical percolation function (Zeng et al., 2012), which is shown as $f(\phi, N_0, D)$, where φ is the porosity (%); N_0 is the initial value of the three-dimensional fractal distribution of the fracture face number (that is, when a cube with side length of L_0 is chosen, N_0 would equal to the number of the fracture faces whose areas are larger than N_0^2), which is chosen as $N_0 = e^0 = 1$ here; and D is the three-dimensional fractal dimension of fracture. When $f(\phi, N_0, D) > 0$, there is no percolation; when $f(\phi, N_0, D) < 0$, percolation is possible; and $f(\phi, N_0, D) = 0$ is the critical condition. The smaller this value, the greater the probability of percolation; that is, the greater the possibility of water forming a mass transfer path in it.

11.2.2.3.2 Relationships between Porous Properties and Threshold Pressure

The threshold pressure is the minimum hydraulic gradient required for a fluid to pass through a saturated capillary pore, i.e., the initial hydraulic gradient at which a saturated porous medium begins to percolate, and flow will not occur at a value less than this hydraulic gradient (Qin & Zhang, 1996).

According to the definition of threshold pressure, the relationship between threshold pressure and capillary force is as follows (Kaptay, 2008):

$$p_{th} = -p_c \tag{11.11}$$

where p_{th} is the threshold pressure (Pa) and p_c is the capillary pressure (Pa).

From the pore size and interfacial hydrophilic analysis, the capillary force is calculated as follows (Yu, 2011; Zhang et al., 2012):

$$p_c = \frac{2\sigma \cos\theta}{r_0} \tag{11.12}$$

where σ is the surface tension of liquid (*N/m*), θ is the contact angle (°) between liquid and the surface of the capillary, and r_0 is the capillary radius (*m*).

From the analysis of the specific surface area and the surface tension of the liquid, the capillary force can also be calculated by the following equation (Kaptay, 2008):

$$p_c = \sigma \times S_v \tag{11.13}$$

where S_v is the mass-specific surface area on the volume of the capillary (m^2/m^3).

From the above formula, it is known that the parameters affecting the threshold pressure include the surface tension of the fluid, the interfacial contact angle of the fluid, the pore size distribution of the porous medium and the surface area of the porous medium.

11.2.2.3.3 Relationships between Porous Properties and Seepage

Seepage is the process of flow of a saturated liquid in a permeable medium, which is different from diffusion. Diffusion is the movement of a solute in a solvent from a region of high concentration to a region of low concentration. The rate of percolation can be calculated by the Darcy equation:

$$Q = \left(\frac{k}{\mu}\right) A_s \Delta P / L \tag{11.14}$$

Where Q is the volume flow of liquid in the porous medium (m^3/s), k is the permeability of the porous medium (m^2), μ is the fluid viscosity (Pa s), A_s is the cross-sectional area to flow (m^2) and $\Delta P / L$ is the hydraulic pressure gradient of fluid in the porous medium (Pa/*m*).

According to the Kozeny equation, the equation of permeability can be expressed as follows:

$$k = c\varphi^3 / \left(TS_v^2\right) \tag{11.15}$$

where c is the shape factor and T is the tortuosity of the porous medium. S_v is calculated as follows:

$$S_v = S_g \times \rho \times 10^6 \tag{11.16}$$

where S_v is the specific surface area based on weight (m^2/g) and ρ is the bulk density of the porous medium (g/mL).

11.2.2.3.4 Relationships between Porous Properties and Enzymatic Hydrolysis Efficiency

According to the mass transfer phenomena in capillary channels of porous media (Lago & Araujo, 2003), the initial percolation process of the enzyme-containing buffer during enzymatic hydrolysis is the process of repelling the air phase in the capillary of the lignocellulosic substrate, which acts as a wetting agent for the material. Therefore, the capillary forces between the air and the capillary and between the enzyme-containing buffer and the capillary plus the external pressure are the main influencing factors. After reaching saturation, the percolation process turns into the seepage process. As the number of contact sites in a capillary decreases, the original solution in the capillary needs to be replaced by a new one for the cellulase molecules to reach the next enzymatic site and for the degradation products to move away from the cellulase in a reasonable time so that the threshold pressure value again becomes the main parameter affecting the enzymatic hydrolysis. If in the process of percolation, it mainly depends on the wetting phase breaking through the capillary force formed between the non-wetting phase and the capillary, then in the process of seepage, since the wetting phase is the same, it primarily relies on the action of external forces breaking through the threshold pressure. It can be seen that the higher the threshold pressure value, the lower the efficiency of enzymatic hydrolysis. This is the same as the conclusion from studies on mass transfer in porous media, that is, the most important factor affecting seepage is the capillary threshold pressure (Lago & Araujo, 2001; Chalbaud et al., 2009; Kaptay, 2008). Therefore, it is scientific to consider the mass transfer process of enzymatic hydrolysis from the perspective of porous media. Based on this analysis, Zhao and Chen (2013) proposed the concept of seepage recalcitrance for enzymatic hydrolysis of lignocellulosic biomass based on the resistance to mass transfer generated by porous media. The seepage recalcitrance can be quantitatively characterized by the threshold pressure value, and the larger the threshold pressure value, the larger the seepage recalcitrance. However, due to the large number of lignocellulosic species and individualized differences, they all play an important role in the formation of lignocellulosic porous structures at the tissue level, cellular level and molecular level, resulting in different shapes and different orientations of the pore structures. Meanwhile, the porous structure changes accordingly with the reaction process during the enzymatic hydrolysis. Therefore, the source of threshold pressure values caused by the porous structure is complex and dynamic.

To break the bottleneck, high-solids enzymatic hydrolysis needs to adopt corresponding measures. Based on the aforementioned theory, in the process of enzyme hydrolysis, if stirring is used, at a certain liquid–solid ratio, if the initial shear force has a normal directional force on the capillary, then as the enzyme hydrolysis proceeds, the particles become smaller and smaller, in which the capillary will be arranged in the tangential direction of the shear force under the action of the stirring field, at which time a relatively strong shear force is required to achieve the later improved enzyme hydrolysis efficiency.

According to the formula of capillary force (Kaptay, 2008; Yu, 2011; Zhang et al., 2012), the liquid surface tension, the pore size of porous media and the contact angle between liquid and solid directly affect the magnitude of capillary force. It has also been shown that there is a relationship between capillary tortuosity, fractal dimension, and specific surface area and capillary force (Lago & Araujo, 2003; Kaptay, 2008). It is thus clear that the porous medium properties of the enzymatic substrate affect enzymatic hydrolysis by influencing the threshold pressure value during the enzymatic hydrolysis process. To improve the efficiency of enzymatic hydrolysis, strengthening measures can be taken in two aspects: first, the porous structure parameters of lignocellulose can be improved by pretreatment, thus reducing the threshold pressure and improving the efficiency of enzymatic hydrolysis; second, during the enzymatic hydrolysis, a medium with lower surface tension can be used to reduce the threshold pressure value, but cellulase requires a special buffer system. Meanwhile, external reinforcement can be used to promote mass transfer in the capillary, especially in the early stage of enzyme hydrolysis. Promoting the repulsion of gas in the capillary by enzyme-containing buffer solution is the key to improving the rate of enzymatic hydrolysis, while in the late stage of enzyme hydrolysis, the porous structure of lignocellulosic substrate is destroyed and capillary water mass transfer is converted to free water mass transfer.

From the preceding analysis, it can be seen that it is reasonable whether it is divided into two stages (reactor scale and pore scale), or three stages (solid stage, transition stage and liquid stage) during the high-solids enzymatic hydrolysis process. The mixing process and enzymatic hydrolysis process promote each other, and in the high-solids system, mass transfer is a prerequisite for enzymatic hydrolysis, and enzymatic hydrolysis further promotes effective mass transfer.

11.3 DEVELOPMENT OF HIGH-SOLIDS ENZYMATIC HYDROLYSIS INTENSIFICATION MEASURES

In the traditional chemical industry, the mass and heat transfer of solid substrates is generally solved by mechanical stirring with the aim of mixing water and solid particles well and accelerating the reaction. However, as the interior of the particles is not affected by macroscopic stirring, mechanical stirring has almost no effect on the microscopic scale (Chen, 2018). In addition, mechanical stirring consumes more energy and affects enzyme activity (Zhang et al., 2020; Wang et al., 2022). Pressure intensification as an external field force is usually effective for reaction systems where mass transfer is limited. For example, the stomach and intestine can efficiently mix food and digestive fluids (mainly digestive enzymes) by periodic expansion and contraction, which is known as periodic peristalsis. Periodic peristalsis differs from conventional agitation in that it enhances mass transfer primarily through high turbulence intensity rather than high velocity (Chen, 2018). For lignocellulose with pores, periodic force enhancing mass transfer in high-solids enzyme hydrolysis is an effective measure.

11.3.1 Periodic Force Improves High-Solids Enzymatic Hydrolysis Efficiency in the Premixing Stage

Since mixing of trace enzyme and bulk substrate is the rate-limiting step for mass transfer at high-solids loading, it is more effective to place mixing as an operational unit before the enzymatic hydrolysis process, so that the enzyme and substrate are well mixed before enzymatic hydrolysis. The necessity of premixing operation was examined by Shi et al. (2022), who found that the dosage of the enzyme used was saved by 87.5% when the same enzymatic efficiency was achieved at 20% solids loading after sufficient premixing compared to that without premixing. The lower the enzyme dosage used, the more necessary the sufficient premixing. When the enzyme dosage was 5 FPU/g DM (dry matter), the efficiency of enzymatic hydrolysis can be increased by nearly 200% with sufficient premixing. At the same time, the liquefaction rate of the high-solids system was accelerated after sufficient premixing, and the inflection point was reached after 24 hours.

Shi et al. (2022) developed a measure to enhance high-solids enzymatic hydrolysis efficiency by periodic high-frequency vibratory premixing. The higher the solids loading, the higher the improvement of high-solids enzymatic hydrolysis efficiency by periodic vibration premixing. In contrast, premixing by shaking was not effective in enhancing high-solids enzymatic hydrolysis efficiency and was almost ineffective. This is because being in a fluid with oscillatory flow in a porous medium, mass transfer is greatly improved (Dey & Raja Sekhar, 2014). Substrates can roll rapidly and collide with each other under periodic vibrational forces, and trace enzyme and bulk substrate mix better. However, the main force of shaker premixing is shear force, which has little utility for substrates with high viscosity, and under this force, capillaries are aligned in their tangential direction, which does not enhance capillary mass transfer, and trace enzymes can only diffuse gradually. The diffusion is very slow due to the porous nature of the substrate.

The periodic force can enhance mass transfer not only in the premixing stage but also in the high-solids enzymatic hydrolysis process by vibration or peristaltic motion to enhance the enzymatic efficiency.

11.3.2 Periodic Force Improves High-Solids Enzymatic Hydrolysis Efficiency in the Intensification Stage

11.3.2.1 Periodic Peristalsis

Liu and Chen (2016) and Wang et al. (2022) have demonstrated that periodic peristalsis can enhance mass transfer and promote high-solids enzymatic hydrolysis efficiency.

11.3.2.1.1 Periodic Peristalsis Has Little Effect on Enzymatic Activity

One of the important issues in the coupling of mixing and enzymatic processes in high-solids systems is the possibility of mixing measures to lower the enzymatic activity. Liu and Chen (2016) confirmed that mechanical stirring significantly reduced the enzymatic activity, which further decreased with the increase in stirring rate and mixing time. The reason for this may be, on the one hand, that stirring enhanced the interaction between enzyme and air, and the nature of the enzyme interface changed, which affected its spatial structure and consequently the activity (Cao & Tan, 2004; Kaya et al., 1994). On the other hand, stirring may have changed the thermal denaturation of the enzyme, producing an aggregation effect (Bose et al., 2012; Kudou et al., 2014). However, the higher the enzyme loading, the less the shear force generated by stirring affected the enzymatic activity (Cao & Tan, 2004). It can be seen that shear forces can disrupt the enzyme structure and lead to enzyme inactivation. In contrast, the periodic force generated by periodic peristalsis has less effect on the enzyme activity and favorably guarantees the high-solids enzymatic hydrolysis.

11.3.2.1.2 Periodic Peristalsis Causes a Rapid Decrease in Substrate Particle Size, Viscosity and Yield Stress

The substrate particle size reflects the degree of lignocellulose fragmentation, substrate homogeneity and viscosity. The substrate particle size gradually becomes smaller as the enzymatic hydrolysis proceeds, and the smaller the particle size the more homogeneous the system and the better the mass transfer, and thus the higher the enzymatic hydrolysis efficiency. Wang et al. (2022) found that the trends of particle size changes were very similar in the static state and under stirring, indicating that the shear force generated by stirring had little effect on the particle size changes. However, periodic peristalsis accelerated the particle size reduction, showing that periodic peristalsis can significantly increase the rate of fiber breakage and particle fragmentation. During the enzymatic hydrolysis, the decrease rate of D50 under periodic peristalsis could reach the maximum at 2 h, which was 1.85 times in the static state and under stirring.

The chemical composition, physical properties (e.g., particle size and porosity) and solids loading of lignocellulose have important effects on the enzymatic hydrolysis of the system (Kristensen et al., 2009; Rosgaard et al., 2007a; Viamajala et al., 2009). Studies by Liu and Chen (2016) and Wang et al. (2022) showed that the viscosity of the system increased with increasing solids loading and that the reduction of viscosity at high-solids loading was later than that at low-solids loading. This indicates that the high-solids loading limits the mass transfer efficiency. As enzymatic hydrolysis proceeded, both the viscosity and yield stress of the system decreased gradually. The viscosity and yield stress of the system decreased more rapidly with cyclic creep than with stirring, shaking and resting conditions. The viscosity at 1 h of periodic peristalsis was lower than that at 10 h in the stirred and static states. The viscosity of the system after 1 h was almost the same as 90% glucose slurry viscosity (Schellart, 2011). After 2 h, the viscosity under periodic peristalsis no longer changed significantly, and the system was basically completely liquefied and became pumpable. Liu and Chen (2016) found that the trend of decreasing the viscosity was consistent with the trend of increasing the glucan conversion, so the system viscosity could characterize the change in enzymatic efficiency to some extent.

11.3.2.1.3 Periodic Peristalsis Reduces Energy Consumption for High-Solids Enzyme Hydrolysis

During the high-solids enzymatic hydrolysis of lignocellulose, the energy consumption of the equipment that promotes enzyme and substrate mixing and enzymatic hydrolysis should be given priority attention, which is related to the technical economy of the whole high-solids enzymatic hydrolysis process. Generally speaking, stirring is the most widely used process in high-solids enzymatic hydrolysis. The stirring power is mainly considered in two aspects (Chen, 2018): the power required to maintain the normal operation of the stirrer, i.e., the stirring operation power; and the power required to perform the stirring operation according to the state of the high-solids system, i.e., the stirrer power. The two powers do not exist independently and cross each other. In the actual evaluation of energy consumption, the energy consumption is usually calculated mainly by the stirrer power, which is also more reasonable because the stirrer power is related to many elements of the reaction system and is used as the basis for energy consumption evaluation. In the case of high-solids systems, all the factors affecting the rheological properties are influencing factors of the stirrer power. The most direct influencing factors are the viscosity and density of the stirred medium, followed by the operating parameters of the stirrer (e.g., rotational speed) and its geometric parameters (e.g., shape and size of the blades, number of blades, mounting height of the blades), as well as the geometric parameters of the stirred tank (e.g., inner diameter of the tank, width of the baffle) (Chen, 2018). As can be seen, the power calculation of the stirrer is extremely complex and requires a combination of theoretical calculations and experimental tests. Wang et al. 92022) compared the power of two processes—stirrer and periodic peristalsis—in high-solids enzymatic hydrolysis. The stirring power can be expressed by the simplified Navier–Stokes equation:

$$P_s = \varnothing \rho n^3 d^5 \left(\frac{n^2 d}{g}\right)^{\left(\frac{a - \log Re}{\beta}\right)} \tag{11.17}$$

$$Re = \frac{\rho n d^2}{\mu} \tag{11.18}$$

where P_s is the stirring power (kgf·m/s); φ is the power coefficient; ρ is the density (kg/m^3); n is the stirring speed; d is the impeller diameter (m); μ is the viscosity (Pa·s); g is the gravitational acceleration (m/s^2); α and β are the coefficient (constant); and Re (Reynolds number) is a dimensionless number that can be used to characterize the flow of a fluid.

As the geometrical parameters of the paddle and tank, and the operating parameters of the paddle were fixed in the high-solids system, Equation 11.18 is substituted into Equation 11.17, and the relation between stirring power and viscosity can be expressed as Equation 11.19:

$$P_s = k_1 \left(k_2\right)^{\log \mu} \tag{11.19}$$

where k_1, k_2 were constant and $k_2 > 1$.

The periodic peristalsis power can be expressed

$$P_p = \delta V \quad . \tag{11.20}$$

where P_p is the periodic peristalsis power (kgf·m/s), δ is the normal stress (kPa) and V is the velocity in the direction of normal stress (m/s).

Wang et al. (2022) went on to experimentally determine the relationship between solids loading and viscosity and between normal stress and viscosity. The stirring power increased exponentially with increasing viscosity, while the viscosity increased exponentially with increasing solids loading. The viscosity and stirring power of a system with 20% solids loading were about 1600 and 200 times higher than those of 1% solids loading, respectively. Zhang et al. (2010) also noted that the stirring energy consumption of high-solids systems accounted for 9%–59% of the energy consumption of ethanol distillation in simultaneous saccharification fermentation. Some researchers pointed out that stirring had little positive effect on slurry mixing and enzymatic hydrolysis when solids loading exceeded 10% (Jørgensen et al., 2007; Baig, 2020). It was evident that stirring did not effectively enhance the high-solids enzymatic hydrolysis process and also required a large amount of energy. Compared to stirring, periodic peristalsis reduced the energy consumption of enzymatic hydrolysis by more than nine times and improved the efficiency of high-solids enzymatic hydrolysis.

11.3.2.2 Periodic High-Frequency Vibration Coupling Rigid Spheres

Zhang et al. (2020) enhanced the mixing process of enzyme and substrate by periodic high-frequency vibration coupling rigid spheres, which significantly accelerated the rate of fiber breakage and particle fragmentation, reduced the crystallinity and viscosity of the substrate, and attenuated the water-binding effect of the substrate, thus improving the efficiency of high-solids enzymatic hydrolysis.

The vibration interval of periodic vibration is an important parameter affecting the mass transfer efficiency of the reaction system. Zhang et al. (2020) found that the glucan and xylan conversions of steam-exploded corn stover after 72 h of high-solids enzymatic hydrolysis were higher at a vibration interval of 10 min than at both a vibration interval of 5 min and 20 min. The efficiency of high-solids enzymatic hydrolysis gradually decreased with increasing the radius of the rigid sphere, which might be because the larger rigid sphere hindered the relative motion of particles and reduced the macroscopic mixing efficiency of the system at the later stage. The addition of different numbers of rigid spheres had little effect on the high-solids enzymatic hydrolysis efficiency. However, periodic high-frequency vibration coupling rigid spheres also reduced the cellulase activity due to the shear force as the enzymatic hydrolysis process proceeded. However, it was shown that compared to the static enzymatic hydrolysis system, the conversion of glucan and xylan was increased by 8.3%–35.0% and 5.0%–33.5%, respectively, using the periodic high-frequency vibration coupling rigid spheres at 15%–25% solids loading, which showed that the importance of efficient mass transfer is much more important than the negative effect from the reduction of enzyme activity in response to the shear force. It was further demonstrated by time-domain nuclear magnetic resonance (TD-NMR) that periodic high-frequency vibration coupling rigid spheres improved high-solids enzymatic hydrolysis efficiency by releasing bound water at the initial stage. Based on the division of high-solids enzymatic hydrolysis stages, it was shown that the efficiency of high-solids enzymatic hydrolysis could only be improved by simultaneously enhancing micro- and macro-mixing in all stages. It was necessary to enhance the high-solids enzymatic hydrolysis efficiency by using different effective measures. In the early stage of enzymatic hydrolysis, periodic high-frequency vibration coupling rigid spheres is more effective for mass transfer in the solid stage; while in the late stage of enzymatic hydrolysis, stirring is more effective for mass transfer in the liquid stage.

11.4 CONCLUSION AND PERSPECTIVES

This chapter cognizes the characteristics of mixing and enzymatic hydrolysis stages at high-solids loadings and elucidates the importance of the rheological and porous properties of substrates and water properties of the system in high-solids enzymatic hydrolysis. Periodic forces such as premixing and process intensification measures break the water binding of the substrate and improve the mass transfer efficiency of the system, which in turn improves the mixing and enzymatic hydrolysis efficiency of the system. Therefore, it is a new strategy for the enhancement of periodic forces in high-solids reaction systems of bioresources.

REFERENCES

Arantes, V. & Saddler, J. N. 2011. Cellulose accessibility limits the effectiveness of minimum cellulase loading on the efficient hydrolysis of pretreated lignocellulosic substrates. *Biotechnology for Biofuels*, 4, 3.

Ashton, M. & Valentin, F. 1966. The mixing of powders and particles in industrial mixers. *Transactions of the Institution of Chemical Engineers*, 44, 166–188.

Asmar, B. N., Langston, P. A. & Matchett, A. J. 2002. A generalised mixing index in distinct element method simulation of vibrated particulate beds. *Granular Matter*, 4(3), 129–138.

Baig, K. S. 2020. Interaction of enzymes with lignocellulosic materials: Causes, mechanism and influencing factors. *Bioresources and Bioprocessing*, 7(1), 21.

Balami, A., Adgidzi, D. & Mua'zu, A. 2013. Development and testing of an animal feed mixing machine. *International Journal of Basics and Applied Sciences*, 1(3), 491–503.

Bánfai, B., Ganzler, K. & Kemény, S. 2007. Content uniformity and assay requirements in current regulations. *Journal of Chromatography A*, 1156(1–2), 206–212.

Bose, S., Barnes, C. A. & Petrich, J. W. 2012. Enhanced stability and activity of cellulase in an ionic liquid and the effect of pretreatment on cellulose hydrolysis. *Biotechnology and Bioengineering*, 109(2), 434–443.

Cao, Y. & Tan, H. 2004. The effect of shear field on the hydrolysis of cellulose. *Journal of Macromolecular Science, Part B*, 43(6), 1115–1121.

Chalbaud, C., Robin, M., Lombard, J., Martin, F., Egermann, P. & Bertin, H. 2009. Interfacial tension measurements and wettability evaluation for geological CO_2 storage. *Advances in Water Resources*, 32(1), 98–109.

Chandratilleke, G. R., Yu, A. B., Bridgwater, J. & Shinohara, K. 2012. A particle-scale index in the quantification of mixing of particles. *AIChE Journal*, 58(4), 1099–1118.

Chen, H. 2018. Design and scale-up of high-solid and multi-phase bioprocess. In *High-Solid and Multi-Phase Bioprocess Engineering: Theory and Practice*, 243–293.

Cho, M., Dutta, P. & Shim, J. 2017. A non-sampling mixing index for multicomponent mixtures. *Powder Technology*, 319, 434–444.

Chou, S.-H., Song, Y.-L. & Hsiau, S.-S. 2017. A study of the mixing index in solid particles. *KONA Powder and Particle Journal*, 34, 275–281.

Deng, Y. Y., Koper, M., Haigh, M. & Dornburg, V. 2015. Country-level assessment of long-term global bioenergy potential. *Biomass and Bioenergy*, 74, 253–267.

Dey, B. & Raja Sekhar, G. P. 2014. Mass transfer and species separation due to oscillatory flow in a Brinkman medium. *International Journal of Engineering Science*, 74, 35–54.

Ding, S.-Y., Liu, Y.-S., Zeng, Y., Himmel, M. E., Baker, J. O. & Bayer, E. A. 2012. How does plant cell wall nanoscale architecture correlate with enzymatic digestibility? *Science*, 338(6110), 1055–1060.

Du, J., Cao, Y., Liu, G., Zhao, J., Li, X. & Qu, Y. 2017. Identifying and overcoming the effect of mass transfer limitation on decreased yield in enzymatic hydrolysis of lignocellulose at high solid concentrations. *Bioresource Technology*, 229, 88–95.

Dutta, S. K. & Chakraborty, S. 2019. Multiscale dynamics of hemicellulose hydrolysis for biofuel production. *Industrial and Engineering Chemistry Research*, 58(21), 8963–8978.

Egermann, H., Kemptner, I. & Pichler, E. 1985. Effects of interparticulate interactions on mixing homogeneity. *Drug Development and Industrial Pharmacy*, 11(2–3), 663–676.

Felby, C., Thygesen, L. G., Kristensen, J. B., Jørgensen, H. & Elder, T. 2008. Cellulose–water interactions during enzymatic hydrolysis as studied by time domain NMR. *Cellulose*, 15(5), 703–710.

Feng, Y. Q., Xu, B. H., Zhang, S. J., Yu, A. B. & Zulli, P. 2004. Discrete particle simulation of gas fluidization of particle mixtures. *AIChE Journal*, 50(8), 1713–1728.

Foreman, P. K., Brown, D., Dankmeyer, L., Dean, R., Diener, S., Dunn-Coleman, N. S., Goedegebuur, F., Houfek, T. D., England, G. J. & Kelley, A. S. 2003. Transcriptional regulation of biomass-degrading enzymes in the filamentous fungus Trichoderma reesei. *Journal of Biological Chemistry*, 278(34), 31988–31997.

Freeman, M. K., White, W. & Iranikhah, M. 2012. Tablet splitting: A review of weight and content uniformity part 1 of a 2-part series. Next month: Table splitting—A review of the clinical and economic outcomes and patient acceptance. *The Consultant Pharmacist®*, 27(5), 341–352.

Himmel, M. E., Ding, S.-Y., Johnson, D. K., Adney, W. S., Nimlos, M. R., Brady, J. W. & Foust, T. D. 2007. Biomass recalcitrance: Engineering plants and enzymes for biofuels production. *Science*, 315(5813), 804–807.

Holdich, R. 2020. *Fundamentals of Particle Technology*. Leicestershire: MidlandIT.

Jørgensen, H., Vibe-Pedersen, J., Larsen, J. & Felby, C. 2007. Liquefaction of lignocellulose at high-solids concentrations. *Biotechnology and Bioengineering*, 96(5), 862–870.

Kaptay, G. 2008. The threshold pressure of infiltration into fibrous preforms normal to the fibers' axes. *Composites Science and Technology*, 68(1), 228–237.

Kaya, F., Heitmann, J. A. & Joyce, T. W. 1994. Cellulase binding to cellulose fibers in high shear fields. *Journal of Biotechnology*, 36(1), 1–10.

Kramer, H. A. 1968. Effect of grain velocity and flow rate upon the performance of a diverter-type sampler. Washington, DC: US Dept. of Agriculture[J]. Agricultural Research Service.

Kristensen, J. B., Felby, C. & Jørgensen, H. 2009. Yield-determining factors in high-solids enzymatic hydrolysis of lignocellulose. *Biotechnology for Biofuels*, 2(1), 11.

Kudou, M., Kubota, Y., Nakashima, N., Okazaki, F., Nakashima, K., Ogino, C. & Kondo, A. 2014. Improvement of enzymatic activity of β-glucosidase from Thermotoga maritima by 1-butyl-3-methylimidazolium acetate. *Journal of Molecular Catalysis B: Enzymatic*, 104, 17–22.

Lacey, P. M. C. 1954. Developments in the theory of particle mixing. *Journal of Applied Chemistry*, 4(5), 257–268.

Lago, M. & Araujo, M. 2001. Threshold pressure in capillaries with polygonal cross section. *Journal of Colloid and Interface Science*, 243(1), 219–226.

Lago, M. & Araujo, M. 2003. Threshold capillary pressure in capillaries with curved sides. *Physica A: Statistical Mechanics and Its Applications*, 319, 175–187.

Larsen, J., Østergaard Petersen, M., Thirup, L., Wen, L. I. H. & Krogh Iversen, F. 2008. The IBUS process–lignocellulosic bioethanol close to a commercial reality. *Chemical Engineering and Technology*, 31(5), 765–772.

Lavenson, D. M., Tozzi, E. J., Mccarthy, M. J., Powell, R. L. & Jeoh, T. 2011. Investigating adsorption of bovine serum albumin on cellulosic substrates using magnetic resonance imaging. *Cellulose*, 18(6), 1543–1554.

Liu, Z.-H. & Chen, H.-Z. 2016. Periodic peristalsis enhancing the high solids enzymatic hydrolysis performance of steam exploded corn stover biomass. *Biomass and Bioenergy*, 93, 13–24.

Lou, H., Wu, S., Li, X., Lan, T., Yang, D., Pang, Y., Qiu, X., Li, X. & Huang, J. 2014. Improving rheology and enzymatic hydrolysis of high-solid corncob slurries by adding lignosulfonate and long-chain fatty alcohols. *Journal of Agricultural and Food Chemistry*, 62(33), 8430–8436.

Lu, Y., Wang, Y., Xu, G., Chu, J., Zhuang, Y. & Zhang, S. 2010. Influence of high solid concentration on enzymatic hydrolysis and fermentation of steam-exploded corn Stover biomass. *Applied Biochemistry and Biotechnology*, 160(2), 360–369.

Luterbacher, J. S., Parlange, J.-Y. & Walker, L. P. 2013. A pore-hindered diffusion and reaction model can help explain the importance of pore size distribution in enzymatic hydrolysis of biomass. *Biotechnology and Bioengineering*, 110(1), 127–136.

McFarland, A. R., Gupta, R. & Anand, N. K. 1999. Suitability of air sampling locations downstream of bends and static mixing elements. *Health Physics*, 77(6), 703–712.

Modenbach, A. A. & Nokes, S. E. 2013. Enzymatic hydrolysis of biomass at high-solids loadings – A review. *Biomass and Bioenergy*, 56, 526–544.

Nicholls, D. 2015. Bioenergy from forests: The power potential of woody biomass. *Science Findings-Pacific Northwest Research Station, USDA Forest Service.*

Portillo, P. M., Ierapetritou, M. G. & Muzzio, F. J. 2008. Characterization of continuous convective powder mixing processes. *Powder Technology*, 182(3), 368–378.

Qin, Y. L. & Zhang, J. F. 1996. Recognization and comparison of threshold pressure in porous media. *Foreign Oil Filed Engineering*, 1, 5.

Ramos, L. P., Da Silva, L., Ballem, A. C., Pitarelo, A. P., Chiarello, L. M. & Silveira, M. H. L. 2015. Enzymatic hydrolysis of steam-exploded sugarcane bagasse using high total solids and low enzyme loadings. *Bioresource Technology*, 175, 195–202.

Roberts, K. M., Lavenson, D. M., Tozzi, E. J., McCarthy, M. J. & Jeoh, T. 2011. The effects of water interactions in cellulose suspensions on mass transfer and saccharification efficiency at high solids loadings. *Cellulose*, 18(3), 759–773.

Roche, C. M., Dibble, C. J., Knutsen, J. S., Stickel, J. J. & Liberatore, M. W. 2009. Particle concentration and yield stress of biomass slurries during enzymatic hydrolysis at high-solids loadings. *Biotechnology and Bioengineering*, 104(2), 290–300.

Rosgaard, L., Andric, P., Dam-Johansen, K., Pedersen, S. & Meyer, A. S. 2007a. Effects of substrate loading on enzymatic hydrolysis and viscosity of pretreated barley straw. *Applied Biochemistry and Biotechnology*, 143(1), 27–40.

Rosgaard, L., Pedersen, S., Langston, J., Akerhielm, D., Cherry, J. R. & Meyer, A. S. 2007b. Evaluation of minimal Trichoderma reesei cellulase mixtures on differently pretreated barley straw substrates. *Biotechnology Progress*, 23(6), 1270–1276.
Russ, D. C., Thomas, J. M. D., Miller, Q. S. & Berson, R. E. 2015. Predicting power for a scaled-up non-Newtonian biomass slurry. *Chemical Engineering and Technology*, 38(1), 53–60.
Sahin, H. & Ozdemir, F. 2004. Effect of some hydrocolloids on the rheological properties of different formulated ketchups. *Food Hydrocolloids*, 18(6), 1015–1022.
Schellart, W. P. 2011. Rheology and density of glucose syrup and honey: Determining their suitability for usage in analogue and fluid dynamic models of geological processes. *Journal of Structural Geology*, 33(6), 1079–1088.
Selig, M. J., Thygesen, L. G. & Felby, C. 2014. Correlating the ability of lignocellulosic polymers to constrain water with the potential to inhibit cellulose saccharification. *Biotechnology for Biofuels*, 7(1), 159.
Sharoba, A. M., Senge, B., El-Mansy, H. A., Bahlol, H. E. & Blochwitz, R. 2005. Chemical, sensory and rheological properties of some commercial German and Egyptian tomato ketchups. *European Food Research and Technology*, 220(2), 142–151.
Shi, X., Wang, L., Sun, L. & Chen, H. 2022. Sufficient premixing enhances enzymatic hydrolysis efficiency of lignocellulose at high-solids loading. *Chemical Engineering Journal*, 444, 136612.
Singh, K., Jung, M., Brinkmann, M. & Seemann, R. 2019. Capillary-dominated fluid displacement in porous media. *Annual Review of Fluid Mechanics*, 51(1), 429–449.
Siraj, M. S., Radl, S., Glasser, B. J. & Khinast, J. G. 2011. Effect of blade angle and particle size on powder mixing performance in a rectangular box. *Powder Technology*, 211(1), 100–113.
Staniforth, J. N. & Rees, J. E. 2011. Effect of vibration time, frequency and acceleration on drug content uniformity. *Journal of Pharmacy and Pharmacology*, 34(11), 700–706.
Van Leeuwen, M. L. J., Bruinsma, O. S. L. & Van Rosmalen, G. M. 1996. Influence of mixing on the product quality in precipitation. *Chemical Engineering Science*, 51(11), 2595–2600.
Viamajala, S., Mcmillan, J. D., Schell, D. J. & Elander, R. T. 2009. Rheology of corn stover slurries at high solids concentrations – Effects of saccharification and particle size. *Bioresource Technology*, 100(2), 925–934.
Vipulanandan, C. & Mohammed, A. S. 2014. Hyperbolic rheological model with shear stress limit for acrylamide polymer modified bentonite drilling muds. *Journal of Petroleum Science and Engineering*, 122, 38–47.
Wang, L., Feng, X. Y., Zhang, Y. Z. & Chen, H. Z. 2022. Lignocellulose particle size and rheological properties changes in periodic peristalsis enzymatic hydrolysis at high solids. *Biochemical Engineering Journal*, 178, 108284.
Wildemuth, C. R. & Williams, M. C. 1984. Viscosity of suspensions modeled with a shear-dependent maximum packing fraction. *Rheologica Acta*, 23(6), 627–635.
Williams, J. C. 1968. The mixing of dry powders. *Powder Technology*, 2(1), 13–20.
Yu, C. M. 2011. *Numerical Analysis of Heat and Mass Transfer for Porous Materials (A Theory of Drying)*. Beijing: Tsinghua University Press.
Zacchi, G. & Axelsson, A. 1989. Economic evaluation of preconcentration in production of ethanol from dilute sugar solutions. *Biotechnology and Bioengineering*, 34(2), 223–233.
Zeng, M., Ximenes, E., Ladisch, M. R., Mosier, N. S., Vermerris, W., Huang, C.-P. & Sherman, D. M. 2012. Tissue-specific biomass recalcitrance in corn stover pretreated with liquid hot-water: Enzymatic hydrolysis (part 1). *Biotechnology and Bioengineering*, 109(2), 390–397.
Zhang, J., Chu, D., Huang, J., Yu, Z., Dai, G. & Bao, J. 2010. Simultaneous saccharification and ethanol fermentation at high corn stover solids loading in a helical stirring bioreactor. *Biotechnology and Bioengineering*, 105(4), 718–728.
Zhang, J. T., Wang, L. & Chen, H. Z. 2020. Effect of periodic high-frequency vibration with rigid spheres added on high solids enzymatic hydrolysis of steam-exploded corn straw. *Process Biochemistry*, 94, 99–109.
Zhang, Y., Li, H. & Yang, D. 2012. Simultaneous estimation of relative permeability and capillary pressure using ensemble-based history matching techniques. *Transport in Porous Media*, 94(1), 259–276.
Zhao, J. & Chen, H. 2013. Correlation of porous structure, mass transfer and enzymatic hydrolysis of steam exploded corn stover. *Chemical Engineering Science*, 104, 1036–1044.

12 Enzymatic Hydrolysis of Microalgal Biomass

Luciane Maria Colla, Viviane Simon,
João Felipe Freitag, Victória Dutra Fagundes,
Alan Rempel, and Julia Catiane Arenhart Braun

12.1 INTRODUCTION

Fossil fuels are still considered one of the main sources of energy in the world. However, they are characterized as an exhaustible reserve. Furthermore, the forms of extraction and processing bring harm to the environment, which reinforces the need for further development of new sustainable energy sources (Neves et al., 2021). Another current factor that puts pressure on this energy matrix is the global geopolitical crisis, triggering price increases and reducing fossil raw material supplies. In this way, new technologies are being studied, aiming at the search for new energy matrices, from renewable raw materials, with processing processes that result in less environmental impact (Mota et al., 2022). In recent years, the production of biofuels from different industrial and agro-industrial residues, as well as plant and microbiological biomass, has not been reported as promising in the production of bioethanol, biogas, and biodiesel, in addition to other products with higher added value (Junior et al., 2022).

Within these new technologies, the use of microalgal biomass has been studied to obtain different products and applications, such as the production of biodiesel, biomethane, biohydrogen, biofertilizers, bioethanol, pharmaceuticals, and cosmetics (Cheirsilp & Maneechote, 2022). In addition to the range of products produced via the conversion of these microalgal biomasses, they do not require arable land, have advantages linked to the form of cultivation, and agro-industrial waste can be used resulting in its treatment and nutrient recovery, thus adding greater sustainability and the circular bioeconomy to the processes (Khan et al., 2018). In addition, microalgae perform carbon sequestration, have high photosynthetic efficiency, and accumulate lipids, proteins, and carbohydrates. These intracellular compounds vary according to culture strategies and microalgal species. These compounds, after going through several stages of cell disruption, hydrolysis, and fermentation, give rise to the aforementioned products (Mota et al., 2022).

The cell disruption step is the first process, in which through several strategies the cell wall of the biomasses is broken and the compounds are exposed for the next steps. The cell walls of microalgae vary by species, formed by multilayered polymers of glucans rich in amino acids. Still, there are species whose cellular structure is formed of more resistant materials, formed of insoluble cellulose biopolymers, which results in greater rigidity, hindering their rupture and, consequently, the access to intracellular compounds (Voigt et al., 2014).

Therefore, pretreatment technologies must be employed to break down these cell walls, thus releasing these intracellular compounds (Maity & Mallick, 2022), which will be later hydrolyzed and transformed into less complex structures, thus being able to carry out their transformation into the desired products, via fermentation processes. There are several methods reported in the literature to rupture the cell walls of microalgal biomass, including chemical hydrolysis (acid or alkaline), enzymatic hydrolysis, or mechanical processes (Maity & Mallick, 2022).

 DOI: 10.1201/9781003335313-14

As it is a first process for the release of these compounds, it is extremely important to evaluate methods that have high rupture efficiencies and do not generate reaction by-products. The use of enzymes tends to have advantages, since it is performed under milder conditions of pH and temperature compared to chemical hydrolysis. Another factor is that it is possible to use different enzymes so that all the cell wall compounds are accessed (Bahlawan et al., 2022).

After the rupture of the cell walls, the use of enzymes can also be used to carry out the hydrolysis of the released compounds (carbohydrates, proteins, lipids) presenting advantages, in addition to the milder pH and temperature conditions, such as the non-generation of by-products, that can be obtained when chemical hydrolysis is employed. According to the compound to be hydrolyzed, enzymes such as amylases, cellulases, xylanases, pectinases, and proteases can be used. Such enzymes can be commercial or produced via fermentation processes using different residues, as a nutritional source for specific producing microorganisms, which adds more sustainability to the process (Rodrigues et al., 2017).

The use of enzymes in cell disruption and hydrolysis of intracellular compounds has emerged as a promising strategy in the generation of bioproducts using microalgal biomass. However, it is extremely important to carry out studies on the composition of the cell walls of the microalgae species used, in addition to evaluating optimal pH and temperature ranges for enzyme action, as well as the dosages to be added to the processes.

This chapter aims to present the use of different enzymes that can be used in the rupture of cell walls and hydrolysis of intracellular compounds of different species of microalgae, demonstrating differences between the structures of the cell walls and enzymes. It also presents studies with optimal pH and temperature ranges of these enzymes and products generated after hydrolysis.

12.2 MICROALGAE PRODUCTION AND APPLICATIONS

Microalgal biomass is considered a sustainable raw material, with numerous advantages over other sources of biomass, such as the lack of arable land for its growth and the possibility of using alternative sources of water (waste, brackish and marine waters) for its cultivation, thus reducing dependence on freshwater sources. In addition, it has higher photosynthetic efficiency (3%–8%), biomass yield (more than 50%), and CO_2 capture (10–50 times) when compared to terrestrial plants (Sankaran et al., 2020).

Microalgae are microscopic and unicellular photosynthetic organisms, prokaryotic (cyanobacteria) or eukaryotic (green algae, diatoms), and have diverse forms, such as filamentous to coccoid. Microalgae can be grown in different growing media, including wastewater, with high growth rates (Udaiyappan et al., 2017). The algal biomasses, as well as the bioproducts obtained from them, can be used in pharmaceutical, food, agricultural, environmental, and bioenergy areas, among others (Siddiki et al., 2022).

Microalgae cultures can be carried out in open (open ponds) or closed (photobioreactors) systems. Cultures in photobioreactors have the advantage of having the most controllable cultivation conditions, such as temperature and contamination. On the other hand, open systems allow a greater scale of application resulting in greater final biomass. Open ponds can be natural or artificial, and because they are open they can have greater external interference, such as different weather conditions, which result in different yields throughout the year. This system is less expensive but requires a larger area for cultivation (Chew et al., 2018).

Photobioreactors differ from each other, being tubular (horizontal or vertical), flat panel, or stirred tank. In addition, these systems allow greater control of luminosity, temperature, and agitation (mechanical or by air bubbles). Due to the greater control of conditions, different species of microalgae can be cultivated in this type of cultivation. However, this system can become more costly to operate. Other fundamental factors for the cultivation of microalgae are the pH and the presence of the nutrients necessary for the development of the strains (Chew et al., 2018).

The carbon source used in the cultivation can be from different sources depending on the nature of the microalgae (autotrophic, heterotrophic, or mixotrophic). The autotrophic system consists of obtaining carbon sources from inorganic compounds, such as by absorbing light energy and CO_2 through photosynthesis processes. In the heterotrophic system, microalgae obtain their carbon sources for growth through organic compounds (Dragone, 2022).

Mixotrophic cultivation consists of obtaining carbon from organic and inorganic sources. This culture medium is used for the valuation of effluents and to increase biomass concentrations, increasing up to two to three times the amount of biomass when compared to the autotrophic system (Roostaei et al., 2018), facilitating the adaptation of microalgae in different culture media. The different cultivation conditions in the mixotrophic system can result in a biochemical composition of interest for the production of bioproducts. With the combination of different light intensities and organic and inorganic carbon sources, microalgae can accumulate lipids, proteins, carbohydrates, and other value-added bioproducts in their cells. In addition, the possibility of using algal biomass to obtain several bioproducts simultaneously makes the system more viable and inserts microalgal biomass within the concept of integrated biorefineries (Cheng et al., 2022).

Microalgal biomass can be used in the development of biofuels, such as bioethanol, biodiesel, biomethane, and biohydrogen. It is also a raw material for the development of biofertilizers, food products, pharmaceuticals, and cosmetics. Figure 12.1 presents the potential uses of microalgal biomass in industry (Ferreira et al., 2019). The application of microalgal biomass depends on the intracellular composition and the cell walls of the various existing species. High levels of carbohydrates in microalgal biomass are of interest for the production of bioethanol, and high levels of lipids are important for the production of biodiesel (Constantino et al., 2021).

The microalgae *Chlorella homosphaera* has great biotechnological potential, mainly in the production of biodiesel due to its high levels of lipids, and still can be used as a heavy metal adsorbent (Kumar et al., 2015). In general, the microalgae cell wall is composed of polysaccharides formed

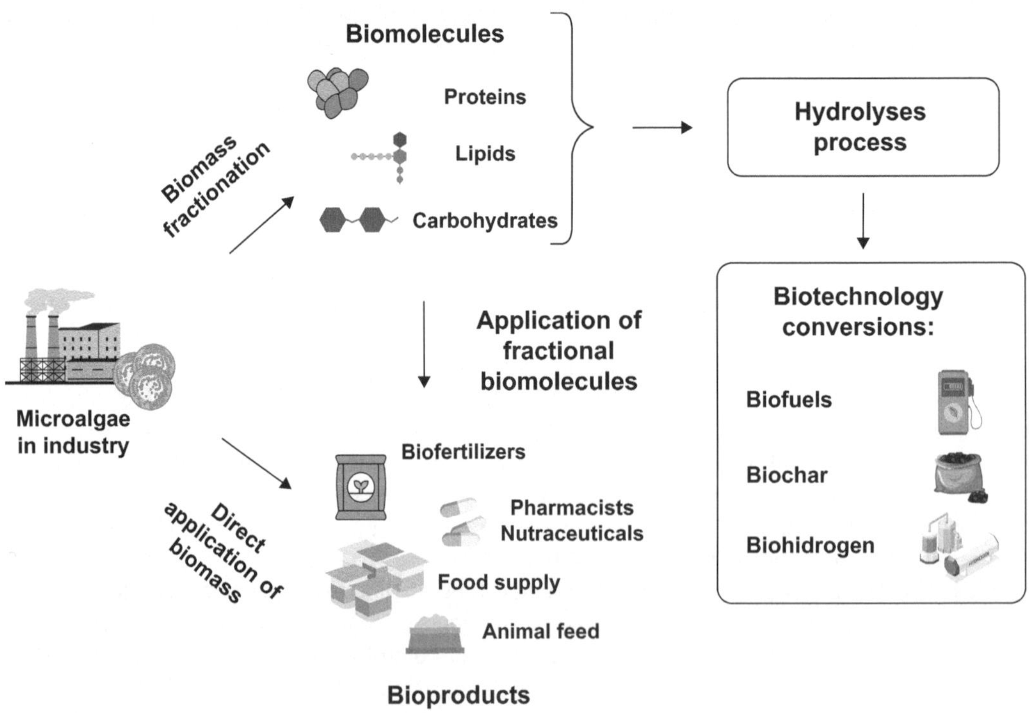

FIGURE 12.1 Applications of microalgae in industry.

by glucose monomers linked by α-1,4 glycosidic bonds. However, it is important to verify the differences between eukaryotic and prokaryotic cells to define potential biotechnological applications, such as differences in the cell wall and the presence of intracellular organelles (Russel et al., 2021). Next, characteristics of prokaryotic and eukaryotic cells, related to the enzymatic hydrolysis process, will be discussed.

12.3 COMPOSITION CHARACTERISTICS OF MICROALGAE CELL WALLS

12.3.1 Prokaryotic Microalgae

Prokaryotic microalgae, also called cyanobacteria, have morphological properties similar to bacteria (Udaiyappan et al., 2017). They belong to the phylum Cyanobacteria. Several microalgae are known and studied, including *Phormidium*, *Anabaena*, *Nostoc*, *Oscillatoria*, *Synechocystis*, *Microcystis*, *Cyanobacterium*, *Lyngbya*, *Synechococcus*, and *Spirulina* (*Arthrospira*). However, although belonging to the same phylum, microalgae have morphological differences between them.

The genus *Phormidium* is reported with approximately 170 identified microorganisms. Although there are differences within the genus itself, these microalgae are characterized by having curved trichomes measuring 2.5–11 μm in width, approximately isodiametric with a pointed or rounded apical cell (Mccallister et al., 2016). The cell wall of this microalgae is composed of about 52% dry weight of amino acids such as glucosamine, muramic acid, alanine, glutamic acid, and diaminopimelic acid. This composition comprises a layer of mucopolymers that form a rigid backbone that provides most of the cell's mechanical strength.

The cell wall of the *Anabaena* is formed by an outer membrane and a meshwork of peptidoglycans. It is also composed of proteins at the septal junction that unites the cells; these are important for the integrity of the filament. About 550 morphospecies of microalgae of the genus *Anabaena* are reported, being benthic and planktonic. This is a filamentous microalga that has the ability to float on water due to the presence of a gas vesicle in its cells (Li et al., 2016).

The *Spirulina* is highlighted since it is widely used around the world and with a great biotechnological application, with approximately 35 different species being used in the industry. This genus is helical or cylindrical, with an average diameter of 2 to 12 μm and can reach 16 μm in length. *Spirulina* can present high levels of proteins in its biomass when grown in media full of nutrients (especially N), which characterizes its main application in the food and nutraceutical industry (Zhao et al., 2022). *Spirulina platensis* has a cell wall composed of multilayers of glucan and peptidoglycan polymers that form a layer composed of sugars and amino acids linked by covalent bonds (Cheng et al., 2022), fibrillar (one of them β-1,2-glucan), one of peptidoglycans and one of acidic polysaccharides. Glucan and peptidoglycan polymers are formed by chains of molecules (sugars and/or amino acids) linked by stable covalent bonds, and these chains of molecules are connected by hydrogen bonds and Van der Waals forces (Cheng et al., 2022).

Among the intracellular compounds of interest, carbohydrates are highlighted. They are used in the development of biofuels. These polysaccharides can be similar to starch and laminarin. In laminarin-type carbohydrates, β-1,3 and β-1,6 glycosidic bonds are found between glucose monomers, and in starch-type carbohydrates, which consist of amylopectin (branched structure with α-1,4 and α-1,6) and amylose (α-1,4 glycosidic bonds). In cyanobacteria, glycogen is found mainly as a polysaccharide similar to starch but smaller, facilitating fermentation (Maity & Mallick, 2022). Thus, when enzymatic hydrolysis is performed, it is important to know and evaluate the type of bonds that each enzyme can break to bring greater efficiency to the process.

12.3.2 Cell Walls of Eukaryotic Microalgae

Eukaryotic microalgae belong to several taxonomic groups, and among these, the main phyla of microalgae are Rhodophyta (red algae), Chlorophyta (green algae), Bacillariophyta (diatoms), and

Chrysophyta (golden algae). Also, among these phyla, different classes and orders are reported as biomasses with industrial use. Some eukaryotic microalgae reported are *Desmodesmus*, *Dunaliella*, *Scenedesmus*, *Botryococcus*, *Haematococcus*, *Parachlorella*, *Chlamydomonas*, *Tetradesmus*, and *Chlorella* (Russel et al., 2021; Udaiyappan et al., 2017).

The intracellular compositions of proteins, carbohydrates, and lipids vary among themselves. Protein concentrations can range from 28% to 68% for *Scenedesmus subspicatus* and *Haematococcus pluvialis*, respectively. The lipid levels can have values that reach 33% for *Botryococcus braunii*, but there are lower levels that vary within the genus itself, as in the case of *Chlorella* (2 to 22%) (Russel et al., 2021).

The eukaryotic microalgae of the Chlorococcales polyphyletic group, belonging to the Chlorophyceae class, such as *Scenedesmus*, feature a resistant cell wall, formed by a trilaminar structure that may contain insoluble biopolymers such as cellulose on the inside and algaenan on the outside that also contribute to the rigidity of the structure. These microalgae may also have cross-linked glycoproteins in their cell wall that contain glucosamine and hydroxyproline (Voigt et al., 2014).

Chlorella microalgae also have a rigid, stable cell wall with a complex structure. It is formed by two cell layers, and in its external part, it has a polysaccharide matrix, containing pectin, agar, alginate, and biopolymers. In its inner layer, it has hemicellulose and glycoproteins in a matrix of cellulose microfibers (Kusmiyati et al., 2020).

The cell wall of the eukaryotic microalgae *Chlorella homosphaera* has a rigid fibrillar matrix inside, with high levels of cellulose and hemicellulose, which can be degraded by cellulolytic enzymes. This microalga has chitin-like polysaccharides in its inner and outer cell walls. After being hydrolyzed, it forms monosaccharides of rhamnose, mannose, glucose, and galactose (Monjed et al., 2021).

In contrast, the eukaryotic microalgae *Dunaliella* does not have a very rigid cell wall. Its cells are surrounded only by a plasma membrane that can undergo osmotic shocks that bring changes in volume and shape. After osmotic equilibrium by the release of glycerol, it can return to normal size, which makes it a halotolerant microalga. Osmoregulation and ATPase activation in the plasma membrane allow this microalga to be tolerant in saline environments and with the presence of metals, making it possible to eliminate these compounds and making it sometimes less susceptible than microalgae with complex cell walls (Sendra et al., 2018).

12.4 ENZYMATIC HYDROLYSIS OF MICROALGAE

Hydrolysis processes are generally used to convert biomass into high value-added bioproducts. The main applications of hydrolytic methods involve the extraction or conversion of biomolecules such as carbohydrates, lipids, proteins, and bioactive compounds (Juárez et al., 2016; Zhao et al., 2022). These processes can be used as pretreatment of subsequent steps, or individually, for the rupture of the microalgal cell wall and release of intracellular biocompounds. Physical–chemical or biological processes are commonly used as pretreatment to increase biomass conversion efficiency results (Velazquez-Lúcio et al., 2018). Among the most used pretreatments, the use of ultrasound, microwaves, freezing and thawing, heat treatments, and acid or alkaline hydrolysis stand out.

The use of pretreatments does not necessarily mean greater conversion efficiency. In their study, Lee et al. (2013) performed the chemical pretreatment of microalgal biomass, using hydrochloric acid (acid hydrolysis) to release reducing sugars. In enzymatic hydrolysis, amylolytic (AMG 300L) and cellulolytic (Celluclast 1.5 L, Novoprime B957) enzymes were used, in addition to a multienzyme complex (Viscozyme L) composed of arabinase, cellulase, β-glucanase, hemicellulose, and xylanase. The best hydrolysis efficiencies were obtained when using AMG 300 L enzymes (80.9%), followed by the multienzymatic complex, and when combined, the hydrolysis efficiency did not increase significantly (81.9%). Furthermore, the study emphasizes that the use of acid hydrolysis proved to be unnecessary when combined with enzymatic hydrolysis.

The use of enzymes appears to be promising, as it presents efficiency in the rupture of the microalgal cell wall, due to the high specificity in relation to substrates. Furthermore, the methodology has advantages compared to the use of chemical methods, as it works in milder conditions of temperature and pH, bringing greater safety to the process (Bahlawan et al., 2022). In addition to the use of commercial enzymes, *homemade* microbial enzymes are commonly produced. The process may involve the valuation of by-products, such as agro-industrial residues, for example, and through the use of microorganisms and specific culture media, enzymes or enzyme complexes with hydrolysis capacity can be obtained, resulting in a more sustainable and cost-effective process if the purification conditions are optimized (Rodrigues et al., 2017).

According to Monjed et al. (2021), enzymes of fungal origin have great potential for the hydrolysis of eukaryotic microalgae, such as *Chlorella vulgaris*, especially cellulases, hemicellulases, glucanases, pectinases, and chitinases. In the study, the enzymes were evaluated by the release of reducing sugars, mainly glucose, without detecting sugars such as N-acetyl-glucosamine and other oligosaccharides from the degraded biomass.

Among the factors that most interfere in the enzymatic hydrolysis process, the temperature and pH of the medium stand out (Liu et al., 2020) to obtain greater efficiencies in the conversion processes. Rempel et al. (2018) performed the characterization of optimal temperatures and pHs of α-amylase and amyloglucosidase enzymes (Novozymes), using as substrates starch and biomass of the microalgae *Spirulina platensis*. The enzymes were evaluated for the production of reducing sugars, with higher production at 50°C and pH of 5.5 and 6.0, for α-amylase and amyloglucosidase, respectively.

Another important factor in enzymatic hydrolysis processes is the correct choice of enzymes, as the composition of the microalgal cell wall may vary according to species. In this sense, according to the composition of the microalgal cell wall, associations of enzymes can be used in order to obtain a greater hydrolysis yield (García et al., 2012). The enzymes most used in the hydrolysis of microalgal biomass are amylases (Souza et al., 2020), however, in the case of eukaryotic microalgae, the cellulose composition can reach up to 33% of intracellular polysaccharides, requiring the application of specific enzymes. Cellulases (endoglucanases, exoglucanases, and cellobiases) are already widely used industrially.

In the study by Sriyod et al. (2021), the optimization of the enzymatic hydrolysis of the microalgae *Chlorella* sp., using together the enzymes Termamyl SC (alpha-amylase), Dextrozyme GA (amyloglucosidase), and Cellic CTec2 (cellulase), was used to convert fermentable sugars into biohydrogen. With a substrate concentration of 100 g volatile solids VS L^{-1}, the optimal enzyme concentrations were 160 U g^{-1} VS, 320 U g^{-1} VS, and 40 U of filter paper g^{-1} VS. With the association of enzymes, hydrolysis efficiencies of approximately 50% were obtained.

Hosseini et al. (2020) performed the enzymatic hydrolysis of *Haematococcus pluvialis* for the production of bioethanol. The hydrolysis of α-β-glucan was performed using different commercial enzymes, starting with the use of a mixture of Cellic (Ctec2 and Cellic Htec2) at 45°C for 48 h (with activity of 50 FPU [filter paper per units]). Then, 10 µL of α-amylase enzyme were added to the reaction solution, which was incubated at 90°C in a water bath with shaking for 2 h. Finally, the temperature of the mixture was reduced to 60°C, and 10 µL of glucoamylase enzyme solution was added to the reaction mixture, which was incubated in a water bath at 60°C for 22 h.

Next, the most commonly used enzymes in the hydrolysis of microalgal biomass will be discussed individually, with emphasis on amylases, cellulases, xylanases, pectinases, proteases, and lipases (Alavijeh et al., 2020; Kim et al., 2014; Liu et al., 2021; Rempel et al., 2018).

12.4.1 Amylases

Starch is one of the most abundant carbohydrates in microalgae. It is a polysaccharide composed of several glucose molecules, linked by glycosidic bonds. Starch is composed of a linear fraction, amylose, which is a linear polymer with α-1,4 linkage chains, and a branched fraction, amylopectin, which in addition to α-1,4 bonds has α-linkages (Jiang et al., 2017). Amylases are enzymes that act in the hydrolysis of starch and can be produced by various microorganisms, plants, and humans. According to the mechanism that breaks down starch, amylases are classified into four types:

- α-amylases, which randomly break the bonds of starch in α-1,4 from the inside of the starch chain, producing glucose compounds with different chain sizes;
- β-amylases, which produce maltose, act by breaking glucose–glucose bonds (α-1,4-D-glucan, maltohydrolase) from outside the starch chain;
- amyloglucosidases, which successively break bonds at the non-reducing ends (α-1,4 and α-1,6), producing free glucose; and
- pullulanases, which are debranching enzymes that hydrolyze the α-1,6 bonds of α-glycans (Rodrigues et al., 2021).

Thus, in the hydrolysis process, starches present in microalgal biomass are converted into simpler sugars, since starches are extensive complex structures (Rempel et al., 2018). In a study carried out by Souza et al. (2020), enzymatic hydrolysis of the biomass of the microalgae *Chlorella sorokiniana* was performed using amylolytic enzymes with activity of 4.8 U (*Aspergillus awamori*) and 540 U (*Trichoderma reesei*). The experiments were carried out in 1.5 mL tubes containing 1.25 g of a 1% (m m^{-1}) mixture of *Chlorella sorokiniana*, enzymes, and 0.05 M citrate buffer solution pH 4.8. The tubes were incubated for 24 h at 50°C and 200 rpm. Starch from microalgal cells was hydrolyzed by *homemade* fungal amylases, with a yield and final glucose concentration of 99% and 16 g L^{-1}, respectively, in a period of 5 h.

For the production of bioethanol from the microalgae *Chlorella* sp., Ngamsirisomsakul et al. (2019) used α-amylase (SPEZYME® ALPHA, DuPont) and glucoamylase (GC 147, DuPont) enzymes in acid-pretreated biomass (98% H_2SO_4). The α-amylase enzyme was used at a temperature of 85°C and pH 5.7, while the glucoamylase was used at 65°C and pH 4.5. The best results were observed using only glucoamylase in the pretreated biomass, with 25 U glucoamylase/g of microalgae. The yield of bioethanol produced from the hydrolyzate was 10.58 ± 0.34 g L^{-1} or 52.88 ± 1.69 mg g^{-1} of microalgae. The results were 7.1 times higher in the production of bioethanol compared to the use of biomass without pretreatment.

In the study by Constantino et al. (2021), chemo-enzymatic hydrolysis of biomass from different microalgae was performed. The microalgae *Tetraselmis striata* (C4), *Tetraselmis* sp. (C6), *Cylindrotheca fusiformis* (S3), *Nanofrustulum* sp. (S4), *Picochlorum maculatum* (C2), *Phaeodactylum trichomutum* (S6), *Chlorella sorokiniana* 211-32 (C1), and *Chlamydomonas reinhardtii* (C3) were considered. Acid hydrolysis was used as a pretreatment and was carried out with a 4% sulfuric acid solution and biomass concentration of 50 g L^{-1}, and the mixture was placed in an autoclave at 121°C for 30 min. Enzymatic hydrolysis was carried out in sequence using α-amylase (8000 U g^{-1} of biomass, pH 5.5, and 95°C) and amyloglucosidase (600 U g^{-1} of biomass, pH 4.5, and 55°C). The enzymes were added in the reverse order to the conventional one, that is, first the amyloglucosidase was added and then (after 3 h) the α-amylase was added. The assays were conducted for 6 h, with samples taken every hour for analysis of reducing sugars. Using the proposed strategy, a one-third increase in reducing sugar yields was observed when compared to the conventional strategy. In addition, 34.0 ± 1.0 g of total reducing sugar $(100\ g)^{-1}$ of dry biomass were obtained.

12.4.2 Cellulases

Cellulose is highly stable and composed of glucose molecules in linear chains of up to 12,000 monomers that are joined by β-1,4 glycosidic bonds. Microalgal biomass can contain up to 75% carbohydrates, depending on the species, which are held together by hydrogen bonds in the native state, but they have a strong tendency to form hydrogen bonds, which increases the rigidity of the cellulose and makes it highly insoluble and resistant for most organic solvents (Hamed et al., 2016).

The complete hydrolysis of cellulose is mediated by a combination of three main types of cellulases. The classification of cellulases, according to Lynd et al. (2002), is dependent on their site of action on the cellulosic substrate: endoglucanases (EnG), which cleave bonds of the cellulosic fiber; exoglucanases (ExG), which act on the external region of cellulose; and β-glucosidases (BG), which hydrolyze glucose-soluble oligosaccharides.

In the study carried out by Maffei et al. (2018), the effects of an enzymatic treatment with cellulase and mannase on the properties of the marine microalgae *Nannochloropsis* sp. Cellulyve 50LC® (CEL) and Feedlyve GMA® (GMA) enzymes were used at concentrations of 13.8 mg L^{-1} and 1.5 mg L^{-1}, respectively. The tests were carried out under optimized conditions with 0.2 g of biomass for 10 mL of enzyme solution at the concentrations described, pH 4.4 and 50°C, for 24 h. At the end of the tests, the biomass was centrifuged for Fourier-transform infrared (FTIR), powder x-ray diffraction (XRD), and scanning electron microscope (SEM) image analysis. The combined use of these enzymes synergistically promoted lipid recovery from microalgae, increasing the extraction yield from 40.8% to over 73%. The use of enzymes increased the ratio of crystalline to amorphous cellulose. Dramatic changes in cell morphology, extensive cell damage, and release of intracellular material were also observed. The results obtained indicate that the enzymes used are capable of breaking the cell wall of microalgae.

Bioethanol production from the biomass of the microalgae *Dunaliella tertiolecta* was investigated by Lee et al. (2013) using chemo-enzymatic saccharification. HCl-catalyzed saccharification of residual biomass at 121°C for 15 min produced reducing sugars in a yield of 29.5% (w w^{-1}) based on the dry weight of residual biomass. Initially, the biomass was pretreated (5% m v^{-1}) with HCl or H_2SO_4 of different molarities, and the mixture was taken to an autoclave at 121°C for 15 min. Then, the saccharification of the biomass (5% w v^{-1}) was carried out at temperatures and pHs of 35–55°C and 3.5–6.5, respectively, using the enzymes Celluclast® 1.5 L, AMG® 300L, and/or Viscozyme® L (contains arabinase, cellulase, β-glucanase, hemicellulose, and xylanase) (Novozymes) with concentrations ranging from 0.1 to 1.0 mg g^{-1} of dry microalgae biomass. AMG 300L and Viscozyme L showed better hydrolysis efficiency when compared to the other enzymes. At pH 5.5 and temperature 55°C, enzymatic saccharification yields were 21.9% and 19.2% (w/w based on residual biomass) for AMG 300L and Viscozyme L, respectively.

12.4.3 Pectinases, Xylanases, and Proteases

Pectinases, xylanases, and proteases, in general, are used in combination with amylolytic and cellulolytic enzymes, among others, and not in isolation, in order to increase hydrolysis yields. Thus, they will be addressed in this topic jointly. Pectinases are widely used in the food and industrial production chain, based on the depolymerization of pectin as a result of de-esterification and hydrolysis reactions. More recently, they have been applied to the hydrolysis of microalgal biomass to increase the yields of polysaccharides associated with other enzymes (Zhang et al., 2018). Pectins are known to be classified based on their mechanism of action into three main categories: protopectinases (degrade insoluble protopectins), desterifiers (hydrolyze specific substrate groups), and pectinalyase (hydrolytic removal of pectates) (John et al., 2020). Also, based on their site of action, pectinases can be classified as endopectinases and exopectinases. Endopectinases are characterized by the random cleavage of internal pectin bonds. Exopectinases cleave the outer units of long-chain polysaccharides.

Xylan is the most abundant constituent of hemicellulose and is located at the interface between lignin and cellulose. The main chain of xylan consists mainly of β-(1,4) glycosidic bonds of D-xylopyranose units, with the potential to present a varied degree of polymerization and branching, depending on the plant species. For the degradation of xylan, in turn, due to the complexity and heterogeneity of the chemical nature of xylan, its complete depolymerization is only possible through the synergy of a set of hydrolytic enzymes with different specificities, called the xylanolytic enzyme system (Burlacu et al., 2016).

Endo-1,4-β-D-xylanases are responsible for the endohydrolysis of β-1,4-glycosidic bonds in the xylan main chain. They produce fractions of xylo-oligosaccharides with a higher degree of polymerization, such as xylose, xylobiose, and xylotriose, and are produced by microorganisms, such as the bacteria *Streptomyces* sp. and *Bacillus* sp. and the fungi *Trichoderma* sp. and *Aspergillus* sp. (Burlacu et al., 2016). The current classification of xylanases was proposed based on the comparison of amino acid sequences (Kocabas et al., 2015), divided among the glycosyl hydrolases (GH) system, being a group of enzymes that hydrolyze the glycosidic bond between two or more carbohydrates or between portions of these and non-carbohydrate portions.

Proteins are made up of biomolecules called amino acids, linked by peptide bonds. Protein hydrolysis can be carried out by proteolytic enzymes and is called proteolysis. Proteases are commonly produced by microorganisms such as fungi, yeasts, bacteria, and actinomycetes, and act by breaking the nitrogen and carbon terminal amino acid bonds (exopeptidases), or by breaking the internal bonds of polypeptide chains (endopeptidases). They can also be classified according to catalytic mechanism and specificity (serine, aspartic, threonine, cysteine, glumatic, and metalloproteinase proteases) and structural relationship characteristics (Srivastava et al., 2020).

In proteolysis, specific intracellular proteins are denatured and then the denatured polypeptides are moved into the chamber of a peptidase for protein degradation, producing amino acid molecules. Amino acids, in turn, have a central carbon atom connected with an amine group at one end and a carboxylic acid at the other, with the remaining bonds filled with hydrogen and an R (alkyl) group. During the degradation process, the amine can increase the efficiency of microorganisms involved in the conversion of sugars into alcohol and other biofuel products, such as the extraction of lipids, through the availability of nitrogen. In addition, the carboxylic acid present in the protein can also be used in the production of biofuels through the formation of alcohol (Srivastava et al., 2020).

The study by Bader et al. (2020) sought to identify alternative sources of enzymes for the saccharification of microalgae biomass of the *Chlamydomonas* species. The production of hydrolytic enzymes was carried out from the filamentous fungus *Trichoderma harzianum*. For hydrolysis, temperatures and pHs were used that varied among 40°C, 50°C, or 60°C and pH 5, 7, or 9, respectively. The experiments were carried out with 1% (w v^{-1}) of dry microalgae biomass or 5% (w v^{-1}) of fresh biomass (intact cells) added to 100 μL of fungal medium and incubated at 55°C. The fungus was able to produce pectinolytic, amylolytic, cellulolytic, and proteolytic enzymes and efficiently hydrolyzed the biomasses. The overall efficiency for releasing reducing sugars using fresh biomass was 66% or 74% of its total carbohydrates. When using the same concentration of dry biomass, the efficiency of releasing reducing sugars was approximately 70%. Saccharification of algal biomass yielded up to 22.4 g of L^{-1} reducing sugars.

For the production of bioethanol from the microalgae *Chlorella vulgaris* cultivated under nutritional stress conditions, Kim et al. (2014) investigated the enzymatic hydrolysis of biomass using different enzymes: Celluclast® 1.5 L and Pectinex® SP-L (Novozymes), and xylanase (X2629), β-glucosidase (C6105), amylase (A8220), chitinase (C6137), lysozyme (L6876), and sulfatase (S9626) (Sigma-Aldrich). The cellulase and pectinase enzymatic activities were 0.122 filter paper units (FPU) mg^{-1} protein and 240 international units (IU) mg^{-1} protein, respectively. The activities of other enzymes were 16 FAU mL^{-1} (amylase), 10 U mL^{-1} (β-glycosidase), 100 U mL^{-1} (xylanase), 0.2 U mL^{-1} (chitinase), 4000 U mL^{-1} (lysozyme), and 50 U ml^{-1} (sulfatase). Pectinase was added to the microalgae at concentrations of 0.24–9.4 mg g^{-1} of protein, and the other enzymes were used at concentrations indicated by the manufacturer. Enzymatic hydrolysis was performed with 1% (w v^{-1}) microalgal biomass (w v^{-1}) in buffer (pH 4.8) at 200 rpm and 50°C for 72 h. Pectinase showed the best results with a saccharification yield of 79% after 72 h.

In a study carried out by Zhang et al. (2018), the optimization of the enzymatic extraction of lipids from the microalga *Scenedesmus* sp. was accomplished using the enzymes cellulase, xylanase, and pectinase, at different concentrations (0, 10, 20, 30, 40 mg g^{-1} for cellulase [D1] and 0, 10, 15, 20 mg g^{-1} for xylanase [D2] and pectinase [D3]), temperatures (30°C, 40°C, 50°C, 60°C, 70°C), pHs (2, 3.5, 5, 6.5, 8), and incubation times (60, 120, 180, 240, 300 min). The optimal conditions were established at T = 45°C, pH = 4.4, P = 190 min, D1 = 20 mg g^{-1}, D2 = 14 mg g^{-1}, and D3 = 10 mg g^{-1}. The lipid extraction yields were 13.8 ± 0.4 g 100 g^{-1} with an optimal percentage of 86.4%.

Despite the advantages of using enzymes in hydrolysis and conversion processes of microalgal biomass to obtain biomolecules, their studies are mainly focused on the laboratory scale. Table 12.1 presents different studies that evaluate the use of different enzymes in enzymatic hydrolysis of microalgal biomass, aiming to obtain biomolecules of interest.

The studies presented used a wide range of enzymes in the process of enzymatic hydrolysis of microalgal biomass, highlighting cellulases, amylases, proteases, pectinases, lipases, and others. Among the biomolecules and bioproducts of interest were lipids, proteins and carbohydrates, and bioethanol and biohydrogen. Zhang et al. (2022) used six different commercial enzymes in the enzymatic hydrolysis process to evaluate the production of lipids from the biomass of the eukaryotic microalgae *Scenedesmus* sp. The enzymes were applied individually (with pH ranging from 3.0 to 8.0 and temperature from 35°C to 65°C, as indicated by the manufacturer) and in a mix of enzymes (with pH of 4.5 and temperature of 55°C) with enzyme concentrations of 5, 20, 40, and 80 mg g^{-1}. As a result, they obtained 100% lipid recovery in the assay using an enzyme mix (pectinase, cellulase, and xylanase at concentrations of 10, 20, and 14 mg g^{-1}, respectively). Since cellulase can effectively degrade the cellulose structure in the cell walls of algae, which could promote lipid release from intracellular microalgae, xylanase and pectinase mainly composed of xylanhydrolase and polygalacturonase, respectively, could specifically damage the structure of hemicellulose and pectin in the cell wall of microalgae (Maffei et al., 2018; Liu et al., 2018).

To produce biohydrogen, Liu et al. (2021) evaluated the efficiency of enzymatic hydrolysis of *Chlorella vulgaris* (eukaryotic) biomass using proteolytic and cellulolytic enzymes. The results showed that when using cellulase alone, with pH 7.0, 25.99 mL g^{-1} of biohydrogen (in dry cell weight) were obtained. Using the protease alone with pH 8.0, 16.47 mL g^{-1} of biohydrogen were obtained. The best results were obtained using the enzymes together (cellulase and protease) with pH 7.0, with a production of 27.43 mL g^{-1} of biohydrogen, following the strategy of performing hydrolysis by protease first, followed by hydrolysis with cellulase. For the authors, the advantage of this enzymatic hydrolysis strategy was the solubilization of proteins in the first part of the hydrolysis and the separation of carbohydrates afterward.

Alavijeh et al. (2020) also used the biomass of *Chlorella vulgaris*. However, for the extraction of lipids, proteins, and carbohydrates, the process was carried out by milling the biomass followed by enzymatic hydrolysis carried out with different enzymes (cellulase, lipase, phospholipase, and protease), and the dosage of each enzyme was adjusted to 2% (v w^{-1}) under hydrolysis conditions. The maximum recovery yield for all components was obtained after enzymatic hydrolysis of the biomass by lipases, without the addition of other enzymes, at 37°C and pH 7.4 for 24 h, producing 88% of lipids in the solid phase, while 74% of carbohydrates and 68% of proteins were separated in the liquid phase. Some studies indicate that lipases can act in the release of proteins and carbohydrates in the liquid phase and later purified lipids in the solid phase, mainly due to the potential of lipases to digest the monolayer of polar lipids or membrane proteins that cover the lipid droplets (Alavijeh et al., 2020).

Peron-Schlosser et al. (2021) used commercial enzymes in the enzymatic hydrolysis process, in which the enzymes α-amylase (Liquozyme® Supra 2.2X) and amyloglucosidase (AMG® 300 L) were used simultaneously in the hydrolysis of residues from the process of extracting proteins from biomass of the prokaryotic microalgae *Arthrospira platensis*, for the production of ethanol in the biorefinery concept, the concentration of each enzyme was varied in 0.5, 1.0 and 1.5%, (v v^{-1}), pH from 5.5 to 50°C. Over 12 h, 8.06 g L^{-1} of bioethanol were obtained, with a theoretical yield of 100%. Still, it is important to point out that the medium was not supplemented, which means that the nutrients present in the hydrolysate were sufficient for the full development of the alcoholic fermentation.

Aikawa et al. (2018) carried out the production of bioethanol from the microalgae *Arthrospira platensis* using amylolytic enzymes and $CaCl_2$ supplementation. The authors evaluated the

TABLE 12.1
Use of Enzymes in Enzymatic Hydrolysis Processes to Obtain Biomolecules and/or Bioproducts from Microalgal Biomass

Enzymes	Microalgae	Experimental Conditions	Bioproducts/ Bioprocess	Hydrolysis Efficiency	Reference
Cellulase, xylanase, pectinase, trypsin, and lysozyme from commercial production	*Scenedesmus* sp.	• Enzyme concentrations: 5, 20, 40, and 80 mg g^{-1} of dry biomass. • Temperatures and pH ranged between 30°C and 60°C and 3.5 to 9.0, respectively, according to the manufacturer's information. • Conditions of 55 °C, pH 4.5, and 24 h with cellulase, xylanase, and pectinase enzymes.	Lipid extraction	100% lipid recovery yield (total lipid content) can be achieved with the step-by-step enzymatic hydrolysis sequence: pectinase (10 mg g^{-1} microalgae dry weight), cellulase (20 mg g^{-1}), and xylanase (14 mg g^{-1})	Zhang et al. (2022)
Cellulase and protease	*Chlorella vulgaris*	• Hydrolysis is divided into three groups: group A (cellulase hydrolysis), group B (protease hydrolysis), and group C (mixed enzymatic hydrolysis). • 15% enzyme load. • pH ranging from 5 to 9.	Biohydrogen production	25.99 mL g^{-1} of biohydrogen for cellulase, 16.47 mL g^{-1} using protease, and 27.43 mL g^{-1} using the enzymes together	Liu et al. (2021)
Commercial enzymes: α-amylase (Liquozyme® Supra 2.2X) and amyloglucosidase (AMG® 300L)	*Arthrospira platensis*	• Enzymatic hydrolysis (pH 5.5; 50°C). • Enzymes were added simultaneously. • With concentrations of 0.5, 1.0, and 1.5% (v v^{-1}). • With pH 5.5 and a temperature of 50°C.	Protein extraction for bioethanol production	8.06 g L^{-1} of bioethanol, with a theoretical yield of 100%	Peron-Schlosser et al. (2021)
Commercial enzymes: Endopeptidase (Alcalase 2.4L, Protamex) and cellulase (Celluclast 1.5L)	*Scenedesmus almeriensis, S. acuminatus, S. ecornis, S. quadricaudae, Mychonastes homophaera*	• Experiment A1: Consortium of microalgae (96% of *S. almeriensis* and 4% of *S. acuminatus*) and bacteria in wastewater with pH 6.5. • Experiment A2: 80% *S. almeriensis*, 10% *S. ecornis*, 9% *S. quadricauda*, and 1% *Mychonastes homosphaera* with a pH of 8. • Tests performed with a concentration of microalgae biomass of 5% (w w^{-1}). • Enzymes applied under ideal conditions provided by the manufacturer.	Production of carbohydrates	38.5% carbohydrate yield of the cellulase enzyme in experiment A1	Rojo et al. (2021)

(Continued)

TABLE 12.1 (CONTINUED)
Use of Enzymes in Enzymatic Hydrolysis Processes to Obtain Biomolecules and/or Bioproducts from Microalgal Biomass

Enzymes	Microalgae	Experimental Conditions	Bioproducts/ Bioprocess	Hydrolysis Efficiency	Reference
Commercial enzymes α-amylase (Liquozyme® Supra 2.2X) and amyloglucosidase (AMG® 300 L)	*Spirulina platensis*	• Enzymes were added simultaneously at a concentration of 1% (v v^{-1}), according to Rempel et al. (2018).	Bioethanol production	Bioethanol production from hydrolyzed biomass showed 4,664 kJ kg^{-1}	Rempel et al. (2021)
Lipases, phospholipases, protease, and cellulase	*Chlorella vulgaris*	• The dosage of each enzyme was 2% (v w^{-1}). • Enzymatic hydrolysis by cellulase and protease was carried out at 45°C, while the experiments with lipase and phospholipase A1 were performed at 37°C.	Obtaining lipids, proteins, and carbohydrates	The maximum yield of 88% of lipids in the solid phase, while 74% of carbohydrates and 68% of proteins were separated in the liquid phase	Alavijeh et al. (2020)
Commercial enzymes α-amylase (Liquozyme) and glucoamylase (Dextrozyme), Cellic CTec2, and Cellic HTec2	*Haematococcus pluvialis*	• Four different hydrolysis (α-amylase + glucoamylase, α-amylase + glucoamylase, CTec2 + HTec2, and CTec2 + HTec2 + α-amylase + gl ucoamylase) • Hydrolysis time ranging from 2 h to 49 h. • pH ranging between 4.5 and 6 and temperature from 45°C to 90°C.	Production of glucose and bioethanol	7.2% glucose yield and 88.1% ethanol yield in the combined use of commercial enzymes	Hosseini et al. (2020)
Commercially produced cellulase, α-amylase, β-glucosidase, and amyloglucosidase	*Hindakia tetrachotoma*	• Lyophilized microalgae (10 g L^{-1}) and enzymes with maximum activity. • pH 5.5. • β-glycosidase and cellulase at 65°C and α-amylase at 95°C for 2 h.	Bioethanol production	Enzymatic hydrolysis with β-glucosidase/cellulase + α-amylase showed the highest saccharification values with 92.3 ± 0.9%.	Onay (2019)
Commercially produced cellulase	Mixed microalgae	• Concentration of 25, 50, 75 and 100 g L^{-1} of microalgal biomass. • Concentration of 0.416 mg protein mL^{-1} of cellulase. • Temperature 30°C–50°C and pH 4–6.	Glucose yield for bioethanol production	The highest glucose yield (57%) (microalgae biomass of 50 g L^{-1}, pH 5, and temperature of 50°C)	Shokrkar, Ebrahimi, and Zamani (2018)

(Continued)

TABLE 12.1 (CONTINUED)
Use of Enzymes in Enzymatic Hydrolysis Processes to Obtain Biomolecules and/or Bioproducts from Microalgal Biomass

Enzymes	Microalgae	Experimental Conditions	Bioproducts/ Bioprocess	Hydrolysis Efficiency	Reference
Commercial enzymes: cellulase (Celluclast 1.5 L) and pectinase (Pectinex SP-L)	*Porphyridium cruentum*	• Enzymatic hydrolysis was performed with samples of SPC (seawater *P. cruemtum*) and FPC (freshwater *P. cruemtum*). • 1% (w/v) dry matter, pectinase (4.8 mg/g BW), cellulase (7.2 mg g^{-1} BW), 0.1% (w v^{-1}) yeast extract, peptone at 0. 2% (w v^{-1}) and 50 mM citrate buffer (pH 4.8) at 37°C for 7 h.	Conversion of glucose to bioethanol production	Glucose yield of 89.8% and 85.3% of glucose for SPC and FPC, respectively	Kim, Oh, and Bae (2017)
Commercial enzymes: cellulase (Celluclast 1.5 L) and β-glucosidase (20 CBU g^{-1})	*Scenedesmus obliquus*, *S. quadricauda*, *Nitzschia* sp., *Aphanothece* sp.	• The hydrolysis tests of untreated and pretreated microalgae were carried out at concentrations of 6% w/w of dry solid and a mixture of cellulose (dry basis). • pH was adjusted to 4.9 ± 0.1.	Production of sugars (glucose, xylose, cellobiose, and arabinose)	55%–90% sugar extraction for experiments A1 and A2	Juárez et al. (2016)
Commercially produced cellulase, pectinase, and hemicellulase	*Scenedesmus* sp.	• Biomass concentration between 2.5 and 10g L^{-1}. • Temperature from 30°C to 60°C. • pH of 3.5 to 5.0. • Protein cellulase:pectinase:hemicellulase weight ratio of 1:1:1, 2:1:1, 1:2:1, 1:1:1, and 1:1:2 (w/w/w).	Enzymatic oil extraction	Maximum yield of extracted oil reached 86.1% (algae concentration at 2.5 g L^{-1}, temperature 30/50°C, pH 3.5/4.5) and a cellulase:pectinase: hemicellulase ratio of 1:1 :1 or 1:2:1 (w/w/w)	Huo et al. (2015)
Commercial enzymes Cellulases (Celluclast 1.5 L) and β-glucosidases (Novozyme 188)	*Chlorella vulgaris*	• *C. vulgaris* cells in buffer solution were mixed with 2% (v v^{-1}) Celluclast 1.5 L and 1% (v v^{-1}) Novozyme 188. • The pH varied at 3.8, 4.8, and 5.8 at 50°C. • The temperature varied at 40°C, 50°C, and 60°C at pH 4.8. • The concentration of *C. vulgaris* was varied at 1%, 2%, 3%, 4%, and 5% by weight at pH 4.8 and 50°C.	Lipid extraction	The maximum hydrolysis yield was 85.3% lipid extraction	Cho et al. (2013)

(Continued)

TABLE 12.1 (CONTINUED)
Use of Enzymes in Enzymatic Hydrolysis Processes to Obtain Biomolecules and/or Bioproducts from Microalgal Biomass

Enzymes	Microalgae	Experimental Conditions	Bioproducts/ Bioprocess	Hydrolysis Efficiency	Reference
Commercial enzymes (Celluclast 1.5 L, Novoprime B957, AMG 300 L, and Viscozyme L)	*Dunaliella tertiolecta*	• 5% (w v^{-1}) of residual biomass. • Temperatures (35°C–55°C) and pH (3.5–6.5). • The reaction mixture was incubated with Celluclast 1.5 L, Novoprime B957, AMG 300L, and/or Viscozyme L. • The enzyme in the range of 0.1–1.0 mL g^{-1} was used based on the dry mass of residual biomass.	Lipid extraction for bioethanol production	Saccharification with AMG 300 L produced 21.0 mg mL^{-1} of reducing sugar with a yield of 42.0% (w w^{-1}) based on residual biomass at pH 5.5 and 55°C.	Lee et al. (2013)
Commercial enzymes (Alcalase 2.5 L and Flavourzyme 1000 mL)	*Scenedesmus almeriensis*	• In the first step, the biomass was transferred to the hydrolysis reactor by 2.5 L alkalase with a temperature of 50°C and pH 8.0. • In the second stage, the pH was reduced to 7.0 and then the second Flavourzyme 1000 L enzyme was added. • Optimal enzyme conditions were recommended by enzyme suppliers.	*L*-amino acid production	59% yield in amino acid production	García, Fernández, and Sevilla (2012)

enzymatic hydrolysis process simultaneously with the alcoholic fermentation, using genetic biotechnology, in the studies a recombinant yeast strain S. cerevisiae was used, which was able to maintain the α-amylase of the bacterium *Streptococcus bovis* and also to exhibit glucoamylase from *Rhizopus oryzae*. They tested it on the surface of *A. platensis* to eliminate the need for biomass pretreatment and amylase hydrolysis, with or without $CaCl_2$. However, the ethanol productivity was only increased with the addition of 100 mM of $CaCl_2$, which helped to break the surface layer of polysaccharides of *Arthrospira platensis*. In this way, they directly converted *A. platensis* biomass into ethanol with a yield of 32% (gEtOH g^{-1} biomass).

Simultaneously using commercial enzymes alpha-amylase (Liquozyme® Supra 2.2X) and amyloglycosity (AMG® 300L), both from Novozymes, Rempel et al. (2018) evaluated the saccharification efficiency of the prokaryotic microalgae *S. platensis* biomass for bioethanol production. Enzymes were used freely (1% v v^{-1}) and immobilized (1% v v^{-1}) with a pH of 5.5 and temperature of 50°C, reaching hydrolysis efficiency of 99% and 83%, respectively. Both free and immobilized, the enzymes proved to be efficient in converting polysaccharides into simple sugars, being a promising alternative in the production of biofuels from microalgae.

Rojo et al. (2021) evaluated the release of carbohydrates from microalgae-bacteria biomasses of eukaryotic characteristics, using different enzymes (Alcalase® 2.4L, Protamex®, and Celluclast® 1.5L – Novozymes). The highest carbohydrate release was 38.5%, with the use of a cellulolytic enzyme (Celluclast 1.5 L) at pH 6.5 at 50°C and hydrolysis time of 5 hours. He et al. (2020) used several enzymes to convert the biomass of the eukaryotic microalgae *Nannochloropsis* into biodiesel, including Celluclast® 1.5 L (Novozymes). The enzyme was able to hydrolyze the microalgal cell and approximately 170 mg g^{-1} of lipids and recovery of 35% methyl esters of fatty acids were obtained.

Bader et al. (2020), in turn, studied the saccharification of the biomass of the eukaryotic microalgae *Chlamydomonas reinhardtii* through enzymes released in the growth of the fungus *Trichoderma harzianum*, with pectinolytic, amylolytic, and cellulolytic activities being detected. To optimize the hydrolysis of the biomass, different combinations of temperature (40°C, 50°C, or 60°C) and pH (5, 7, or 9) were analyzed. At the end of the saccharification, 22.4 g L^{-1} reducing sugars were reached, which is promising for the conversion of microalgal strains for the production of bioethanol. Velazquez-Lúcio et al. (2018) discuss that in addition to many high-value molecules, microalgae cells can accumulate neutral lipids (20% to 60% of dry weight) and storage polysaccharides (20% to 50% starch), which can be used for the production of bioethanol.

For the hydrolysis of *Chlorella vulgaris* biomass, Ho et al. (2013) carried out a comparative study between acid hydrolysis and enzymatic hydrolysis (α-amylase and amyloglycosity) for the production of bioethanol. Hydrolysis yields of 87.6% were obtained for acid hydrolysis and 89.3% for enzymatic hydrolysis. A study by Ortega and Galán (2021) proves that, in addition to cellulose and hemicellulose, the microalgae in the study are made up of complex structures, such as pectins, and that, for example, due to the action of pectinase, arabinose release increased because of the degradation of arabinans present in the main structure of pectins. Furthermore, in the Pectinex–Celluclast–Viscozyme enzyme mixture, the authors reported the breakdown of pectins and polygalacturonic acids to improve the interaction between cellulolytic, hemicellulolytic, cellulose, and hemicellulose enzymes present in the residual biomass of the microalgae biomass, increasing the amounts of glucose and final xylose.

In a study carried out by Lee et al. (2013), the enzymes Celluclast® 1.5 L, Novoprime® B957, AMG® 300 L, and/or Viscozyme® L (Novozymes) were used for saccharification of the biomass of the eukaryotic alga *Dunaliella tertiolecta* and production of bioethanol. Biomass was used after lipid extraction to produce biodiesel. The results showed a hydrolysis efficiency of 42% for the residual biomass (pH 5.5 and temperature of 55°C), with the release of 21 mg mL^{-1} of reducing sugars. The bioethanol yield was 0.14 g ethanol g^{-1} of biomass and 0.44 (g of ethanol) (g^{-1} glucose).

Hosseini et al. (2020) implemented a biorefinery approach to produce the antioxidant astaxanthin and the biofuels bioethanol and biogas from the biomass of the eukaryotic microalgae

Haematococcus pluvialis. First, the authors performed the extraction of astaxanthin, for subsequent enzymatic hydrolysis of the biomass to produce bioethanol. Four different hydrolysis processes were used with pH and temperatures varying between 4.5 and 6.5 and 45°C and 90°C. An efficiency of 97.2% was obtained on the total residual glucan, which was then fermented to produce 0.21 g ethanol g^{-1} of residual biomass. The hydrolysis that obtained the best results was performed using α-amylase (Liquezyme) and glucoamylase (Dextrozyme GA), both with an enzyme load of 10 μL g^{-1} of residual biomass, for the hydrolysis of α-glucans, and a mixture of cellulases to hydrolysis of β-glucans (Cellic CTec2 and Cellic HTec2), using 10 FPU g^{-1} of residual biomass.

García et al. (2012) performed enzymatic hydrolysis of *Scenedesmus almeriensis* biomass for amino acid production using commercial enzymes Alcalase 2.5 L and Flavourzyme 1000 mL at concentrations of 2.5 AU-A g^{-1} and 1000 AU-A g^{-1}, respectively, following the recommendations of the manufacturer (Novozymes). The temperature was set at 50°C and the pH was adjusted to 8.0 during the period of action of Alcalase 2.5 L, and then to 7.0 for the addition of Flavourzyme 1000 mL. The study was separated into two hydrolysis steps. The data show that carrying out a second hydrolysis step increases the degree of hydrolysis from 51% to 59%. The authors also address that the concentration of biomass and enzymes during the enzymatic hydrolysis process are important operational parameters and, therefore, need to be optimized, mainly because enzymes are expensive and therefore are used in concentrations as low as possible, which are normally between 1% and 5% in this type of application.

Although much of the genetic resources of microalgae remain unexplored, an increasing number of enzymes with potential applications in the food, pharmaceutical, and chemical industries have been reported, some of the most important challenges will be to increase the concentration of enzymes and reduce the cost of downstream operations (Brasil et al., 2017). In addition, with advancing technologies, more stable and efficient enzymes can be obtained through genetic engineering and the study of enzyme immobilization techniques can considerably affect enzyme pretreatment. Furthermore, synthetic biology can be applied to obtain enzymes with greater efficiency, for example, thermostable enzymes can be exploited to increase biomass hydrolysis (Barati et al., 2021).

The advances, coupled with the use of all algal biomass to produce various bioproducts in a biorefinery strategy, offer promising possibilities to overcome these limitations and achieve cost-effective production of industrial enzymes in microalgae cells in the future. Microalgae biorefinery involves the combination of technologies of harvesting, fractionation, hydrolysis of algal biomass, and conversion into products with the concomitant recovery of intermediate products, thus, a biorefinery approach in microalgae technology is essential from an economic perspective, as the cultivation of microalgae exclusively for the purpose of biofuel production is quite expensive and the cost of the product is not competitive with conventional chemicals available on the market. In addition, microalgae biomass has great potential for a diverse number of products such as food biochemicals, nutraceuticals, cosmetics, and fuel production (Brasil et al., 2017).

12.5 CONCLUSIONS AND PERSPECTIVES

Enzymatic hydrolysis of microalgal biomass to obtain biomolecules of industrial interest, such as carbohydrates, proteins, lipids, and even bioactive compounds, has been the subject of studies for some time. However, some factors need to be considered to optimize the enzymatic hydrolysis process, mainly due to the characteristics of microalgal biomass and enzymes. This is because the way of adding the enzymes varies according to the species of microalgae applied, with amylases and cellulases being the most used enzymes in the enzymatic hydrolysis of microalgae. Still, there is a need to characterize the pH, temperature, and hydrolysis time of the enzymes used, mainly due to the variation in enzymatic activity of each enzyme. It is also important to highlight the importance of pretreatments carried out to break the cell walls of the microalgal biomass, making the subsequent process of enzymatic hydrolysis more efficient.

The optimization of biotechnological processes to obtain products via microalgal biomass can be one of the biggest challenges. Thus, efforts are being made to reach industrial scales of production and increasingly processes that meet the concepts of biorefineries and circular bioeconomy. Much has been reported about new pretreatments, carried out under milder conditions, without the use of temperatures and pressures, or even the use of enzymes that can act on different biochemical compounds of the biomass in pH ranges and temperatures close to each other. However, it is necessary to seek and elucidate technologies and bioprocesses that jointly idealize different stages. An example of this is the use of consolidated bioprocesses. In this case, most of the steps developed together, with the production of enzymes, saccharification, and fermentation to obtain bioethanol. Residues from this process can still be used as biofertilizer.

REFERENCES

Aikawa, S., Inokuma, K., Wakai, S., Sasaki, K., Ogino, C., Chang, J. S., … Kondo, A. (2018). Direct and highly productive conversion of cyanobacteria *Arthrospira platensis* to ethanol with CaCl 2 addition. *Biotechnology for Biofuels, 11*(1), 1–9.

Al Abdallah, Q., Nixon, B. T., & Fortwendel, J. R. (2016). The enzymatic conversion of major algal and cyanobacterial carbohydrates to bioethanol. *Frontiers in Energy Research, 4*, 36.

Alavijeh, R. S., Karimi, K., Wijffels, R. H., van den Berg, C., & Eppink, M. (2020). Combined bead milling and enzymatic hydrolysis for efficient fractionation of lipids, proteins, and carbohydrates of Chlorella vulgaris microalgae. *Bioresource Technology, 309*, 123321

Bader, A. N., Rizza, L. S., Consolo, V. F., & Curatti, L. (2020). Efficient saccharification of microalgal biomass by *Trichoderma harzianum* enzymes for the production of ethanol. *Algal Research, 48*, 101926.

Bahlawan, S., Ash, Z., Damayanti, A., Putri, A., Dewi, R., Triwibowo, B., & Prasetiawan, H. (2022). Comparative study on the various hydrolysis and fermentation methods of *Chlorella vulgaris* biomass for the production of bioethanol. *International Journal of Renewable Energy Development, 11*(2), 515–522.

Barati, B., Zafar, F. F., Rupani, P. F., & Wang, S. (2021). Bacterial pretreatment of microalgae and the potential of novel nature hydrolytic sources. *Environmental Technology and Innovation, 21*, 101362.

Brasil, B. D. S. A. F., de Siqueira, F. G., Salum, T. F. C., Zanette, C. M., & Spier, M. R. (2017). Microalgae and cyanobacteria as enzyme biofactories. *Algal Research, 25*, 76–89.

Burlacu, A., Cornea, C. P., & Israel-Roming, F. (2016). Microbial xylanase: A review. *Scientific Bulletin. Series F. Biotechnologies, 20*, 335–342.

Cheirsilp, B., & Maneechote, W. (2022). Insight on zero waste approach for sustainable microalgae biorefinery: Sequential fractionation, conversion and applications for high-to-low value-added products. *Bioresource Technology Reports*,18, 101003.

Cheng, P., Huang, J., Song, X., Yao, T., Jiang, J., Zhou, C., … Ruan, R. (2022). Heterotrophic and mixotrophic cultivation of microalgae to simultaneously achieve furfural wastewater treatment and lipid production. *Bioresource Technology, 349*, 126888.

Chew, K. W., Chia, S. R., Show, P. L., Yap, Y. J., Ling, T. C., & Chang, J. S. (2018). Effects of water culture medium, cultivation systems and growth modes for microalgae cultivation: A review. *Journal of the Taiwan Institute of Chemical Engineers, 91*, 332–344.

Cho, H. S., Oh, Y. K., Park, S. C., Lee, J. W., & Park, J. Y. (2013). Effects of enzymatic hydrolysis on lipid extraction from *Chlorella vulgaris*. *Renewable Energy, 54*, 156–160.

Constantino, A., Rodrigues, B., Leon, R., Barros, R., & Raposo, S. (2021). Alternative chemo-enzymatic hydrolysis strategy applied to different microalgae species for bioethanol production. *Algal Research, 56*, 102329.

Dragone, G. (2022). Challenges and opportunities to increase economic feasibility and sustainability of mixotrophic cultivation of green microalgae of the genus *Chlorella*. *Renewable and Sustainable Energy Reviews, 160*, 112284.

Ferreira, A., Ribeiro, B., Ferreira, A. F., Tavares, M. L., Vladic, J., Vidović, S., … Gouveia, L. (2019). *Scenedesmus obliquus* microalga-based biorefinery–from brewery effluent to bioactive compounds, biofuels and biofertilizers–aiming at a circular bioeconomy. *Biofuels, Bioproducts and Biorefining, 13*(5), 1169–1186.

García, J. R., Fernández, F. A., & Sevilla, J. F. (2012). Development of a process for the production of l-amino-acids concentrates from microalgae by enzymatic hydrolysis. *Bioresource Technology, 112*, 164–170.

Hamed, S. B., Hamed, M. B. B., Kassouar, S., & Abi Ayad, S. M. E. A. (2016). Physicochemical analysis of cellulose from microalgae *Nannochloropsis gaditana*. *African Journal of Biotechnology, 15*(24), 1201–1206.
He, Y., Zhang, B., Guo, S., Guo, Z., Chen, B., & Wang, M. (2020). Sustainable biodiesel production from the green microalgae *Nannochloropsis*: Novel integrated processes from cultivation to enzyme-assisted extraction and ethanolysis of lipids. *Energy Conversion and Management, 209*, 112618.
Ho, S. H., Huang, S. W., Chen, C. Y., Hasunuma, T., Kondo, A., & Chang, J. S. (2013). Characterization and optimization of carbohydrate production from an indigenous microalga *Chlorella vulgaris* FSP-E. *Bioresource Technology, 135*, 157–165.
Hosseini, A., Jazini, M., Mahdieh, M., & Karimi, K. (2020). Efficient superantioxidant and biofuel production from microalga *Haematococcus pluvialis* via a biorefinery approach. *Bioresource Technology, 306*, 123100.
Huo, S., Wang, Z., Cui, F., Zou, B., Zhao, P., & Yuan, Z. (2015). Enzyme-assisted extraction of oil from wet microalgae *Scenedesmus* sp. G4. *Energies, 8*(8), 8165–8174.
Jiang, J., Yao, C., Cao, X., Liu, Y., & Xue, S. (2017). Characterization of starch phosphorylase from the marine green microalga (*Chlorophyta*) *Tetraselmis subcordiformis* reveals its potential role in starch biosynthesis. *Journal of Plant Physiology, 218*, 84–93.
John, J., Kaimal, K. S., Smith, M. L., Rahman, P. K., & Chellam, P. V. (2020). Advances in upstream and downstream strategies of pectinase bioprocessing: A review. *International Journal of Biological Macromolecules, 162*, 1086–1099.
Juárez, J. M., Hernando, A. L., Torre, R. M., Lanza, S. B., & Rodríguez, S. B. (2016). Saccharification of microalgae biomass obtained from wastewater treatment by enzymatic hydrolysis. Effect of alkaline-peroxide pretreatment. *Bioresource Technology, 218*, 265–271.
Junior, A. D. N. F., Etchebehere, C., Perecin, D., Teixeira, S., & Woods, J. (2022). Advancing anaerobic digestion of sugarcane vinasse: Current development, struggles and future trends on production and end-uses of biogas in Brazil. *Renewable and Sustainable Energy Reviews, 157*, 112045.
Khan, M. I., Shin, J. H., & Kim, J. D. (2018). The promising future of microalgae: Current status, challenges, and optimization of a sustainable and renewable industry for biofuels, feed, and other products. *Microbial Cell Factories, 17*(1), 1–21.
Kim, H. M., Oh, C. H., & Bae, H. J. (2017). Comparison of red microalgae (*Porphyridium cruentum*) culture conditions for bioethanol production. *Bioresource Technology, 233*, 44–50.
Kim, K. H., Choi, I. S., Kim, H. M., Wi, S. G., & Bae, H. J. (2014). Bioethanol production from the nutrient stress-induced microalga *Chlorella vulgaris* by enzymatic hydrolysis and immobilized yeast fermentation. *Bioresource Technology, 153*, 47–54.
Kocabaş, D. S., Güder, S., & Özben, N. (2015). Purification strategies and properties of a low-molecular weight xylanase and its application in agricultural waste biomass hydrolysis. *Journal of Molecular Catalysis B: Enzymatic, 115*, 66–75.
Kumar, K. S., Dahms, H. U., Won, E. J., Lee, J. S., & Shin, K. H. (2015). Microalgae–a promising tool for heavy metal remediation. *Ecotoxicology and Environmental Safety, 113*, 329–352.
Kusmiyati, K., Heratri, A., Kubikazari, S., Hidayat, A., & Hadiyanto, H. (2020). Hydrolysis of microalgae *Spirulina platensis, Chlorella sp.*, and macroalgae *Ulva lactuca* for bioethanol production. *International Energy Journal, 20*(4), 611–620.
Lee, O. K., Kim, A. L., Seong, D. H., Lee, C. G., Jung, Y. T., Lee, J. W., & Lee, E. Y. (2013). Chemo-enzymatic saccharification and bioethanol fermentation of lipid-extracted residual biomass of the microalga, *Dunaliella tertiolecta*. *Bioresource Technology, 132*, 197–201.
Li, X., Dreher, T. W., & Li, R. (2016). An overview of diversity, occurrence, genetics and toxin production of bloom-forming *Dolichospermum* (Anabaena) species. *Harmful Algae, 54*, 54–68.
Liu, H., Ru, G., Zhang, Z., Li, Y., Xia, C., Lu, C., & Zhang, Q. (2021). Experimental study on optimization of initial pH for photo-fermentation bio-hydrogen under different enzymatic hydrolysis of *chlorella vulgaris*. *Bioresource Technology, 338*, 125571.
Liu, H., Zhang, Z., Zhang, H., Lee, D. J., Zhang, Q., Lu, C., & He, C. (2020). Evaluation of hydrogen yield potential from Chlorella by photo-fermentation under diverse substrate concentration and enzyme loading. *Bioresource Technology, 303*, 122956.
Liu, W., Chen, W., Hou, Q., Wang, S., & Liu, F. (2018). Effects of combined pretreatment of dilute acid pre-extraction and chemical-assisted mechanical refining on enzymatic hydrolysis of lignocellulosic biomass. *RSC Advances, 8*(19), 10207–10214.
Lynd, L. R., Weimer, P. J., Van Zyl, W. H., & Pretorius, I. S. (2002). Microbial cellulose utilization: Fundamentals and biotechnology. *Microbiology and Molecular Biology Reviews, 66*(3), 506–577.

Maffei, G., Bracciale, M. P., Broggi, A., Zuorro, A., Santarelli, M. L., & Lavecchia, R. (2018). Effect of an enzymatic treatment with cellulase and mannanase on the structural properties of Nannochloropsis microalgae. *Bioresource Technology*, *249*, 592–598.

Maity, S., & Mallick, N. (2022). Trends and advances in sustainable bioethanol production by marine microalgae: A critical review. *Journal of Cleaner Production*, 345, 131153.

McAllister, T. G., Wood, S. A., & Hawes, I. (2016). The rise of toxic benthic Phormidium proliferations: A review of their taxonomy, distribution, toxin content and factors regulating prevalence and increased severity. *Harmful Algae*, *55*, 282–294.

Monjed, M. K., Achour, B., Robson, G. D., & Pittman, J. K. (2021). Improved saccharification of *Chlorella vulgaris* biomass by fungal secreted enzymes for bioethanol production. *Algal Research*, *58*, 102402.

Mota, G. F., de Sousa, I. G., de Oliveira, A. L. B., Cavalcante, A. L. G., da Silva Moreira, K., Cavalcante, F. T. T., ... dos Santos, J. C. (2022). Biodiesel production from microalgae using lipase-based catalysts: Current challenges and prospects. *Algal Research*, *62*, 102616.

Neves, S. A., & Marques, A. C. (2021). The substitution of fossil fuels in the US transportation energy mix: Are emissions decoupling from economic growth? *Research in Transportation Economics*, *90*, 101036.

Ngamsirisomsakul, M., Reungsang, A., Liao, Q., & Kongkeitkajorn, M. B. (2019). Enhanced bio-ethanol production from Chlorella sp. biomass by hydrothermal pretreatment and enzymatic hydrolysis. *Renewable Energy*, *141*, 482–492.

Onay, M. (2019). Bioethanol production via different saccharification strategies from *H. tetrachotoma* ME03 grown at various concentrations of municipal wastewater in a flat-photobioreactor. *Fuel*, *239*, 1315–1323.

Ortega, J. L. S., & Galán, J. E. L. (2021). Hidrólise enzimática de carboidratos restos na biomassa de microalga *Chlorella vulgaris*. *Revista virtual de Química*, 1, 1–6.

Peron-Schlosser, B., Stobienia, M., Bispo, L. D. O., Colla, L. M., Baraldi, I. J., & Colla, E. (2021). Residual fractions from *Arthrospira platensis* protein extraction as feedstock for ethanol production. *Biocatalysis and Biotransformation*, *39*(5), 378–388.

Rempel, A., Machado, T., Treichel, H., Colla, E., Margarites, A. C., & Colla, L. M. (2018). Saccharification of *Spirulina platensis* biomass using free and immobilized amylolytic enzymes. *Bioresource Technology*, *263*, 163–171.

Rempel, A., Nadal Biolchi, G., Farezin Antunes, A. C., Gutkoski, J. P., Treichel, H., & Colla, L. M. (2021). Cultivation of microalgae in media added of emergent pollutants and effect on growth, chemical composition, and use of biomass to enzymatic hydrolysis. *Bioenergy Research*, *14*(1), 265–277.

Rodrigues, É. F., Ficanha, A. M. M., Dallago, R. M., Treichel, H., Reinehr, C. O., Machado, T. P., ... Colla, L. M. (2017). Production and purification of amylolytic enzymes for saccharification of microalgal biomass. *Bioresource Technology*, *225*, 134–141.

Rodrigues, E. M. G. (2021). Utilização de subproduto da indústria cervejeira como substrato para a produção de amilase por fermentação em estado sólido. *Bioenergia em Revista: Diálogos (ISSN: 2236-9171)*, *11*(1), 46–57.

Rojo, E. M., Piedra, I., González, A. M., Vega, M., & Bolado, S. (2021). Effect of process parameters on the valorization of components from microalgal and microalgal-bacteria biomass by enzymatic hydrolysis. *Bioresource Technology*, *335*, 125256.

Roostaei, J., Zhang, Y., Gopalakrishnan, K., & Ochocki, A. J. (2018). Mixotrophic microalgae biofilm: A novel algae cultivation strategy for improved productivity and cost-efficiency of biofuel feedstock production. *Scientific Reports*, *8*(1), 1–10.

Russell, C., Rodriguez, C., & Yaseen, M. (2021). High-value biochemical products & applications of freshwater eukaryotic microalgae. *Science of the Total Environment*, 809, 151111.

Sankaran, R., Cruz, R. A. P., Pakalapati, H., Show, P. L., Ling, T. C., Chen, W. H., & Tao, Y. (2020). Recent advances in the pretreatment of microalgal and lignocellulosic biomass: A comprehensive review. *Bioresource Technology*, *298*, 122476.

Sendra, M., Blasco, J., & Araújo, C. V. (2018). Is the cell wall of marine phytoplankton a protective barrier or a nanoparticle interaction site? Toxicological responses of *Chlorella autotrophica* and *Dunaliella salina* to Ag and CeO_2 nanoparticles. *Ecological Indicators*, *95*, 1053–1067.

Shokrkar, H., Ebrahimi, S., & Zamani, M. (2018). Enzymatic hydrolysis of microalgal cellulose for bioethanol production, modeling and sensitivity analysis. *Fuel*, *228*, 30–38.

Siddiki, S. Y. A., Mofijur, M., Kumar, P. S., Ahmed, S. F., Inayat, A., Kusumo, F., ... Mahlia, T. M. I. (2022). Microalgae biomass as a sustainable source for biofuel, biochemical and biobased value-added products: An integrated biorefinery concept. *Fuel*, *307*, 121782.

Souza, M. F., Rodrigues, M. A., Freitas, S. P., & da Silva Bon, E. P. (2020). Effect of milling and enzymatic hydrolysis in the production of glucose from starch-rich *Chlorella sorokiniana* biomass. *Algal Research, 50*, 101961.

Srivastava, N., Mishra, P. K., & Upadhyay, S. N. (2020). Industrial enzymes for biofuels production. *Elsevier*, 1, 1–226.

Sriyod, K., Reungsang, A., & Plangklang, P. (2021). Single and combined enzymatic saccharification and biohydrogen production from *Chlorella* sp. *Biomass and Bioenergy Research, 14*(3), 940–953.

Udaiyappan, A. F. M., Hasan, H. A., Takriff, M. S., & Abdullah, S. R. S. (2017). A review of the potentials, challenges and current status of microalgae biomass applications in industrial wastewater treatment. *Journal of Water Process Engineering, 20*, 8–21.

Velazquez-Lucio, J., Rodríguez-Jasso, R. M., Colla, L. M., Sáenz-Galindo, A., Cervantes-Cisneros, D. E., Aguilar, C. N., ... Ruiz, H. A. (2018). Microalgal biomass pretreatment for bioethanol production: A review. *Biofuel Research Journal*, 5(1), 780–791.

Voigt, J., Stolarczyk, A., Zych, M., Malec, P., & Burczyk, J. (2014). The cell-wall glycoproteins of the green alga *Scenedesmus obliquus*. The predominant cell-wall polypeptide of Scenedesmus obliquus is related to the cell-wall glycoprotein gp3 of *Chlamydomonas reinhardtii*. *Plant Science, 215*, 39–47.

Zhang, Y., Kang, X., Zhen, F., Wang, Z., Kong, X., & Sun, Y. (2022). Assessment of enzyme addition strategies on the enhancement of lipid yield from microalgae. *Biochemical Engineering Journal, 177*, 108198.

Zhang, Y., Kong, X., Wang, Z., Sun, Y., Zhu, S., Li, L., & Lv, P. (2018). Optimization of enzymatic hydrolysis for effective lipid extraction from microalgae *Scenedesmus* sp. *Renewable Energy, 125*, 1049–1057.

Zhao, K., Zhang, M., Tian, H., Lei, F., He, D., Zheng, J., & Zhang, L. (2022). Effects of structural and compositional changes of *Nanochloropsis oceania* after enzyme treatment on EPA-rich lipids extraction. *Marine Drugs, 20*(3), 160.

13 Enzymatic Hydrolysis of Macroalgal Biomass

Dominique F. Rocher, Rosemary A. Cripwell, Willem H. van Zyl, and Marinda Viljoen-Bloom

13.1 INTRODUCTION

Macroalgae have been used commercially in several industries, including food, printing and paper, medical, cosmetics, and agriculture (Leandro et al., 2019). Brown macroalgae contain alginate, a polysaccharide that is used as a thickening and gelling agent or a colloid in food, whereas red macroalgae are used in the manufacturing of carrageenan for cosmetics, food processing and other industrial applications. The use of macroalgae for human consumption or as a fertiliser can be attributed to its high nutritional value. Studies have shown that macroalgae contain approximately 27% to 39% carbon, which includes carbohydrates, protein, lipids and polyphenols. A large proportion of the carbon content is structural and storage carbohydrates (Schiener, 2013), whereas an abundance of sulphated polysaccharides includes fucoidans, agars, carrageenans and hemicelluloses.

Different strategies and combinations of processes are required to utilise macroalgae and exploit their valuable compounds efficiently. After harvesting, macroalgae are mechanically milled to reduce the particle size of macroalgae, which increases their surface area and thus the extraction efficiency. The macroalgal biomass can be fractionated into the main components, i.e., proteins, lipids, carbohydrates and pigments (Cikoš et al., 2022). Commercial products such as omega-3 fatty acids or pigments are usually extracted and purified from dried biomass. A selective biorefining with a cascade approach has been suggested for commercial-scale processes, i.e., the extraction of hydrophilic components (proteins and carbohydrates), followed by extraction and fractionation of lipids and, finally, the recovery of insoluble components (Ruiz et al., 2016).

The extraction process can be further assisted by microbial enzymes that can hydrolyse polysaccharides, thus releasing carbohydrates. The hydrolysis of structural polysaccharides would also render the biomass more susceptible to the extraction of lipids, pigments and other unique macroalgal products. Recombinant expression of one or more of these enzymes in a fermenting organism can support a biorefinery concept for macroalgal biomass to produce a range of products, including bioethanol. Brown and red macroalgae are considered potential renewable feedstocks for third-generation biofuels, but this requires microbial enzymes to hydrolyse the polysaccharides to sugars that can be fermented to ethanol.

This chapter discusses the polysaccharides found in macroalgae, in particular mannitol, alginate and laminarin, and the corresponding microbial enzymes (mannitol dehydrogenases, alginate lyases and laminarinases) required for their hydrolysis. The role of these enzymes in substrate conversion is also highlighted by comparing different fermentation approaches for bioethanol production.

13.2 POLYSACCHARIDES IN MACROALGAE

Macroalgae contain high levels of structural polysaccharides but have a low cellulose content, and none or minor quantities of lignin. Other compounds unique to macroalgae that contribute to their commercial value are alginate, carrageenan and agar. The chemical composition of

 DOI: 10.1201/9781003335313-15

macroalgae varies considerably between genera and species. Still, it is possible to group certain compounds under the major macroalgal taxonomic groups, namely green, red and brown macroalgae. Carbohydrates in green algae represent approximately 44.4% (of dry weight) and include ulvan, mannan, starch and cellulose. The red algae typically have a higher carbohydrate content (49.4%), including agar, carrageenan, lignin, and cellulose. Brown algae have the lowest carbohydrate content (43.2%), including alginate, mannitol, laminarin, cellulose and glucan (Dave et al., 2019; Godvin Sharmila et al., 2021).

Macroalgal polysaccharides can be categorised into two groups, namely storage and structural polysaccharides. *Storage polysaccharides* are mostly glucans, and their concentration fluctuates with seasonal and environmental changes. The main storage polysaccharides in brown macroalgae are mannitol and laminarin, with mannitol generally dominating (Horn, 2000; Schiener, 2013). Mannitol is a six-carbon sugar alcohol corresponding to mannose (Figure 13.1A) that can be remobilised from macroalgal tissue to provide energy and carbon for growth. Laminarin is a β-1,3-D-glucan consisting of approximately 25 glucosyl residues that may represent up to 35% of the dry weight in brown macroalgae (Figure 13.1B). Most laminarin molecules terminate with a non-reducing 1-linked D-mannitol residue (resulting in an M-chain), while some molecules terminate with a reducing 3-linked glucose residue (resulting in a G-chain). Moreover, β-1,6-linked branching and β-1,6-interchain links may occur, with different branching ratios observed among brown macroalgae species. The degree of branching affects the solubility of laminarin in water; molecules consisting of only β-1,3-linked residues are water-insoluble, while branched molecules are water-soluble.

The cell wall structures of macroalgae are composed of *structural polysaccharides* that provide mechanical strength and flexibility to enable the algae to endure tidal fluctuations. These compounds are classified into three groups, namely neutral polysaccharides (cellulose), polyuronides (alginate) and sulphated polysaccharides (fucoidans) (Schiener, 2013). In contrast to terrestrial plants, cellulose represents a minor component of macroalgal cell walls (Dave et al., 2019). Cellulose generally comprises β-D-pyranose units linked by β-1,4-glycosidic bonds and occurs in elementary fibrils (Figure 13.1D). Both crystalline and amorphous regions are found within cellulosic material; amorphous regions are less compact and easier to degrade, while the recalcitrant crystalline areas strengthen the cellulose structure. Cellulose can be depolymerised into glucose through the cooperation of several cellulase enzymes, namely endoglucanases, cellobiohydrolases and β-glucosidases.

Alginate is one of the most abundant polymers in the marine environment and is the main structural polysaccharide that provides macroalgae with strength and flexibility. It comprises α-L-guluronic acid and β-D-mannuronic acid, with 1,4-glycosidic bonds linking the pyranose rings of the monosaccharides. The structure of alginate varies according to the position of the monomer on the chain. It can form homopolymeric blocks of poly-1,4-D-mannuronic acid (MM blocks) or poly-1,4-L-guluronic acid (GG blocks), or heteropolymeric blocks with arrangements of both monomers (MG blocks) (Figure 13.1F) (Schiener, 2013).

Fucoidan is a heterogeneous sulphated polysaccharide unique to brown macroalgae and is classified as a mucopolysaccharide (Figure 13.1E) that is extracted for its application in the food and pharmaceutical industries. The heteropolysaccharide mainly consists of sulphated L-fucose residues and is located in the inner boundary of cells or cell walls. Fucoidan may also contain other monosaccharides, uronic acid and proteins. The chemical composition of fucoidan is very complex; it usually includes 1,3-linked α-L-fucopyranose residues and 1,3- and 1,4-linked α-L-fucopyranose residues, with sulphated esters at positions 2 and 4, with rare occurrences at position 3 (Schiener, 2013).

Starch (Figure 13.1C) consists of a mixture of amylose and amylopectin molecules. Amylose is a linear polymer consisting of up to 6 000 glucose units linked by α-1,4 glycosydic bonds, whereas amylopectin is a branched polymer that consists of short linear chains of 10 to 60 glucose units joined by α-1,4 bonds and α-1,6 linked side chains of 15 to 45 glucose units (Den Haan et al, 2021). Other polymers (Figure 13.1G–I) in macroalgae include ulvan, a heteropolymer composed

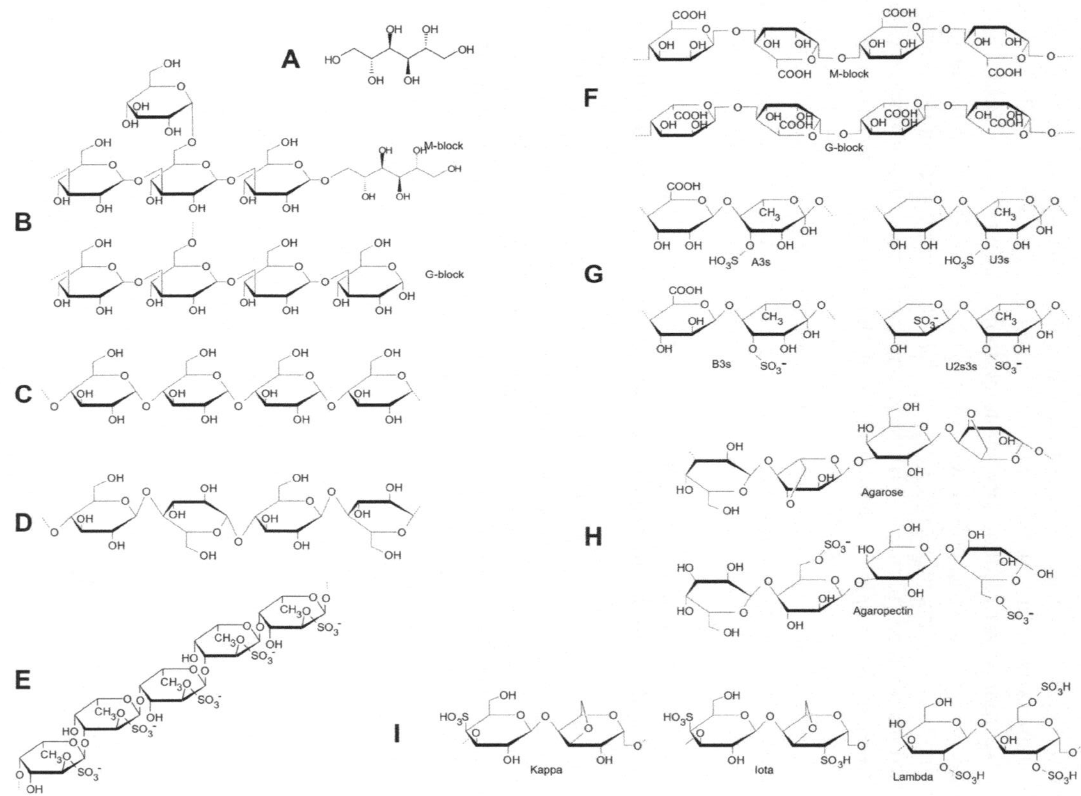

FIGURE 13.1 Chemical structure of different polysaccharides found in macroalgae. (A) Mannitol monomer. (B) Laminarin with α-1,3 bonds between glucose moieties and occasional α-1,6 branches. Laminarin chains end with either a mannitol (M) or glucose (G) block. (C) Starch with α-1,4 bonds and (D) cellulose with β-1,4 bonds between glucose moieties. (E) Fucoidan with α-1,3 bonds or α-1,4 bonds between sulphated fucose moieties. (F) Alginate contains β-1,4 mannuronic acid and/or α-l-guluronic acid moieties. (G) Ulvan is a heteropolymer of ulvanobiouronate A, ulvanobiouronate B and ulvanobiouronate U with sulfate groups on some of the carbon 3 positions. (H) Agarose is composed of 1,3-linked-d-galactopyranose and 1,4-linked 3,6-anhydro-α-l-galactopyranose moieties. (I) Carrageenan is a heteropolymer composed of galactose monomers with different levels of sulphate substitutions.

of α -L-rhamnose-3-sulphate-1,4-β-D-glucuronic acid (ulvanobiouronate A), α-L-rhamnose-3-sulphate-1,4-α-D-iduronic (ulvanobiouronate B), and α-L-rhamnose-3-sulphate-1,4-β-D-xylose (ulvanobiouronate U) with sulphate groups on some of the carbon-3 positions. Agarose is composed of 1,3-linked-d-galactopyranose and 1,4-linked 3,6-anhydro-α-l-galactopyranose moieties. Agaropectin has a similar structure to agarose, but contains sulphate branches on carbon-6 instead of the 3,6-anhydro link. Carrageenan is a heteropolymer composed of galactose monomers with different levels of sulphate substitutions.

The organic waste that remains after the industrial extraction of pigments and lipids mainly contains cellulose and some algal colloids, such as carrageenan and alginate (Dave et al., 2019). Removing alginate from macroalgal waste is essential as it causes various problems during waste treatment and management. Alginate removal could be achieved by converting alginate to alginate oligosaccharides, a high-value product with several bioactivities, including plant growth promotion, antioxidant, antimicrobial, anti-tumour and immunomodulation.

Unique macroalgal chemical components (absent in terrestrial biomass) offer the opportunity for a biorefinery concept where different products of economic value can be derived from the biomass. One of the most valuable products from macroalgae is natural pigments, which vary depending on the macroalgal species and growth conditions. The three main classes of pigments are chlorophylls, carotenoids (carotenes and xanthophylls) and phycobiliproteins (Cikoš et al., 2022). Some macroalgal pigments have strong antioxidant properties, such as the carotenoids astaxanthin and fucoxanthin.

Sulphated oligosaccharides isolated from macroalgae (e.g., carrageenan, agar and fucoidan) are biopolymers with various physicochemical properties and biological activities. They have attracted attention due to their antimicrobial and antiviral properties and are generally produced by either chemical treatment or enzymatic hydrolysis. For example, ulvans in green seaweed cell walls can be depolymerised to oligo-ulvans, whereas oligo-fucans can be obtained by the enzymatic hydrolysis of fucoidan. Macroalgae can also be exploited for rare sugars, which are water-soluble monosaccharides with unique chemical structures that are naturally found in small quantities. They are species-specific and include sugar alcohols, simple sugars, anhydro-sugars and functionalised sugars (Bayu et al., 2021) that can be applied as low-caloric sweeteners (using mannitol) and precursors for valuable chemicals such as L-rhamnose.

13.3 ENZYMATIC HYDROLYSIS OF MACROALGAE CARBOHYDRATES

The macroalgal polysaccharide composition differs from terrestrial biomass in which degradable polysaccharides are mainly targeted for bioconversion. Macroalgae contain neutral sugars and uronic acids similar to terrestrial plants, but also carry sulphate moieties attached to the hydroxyl groups of sugars (Schiener, 2013). The different specific arrangements and linkages of hexose sugars (glucose, galactose and mannose), pentoses (xylose and arabinose) and deoxyhexoses (rhamnose and fucose) in macroalgal polysaccharides contribute to a variety of polysaccharides with different properties. In addition, different structural conformations are attributed to the presence of uronic acids (glucuronic, mannuronic and guluronic acid) and half-ester sulphate groups (Bayu et al., 2021).

Exploiting macroalgae for bioconversion via enzymatic hydrolysis requires identifying and evaluating microbial enzymes that can hydrolyse the most abundant carbohydrates. Glycoside hydrolases and lytic polysaccharide lyases are essential to decompose a variety of algal polysaccharides into simple sugars and are usually found in microorganisms that can degrade complex carbohydrates. However, potential sources for hydrolases are not limited to marine microbes, as various terrestrial microbes produce hydrolases that can degrade the microalgae polysaccharides. The following sections describe the main polysaccharides in brown macroalgae (mannitol, lamanarin and alginate) and the relevant hydrolases, i.e., mannitol dehydrogenases, laminarinases and alginate lyases, as well as potential sources for these enzymes.

13.3.1 Mannitol

Mannitol is a soluble and readily utilisable polysaccharide that is produced from fructose-6-phosphate (F6P) via two steps: F6P is reduced by mannitol-1-phosphate dehydrogenase (M1PDH) to mannitol-1-phosphate (M1P), which is then converted by mannitol-1-phosphatase (M1Pase) to mannitol. The enzymes mannitol-2-dehydrogenase (M2DH) and hexokinase (HK) catalyse the recycling of mannitol back to F6P in macroalgae (Iwamoto & Shiraiwa, 2005).

Mannitol metabolism has been studied extensively in various organisms, including bacteria, fungi, plants and algae, with numerous mannitol dehydrogenases identified (Iwamoto & Shiraiwa, 2005). Mannitol metabolism in the yeast *Saccharomyces cerevisiae* uses the oxidative pathway that converts mannitol to F6P in the presence of mannitol dehydrogenase and hexokinase (KEGG, 2023). Compared to glucose metabolism, an extra NADH is produced, which results in a redox imbalance (Figure 13.2). Yeast cells lack transhydrogenases and thus require oxygen as the final electron acceptor to regenerate NAD^+ during mannitol metabolism (Schiener, 2013). Semi-anaerobic conditions or a separate saccharification and fermentation process are thus required to convert mannitol to ethanol in *S. cerevisiae.*

13.3.2 Laminarin

Laminarin can be hydrolysed to glucose using laminarinase, an enzyme that primarily cleaves the β-1,3-linkages of D-glucan to form oligosaccharides containing two to ten residues (Badur et al., 2020). *Laminarinase* is an endoglucanase that may cleave either (1→3), (1→4), (1→6), or mixed (1→3), (1→4), (1→6) linkages. These endoglucanases can cleave the laminarin–oligosaccharides into glucose units, which can be fermented into ethanol by yeast (Figure 13.2). Laminarinases fall within the group of β-glucanases, but not all β-glucanases can hydrolyse laminarin.

β-Glucanases form part of the glycoside hydrolase (GH) family and are often used as biocatalysts in industrial processes. Various filamentous fungi and yeasts produce exo- and endocellular

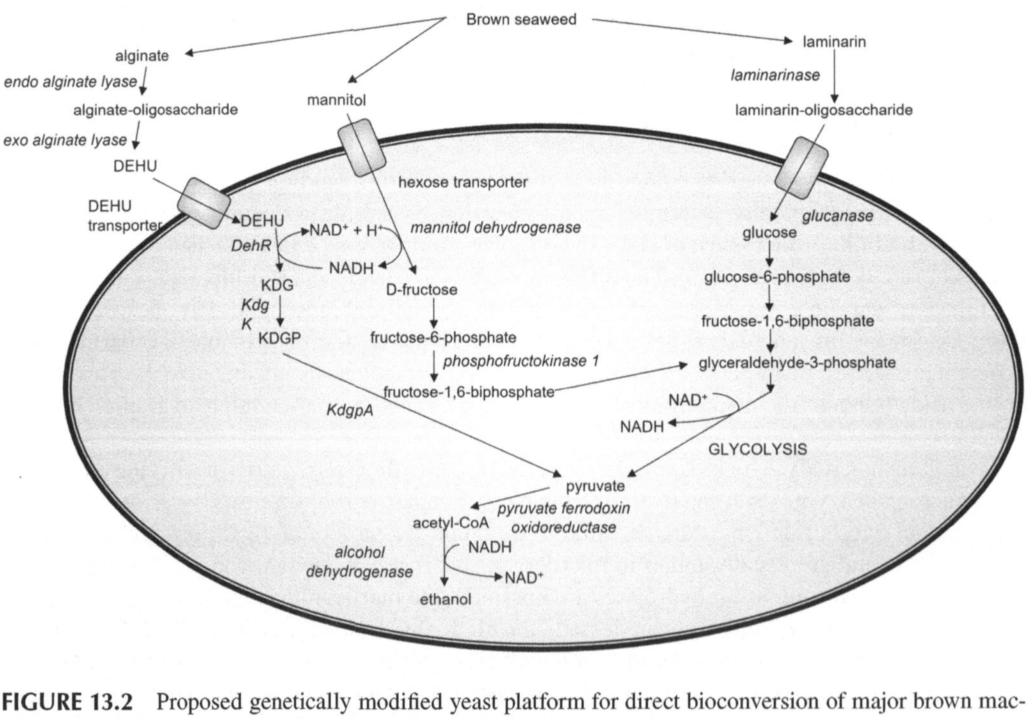

FIGURE 13.2 Proposed genetically modified yeast platform for direct bioconversion of major brown macroalgal components, namely alginate, mannitol, and laminarin.

β-glucanases, known to be the most abundant cell wall hydrolases in yeasts. Soil-borne fungi, such as those from the genus *Trichoderma,* are well known for producing many hydrolytic enzymes when cultured on laminarin, including β-1,3-glucanases, β-1,6-glucanases and β-1,3(6)-glucanases (Nobe et al., 2003).

β-Glucanases capable of cleaving the (1→3) and (1→6) linkages of β-glucan form part of the GH55 family, such as laminarinases identified in *Trichoderma viride* and *Trichoderma reesei* (Nobe et al., 2003) (Table 13.1). Endo-β-1,3(6)-glucanases hydrolysed laminarin more efficiently than endo-β-1,3-glucanases due to the presence of both β-1,6-linked and β-1,3-linked residues in laminarin. However, industrial laminarinases (e.g., from *Trichoderma* sp.) mostly contain either endo-β-1,3-glucanases or endo-β-1,3(4)-glucanases.

β-Glucanases capable of cleaving the (1→3) and (1→4) linkages of β-glucan form part of the GH16 family and were reported in both *Rasamsonia emersonii* (previously named *Talaromyces emersonii*) (You et al., 2016) and *S. cerevisiae*, among others (Table 13.1). In theory, GH55 laminarinases produce higher levels of glucose monomers from laminarin than commercial GH16 laminarinases; microorganisms producing GH55 laminarinases should thus result in higher ethanol yields during the fermentation of brown macroalgae.

13.3.3 Alginate

Alginate is a hetero-polyuronic acid and a major component of the cell wall matrix of brown macroalgae. The abundance of alginate has led to a diverse group of alginolytic organisms, including marine echinoderms and molluscs; bacteria such as *Vibrio* spp., *Pseudomonas* and *Sphingomonas*; as well as fungi such as *Aspergillus*, *Cryptococcus* and *Trichoderma* (Horn, 2000; Schiener, 2013). Alginate hydrolysis is catalysed by alginate lyases, which cleave the alginate polymer to form an unsaturated uronic acid at the new non-reducing end (Horn, 2000). Alginate lyases belong to the PL5 to PL7, PL14 to PL15 and PL17 to PL18 families that may cleave alginate in an endolytic or exolytic manner into oligomers and monomers.

Almost all *alginate lyases* act endolytically on the substrate, yielding unsaturated oligosaccharides or oligoalginates, which are further hydrolysed by cell-bound exolytic alginate lyases. The ultimate product formed during alginate catabolism is the monosaccharide 4-deoxy-L-erythro-5-hexoseulose uronic acid (DEHU), which is converted to 2-keto-3-deoxy-D-gluconic acid (KDG) by DEH reductase (DehR) (Figure 13.3). The enzymes KDG kinase (KdgK) and 2-keto-3-deoxy-6-phosphogluconate aldolase (KdgpA) then convert KDG to pyruvate and glyceraldehyde-3-phosphate, which is fed into the central metabolic pathways.

TABLE 13.1
Foreign Laminarinase and Native Laminarinase-Like Proteins Identified for Expression in *S. cerevisiae*, as well as the Respective Secretion Signals

	Protein			Protein Secretion Signal	
Organism	**Family**	**Size (# aa)**	**Protein Size (kDa)[b]**	**Yes/No**	**Estimated Size (# aa)[a]**
Trichoderma viride	GH55	759	84.36	Yes	22
Trichoderma reesei	GH55	763	84.80	Yes	17
Rasamsonia emersonii	GH16	344	38.29	Yes	19
Saccharomyces cerevisiae	GH16	779	90.84	No	—

GH = Glycoside hydrolase; aa = amino acid(s).

[a] Determined using SignalP 5.0 (www.cbs.dtu.dk/services/SignalP/).

[b] Predicted size (unglycosylated) of the secreted protein (http://www.molbiol.ru/eng/scripts/01_06.html).

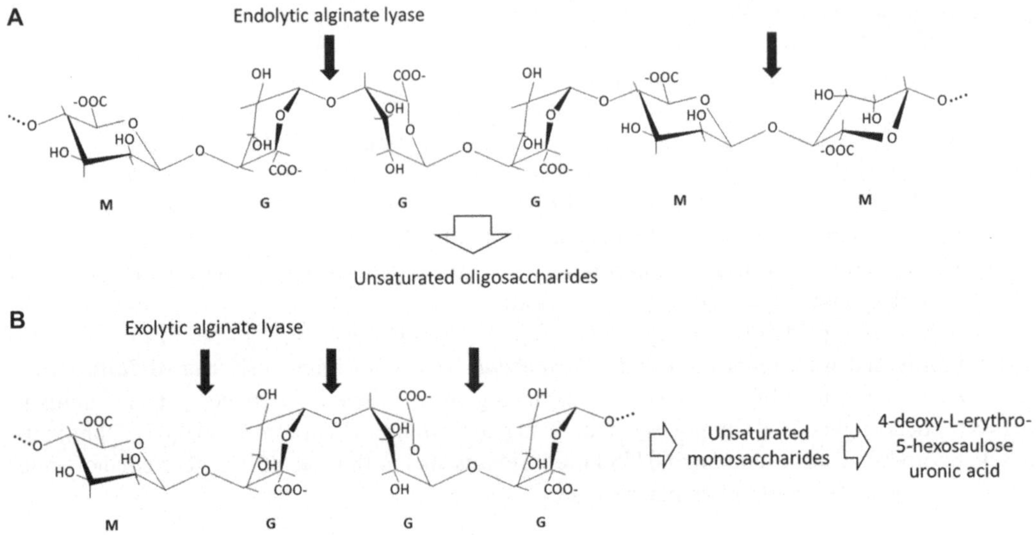

FIGURE 13.3 Alginate hydrolysis by (A) endolytic and (B) exolytic alginate lyases.

13.4 MACROALGAE FOR BIOETHANOL PRODUCTION

The growing global demand for energy production has led to many concerns, such as increased carbon dioxide emissions and the depletion of non-renewable fuel resources. This has paved the way for alternative and sustainable fuel sources such as biofuels derived from biomass or their residuals, which include biogas, biomethanol, biodiesel, bioethers, biohydrogen, bioethanol, synthetic biofuels and vegetable oils. Biofuels are classified as first, second or third generation based on the feedstock used for their production. First-generation biofuels are produced from food crops such as sugarcane, maise and grains, and are preferred as co-fuel sources for the automotive industry (Dave et al., 2019). Second-generation feedstocks are mostly lignocellulosic materials, including wood chips, bagasse, rice straw, organic waste and cotton stalk. Both first- and second-generation biofuels have several limitations, such as potential competition with food production and difficulty processing the lignin in cellulosic plant material. Third-generation biofuel production from algal feedstocks has been identified as an alternative to overcome some of these restrictions.

Carbohydrate-rich algal biomass can serve as feedstock for producing biodiesel, biohydrogen and bioethanol. Algal biomass requires less energy for its production than terrestrial biomass, can take up nutrients across its entire surface, and can be cultivated in three dimensions, resulting in rapid growth (Schiener, 2013). Algal biomass contains little lignin and low cellulose concentrations, making it a simpler substrate for conversion to biofuels than terrestrial biomass (Horn, 2000). Water represents approximately 85% of brown macroalgae biomass, which renders macroalgae a suitable candidate for an aqueous process to produce bioethanol.

Microbial hydrolysis can release valuable compounds from harvested macroalgae or waste discarded after extracting proteins, lipids, pigments, etc., using a staggered biorefinery approach. The sugars (glucose, D-fructose and DEHU/uronic acid) released by these enzymes can be fermented to bioethanol or uronic acid, among others. The high levels of polysaccharides and sugar alcohols in brown macroalgae make it more feasible than the other types of macroalgae to produce liquid biofuels. Technology to cultivate brown macroalgae on a large scale is currently being developed and would support its application as a biofuel feedstock and the production of other high-value products. The fermentation of brown macroalgae may also produce high-value by-products that can improve the economic feasibility of the bioconversion of macroalgae (Schultze-Jena et al., 2022). Several studies have examined ethanol production from various macroalgal species (Godvin Sharmila et al., 2021). Large-scale biofuel production from macroalgae could follow a production pipeline typically used for terrestrial biomass where the biomass is harvested, pretreated (if required) and converted to bioethanol via enzyme-assisted hydrolysis and sugar fermentation.

Various factors affect the solids loading of macroalgae, namely the characteristics of the chosen species/type of biomass, pretreatment conditions, exogenous enzymes and the desired product. A substrate loading of 10%–15% (w/v) is often reported for optimal fermentation for bioethanol production (Dave et al., 2019), but increasing the solids loading can severely affect mixing conditions. Organic waste from the industrial processing of marine macroalgae has also been identified as a cost-effective source for bioethanol production, with an ethanol yield of 0.45 g/g substrate resulting from an 8% slurry.

The yeast *S. cerevisiae* is the most widely used microorganism for industrial bioethanol fermentation due to its high ethanol tolerance and productivity, long industrial history, and well-established genetic background. Varying levels of success have been obtained under fermentation conditions, with an ethanol yield of 43.92 g/L reported for a *Sargassum crassifolium* hydrolysate with *S. cerevisiae* JCM3012 (Dave et al., 2019). Genetically modified yeast strains have also been investigated for their ability to ferment macroalgal biomass to ethanol, including the yeast cell-surface display system to construct *S. cerevisiae* strains able to hydrolyse and ferment mannitol and alginate from macroalgae hydrolysate. The bioconversion of brown macroalgae to ethanol is usually conducted under either anaerobic or semi-anaerobic conditions for two to eight days at 30°C–37°C at pH 4–7 (Dave et al., 2019). Fermentation of laminarin and mannitol in hydrolysate from *Laminaria hyperborea* with *Pichia angophorae* yielded 0.43 g ethanol/g substrate (Horn et al., 2000). Alginate degradation pathways were engineered in *Escherichia coli* for alginate transport and hydrolysis to pyruvate, resulting in an ethanol titre of ±4.7% v/v from the brown macroalgae, *Saccharina japonica* (Wargacki et al., 2012).

Although engineered microbial platforms for bioethanol production from brown macroalgae have been investigated (Leandro et al., 2019), commercial ethanol production from macroalgae has yet to be realised. Short-term projects such as MacroFuels investigated the macroalgae-to-fuel value chain by applying commercial or novel recombinant enzymes to macroalgae biomass (MacroFuels, 2020). Combining commercial cellulase and the bacterium *Clostridium beijerinckii* DSM-6422 effectively produced 7.16 g/L butanol from the brown macroalgae *Laminaria digitata* in batch fermentations (Hou et al., 2017).

Different fermentation processes have been developed to increase the efficiency of ethanol production from biomass. These methods include separate hydrolysis and fermentation (SHF), simultaneous saccharification and fermentation (SSF), simultaneous saccharification and co-fermentation (SSCF), and consolidated bioprocessing (CBP). During SHF, the feedstock is first hydrolysed by exogenous enzymes, followed by the addition of the fermenting microorganism(s) to allow for ethanol production. The SSF method combines the feedstock hydrolysis and fermentation processes in one step, thus reducing the risk of contamination, the time required for ethanol production and the amount of enzyme for efficient substrate hydrolysis. Different enzyme cocktails have been employed during the SSF of macroalgae, with Cellic® Ctec2 commonly used for the pretreatment of brown macroalgae (Dave et al., 2019).

The CBP approach is an energy-efficient process that combines enzyme production, biomass hydrolysis and alcoholic fermentation in one step using a single organism. The CBP approach requires producing key enzymes, namely laminarinase, mannitol dehydrogenase and alginate lyase, to efficiently hydrolyse the biomass (Rocher et al., 2021). In theory, end-product inhibition is minimised since the hydrolysis products are converted to ethanol by the fermenting microorganisms as they are produced. It also eliminates (or reduces) the need for exogenous enzyme addition, which lowers operating costs and represents a more economical and efficient process. However, enzyme supplementation during CBP may still be necessary to improve the substrate conversion rate and allow for higher solids loadings of macroalgal biomass.

Cell-surface engineering has been used by various studies that investigated the bioconversion of selected macroalgal components to ethanol, but pretreatment of the feedstock is required for efficient conversion yield. An alternative approach is to secrete the necessary enzymes into the extracellular environment where they can interact directly with the substrate (Den Haan et al., 2021). Furthermore, accumulative enzyme activity is achieved due to continuous enzyme production by the host, which increases the sugar and ethanol yields.

13.5 CONCLUSIONS AND PERSPECTIVES

Macroalgae are cultured or harvested to produce a range of products, but significant further value can be added to this renewable substrate through enzyme hydrolysis. An attractive process for bioethanol production from macroalgae is CBP, where the fermenting organism is also the host for recombinant expression of the enzymes. Although third-generation biofuels are increasingly being suggested as alternatives to conventional fuels, several factors still require further investigation to fully realise the economic feasibility of utilising algal sources for biofuel production. It would also be necessary to perform a life cycle analysis that considers the environmental impacts and energy returns of such a process.

The composition of macroalgal waste will differ depending on the species, season and upstream processing. It may thus be beneficial to modify the ratio of the enzymes to achieve a more "tailored" enzyme cocktail based on the specific feedstock characteristics. Identifying and constructing metabolic pathways (involving multiple enzymes) in host strains are essential to target different structural and storage polysaccharides in macroalgae. For example, efficient bioconversion of brown macroalgae via CBP would require the bioconversion of mannitol, alginate and laminarin (Figure 13.2). Studies have focused on converting alginate and mannitol to ethanol using a single-engineered organism (Dave et al., 2019). The recombinant co-expression of all three enzymes, i.e., laminarinase, mannitol dehydrogenase and alginate lyase, in *S. cerevisiae* would allow the development of a CBP process that could combine three of the most important enzymes required for the hydrolysis of macroalgae (Rocher et al., 2021).

Macroalgal waste, such as the waste remaining from the plant growth stimulants industry, is rich in carbohydrates and represents a promising substrate for bioethanol production. Utilising macroalgal biomass for bioethanol production may serve as a solution to the energy crisis and assist in diversifying global fuel production. Future research may thus investigate the economic feasibility of utilising this waste in a biorefinery approach for the co-production of ethanol and high-value products. The global economy may also be impacted, which will affect multiple industries and promote the establishment of macroalgal biorefineries. By harnessing the full potential of macroalgae, we can contribute to a more sustainable and diversified energy future.

REFERENCES

Badur, A. H., Ammar, E. M., Yalamanchili, G., Hehemann, J. H., & Rao, C. V. (2020). Characterization of the GH16 and GH17 laminarinases from *Vibrio breoganii* 1C10. *Applied Microbiology and Biotechnology, 104*(1), 161–171. https://doi.org/10.1007/s00253-019-10243-0

Bayu, A., Warsito, M. F., Putra, M. Y., Karnjanakom, S., & Guan, G. (2021). Macroalgae-derived rare sugars: Applications and catalytic synthesis. *Carbon Resources Conversion, 4*, 150–163. https://doi.org/10.1016/j.crcon.2021.04.002

Cikoš, A.-M., Šubarić, D., Roje, M., Babić, J., Jerković, I., & Jokić, S. (2022). Recent advances on macroalgal pigments and their biological activities (2016–2021). *Algal Research, 65*, 102748. https://doi.org/10.1016/j.algal.2022.102748

Dave, N., Selvaraj, R., Varadavenkatsan, T., & Vinayagam, R. (2019). A critical review on production of bioethanol from macroalgal biomass. *Algal Research, 42*, 101606. https://doi.org/10.1016/j.algal.2019.101606

Den Haan, R., Rose, S. H., Cripwell, R. A., Trollope, K. M., Myburgh, M. W., Viljoen-Bloom, M., & van Zyl, W. H. (2021). Heterologous production of cellulose- and starch-degrading hydrolases to expand

Saccharomyces cerevisiae substrate utilization: Lessons learnt. *Biotechnology Advances*, *53*, 107859. https://doi.org/10.1016/j.biotechadv.2021.107859

Godvin Sharmila, V., Dinesh Kumar, M., Pugazhendi, A., Bajhaiya, A. K., Gugulothu, P., & Rajesh Banu, J. (2021). Biofuel production from macroalgae: Present scenario and future scope. *Bioengineered*, *12*(2), 9216–9238. https://doi.org/10.1080/21655979.2021.1996019

Horn, S. J. (2000). Bioenergy production from brown seaweeds. PhD dissertation, Department of Biotechnology, Norwegian University of Science and Technology, Norway.

Horn, S. J., Aasen, I. M., & Østgaard, K. (2000). Ethanol production from seaweed extract. *Journal of Indian Microbiology and Biotechnology*, *25*(5), 249–254. https://doi.org/10.1038/sj.jim.7000065

Hou, X., From, N., Angelidaki, I., Huijgen, W. J. J., & Bjerre, A. (2017). Butanol fermentation of the brown seaweed *Laminaria digitata* by *Clostridium beijerinckii* DSM-6422. *Bioresource Technology*, *238*, 16–21. https://doi.org/10.1016/j.biortech.2017.04.035

Iwamoto, K., & Shiraiwa, Y. (2005). Salt-regulated mannitol metabolism in algae. *Marine Biotechnology*, *7*(5), 407–415. https://doi.org/10.1007/s10126-005-0029-4

KEGG. (2023). Fructose and mannose metabolism – *Saccharomyces cerevisiae* (budding yeast). Retrieved June 14, 2022, from https://www.genome.jp/pathway/sce00051

Leandro, A., Pereira, L., & Goncalves, A. M. M. (2019). Diverse applications of marine macroalgae. *Marine Drugs*, *18*(1), 17. https://doi.org/ 10.3390/md18010017

MacroFuels. (2020). MacroFuels project factsheet. Retrieved June 3, 2022, from http://www.macrofuels.eu/

Nobe, R., Sakakibara, Y., Fukuda, N., Yoshida, N., Ogawa, K., & Suiko, M. (2003). Purification and characterization of laminarin hydrolases from *Trichoderma viride*. *Biosciences, Biotechnology and Biochemistry*, *67*(6), 1349–1357. https://doi.org/10.1271/bbb.67.1349

Rocher, D. F., Cripwell, R. A., & Viljoen-Bloom, M. (2021). Engineered yeast for enzymatic hydrolysis of laminarin from brown macroalgae. *Algal Research*, *54*, 102233. https://doi.org/10.1016/j.algal.2021.102233

Ruiz, J., Olivieri, G., De Vree, J., Bosma, R., Willems, P., Reith, J. H., Eppink, M. H. M., Kleinegris, D. M. M., Wijffels, R. H., & Barbosa, M. J. (2016). Towards industrial products from microalgae. *Energy and Environmental Science*, *9*(10), 3036–3043. https://doi.org/10.1039/c6ee01493c

Schiener, P. (2013). Bioethanol production from macroalgae. PhD dissertation, Department of Molecular and Microbiology, University of the Highlands and Islands, Scotland.

Schultze-Jena, A., Vroon, R. C., Macleod, A. K. A., Hreggviðsson, G. Ó., Adalsteinsson, B. T., Engelen-Smit, N. P. E., de Vrije, T., Budde, M. A. W., van der Wal, H., López-Contreras, A. M., & Boon, M. A. (2022). Production of acetone, butanol, and ethanol by fermentation of *Saccharina latissima*: Cultivation, enzymatic hydrolysis, inhibitor removal, and fermentation. *Algal Research*, *62*, 102618.

Wargacki, A. J., Leonard, E., Win, M. N., Regitsky, D. D., Santos, C. N., Kim, P. B., Cooper, S. R., Raisner, R. M., Herman, A., Sivitz, A. B., Lakshmanaswamy, A., Kashiyama, Y., Baker, D., & Yoshikuni, Y. (2012). An engineered microbial platform for direct biofuel production from brown macroalgae. *Science*, *335*(6066), 308–313. https://doi.org/10.1126/science.121454

You, S., Zhang, L., Wang, Y., Huang, H., Ma, R., Shi, P., Bai, Y., Su, X., Lin, Z., Luo, H., & Yao, B. (2016). Improvement of the thermostability and catalytic efficiency of a highly active β-glucanase from *Talaromyces leycettanus* JCM12802 by optimizing residual charge-charge interactions. *Biotechnology for Biofuels*, *9*, 124. https://doi.org/10.1186/s13068-016-0544-8

Subject Index